POLYFUNCTIONAL STABILIZERS OF POLYMERS

Polyfunctional Stabilizers of Polymers

N.A. Mukmeneva
S.V. Bukharov
G.N. Nugumanova
and
A.M. Kochnev

Nova Science Publishers, Inc.
New York

For permission to use material from this book please contact us:
Telephone 631-231-7269; Fax 631-231-8175
Web Site: http://www.novapublishers.com

LIBRARY OF CONGRESS CATALOGING-IN-PUBLICATION DATA

Polyfunctional stabilizers of polymers / N.A. Mukmeneva ... [et al.].
p. cm.
Includes index.
ISBN 978-1-60692-536-2 (hardcover)
1. Polymers--Additives. 2. Stabilizing agents--Synthesis. I. Mukmeneva, N. A.
TP1142.P64 2009
668.9--dc22

2008047037

Published by Nova Science Publishers, Inc. ✝ New York

CONTENTS

PREFACE

The monograph is of interest to scientists specializing in physical chemistry. The results of research of polyfunctional stabilizers from the point of view of correlation of structure, reactivity and inhibiting efficiency on the basis of general kinetic approach and modern physical, chemical and mathematical methods of identification of mechanisms of their stabilizing action, both in modelling liquid-phase systems and solid polymers and their melts are analyzed. The method of the comparative estimation of inhibiting properties of the polyfunctional compounds based on semiempirical correlations "structure - properties" are considered.

Aspects of chemistry of high-molecular compounds include the results of research of polymer analogous reactions of polyfunctional stabilizers with liable groups, contained in the composition of polymers at various stages of their ageing.

In the final chapters of the monograph the data on ageing and stabilization of various types of polymers (PO, PVC, heterogeneous chain polymers, rubbers, etc.) are considered, some variants of practical realization of intramolecular and mix synergism of antioxidizing action of polyfunctional stabilizers in polymers are presented.

In whole, the book represents the scientific work devoted to the subject of the development of polyfunctional stabilizers, in this work some aspects of organic chemistry, physical chemistry, chemistry of high-molecular compounds, issues of ageing and stabilization of polymers, practical forecasts for the usage of polyfunctional stabilizers are considered. The above said predetermines the importance of the monograph.

FOREWORD

The field of polymer ageing and stabilization includes chemistry and physics of conjugated radical chain, ionic and molecular transformations in complex multicomponent systems which constitute polymeric material. As a rule, polymer loses its valuable initial properties as a result of ageing processes being subjected to destruction or cross-linking and it obtains intensive coloring. Polymer stability to various types of ageing proceeding under processing and operation determines potential limits of its practical application. It is for these reasons that polymer stabilization is one of the urgent issues of chemistry of high-molecular compounds in the long run.

Chemistry and technology of polymeric material stabilizers, mainly formed in 60th–70th of the last century, are covered in the monographs of Foigt J., Emanuel N.M., Buchachenko A.L., Minsker K.S., Denisov E.T., Shlyapintokh V.Y., Piotrovskiy K.P., Tarasova Z.N., Grassy N., Scott J., Ershov V.V. et al, Gorbunov B.N. and Gurvich Y.A., Roginskiy V.A. and others.

In subsequent years representatives of Moscow, Leningrad (St. Petersburg), Nonosibirsk, Kazan Chemical Schools have contributed greatly into the development of new methods and improvement of existing of synthesis methods of known inhibitors of free radical chain oxidative processes. The main directions of these studies are optimization ways of synthesis and development of new polyfunctional stabilizers (Ershov V.V., Nikiforov G.A. et al., Koptyug V.A. et al., Kirpichnikov P.A., Mukmeneva N.A. et al.), synthesis and investigation of oligomer and water-soluble sterically hindered phenol compounds (Domnins N.S. et al.), synthesis and study of ligands and metal complexes with sterically hindered phenol fragments (Milaeva E.R.).

The traditional stabilization of polymers makes provisions for the usage of individual stabilizers containing a reactive center in their structure and acting according to one definite mechanism. However, this monofunctional stabilizer doesn't always provide high degree of polymer protection from ageing. The basic trend developing in the world recently is the application of mixtures of several stabilizers of diverse action conditioning synergetic (strengthening) stabilization effect. The detection of mutual strengthening effect of stabilizing action of multicomponent systems resulted in the idea of developing stabilizers, in molecules of which there are simultaneously several reaction centers capable of performing various functions during polymer stabilization (the so called polyfunctional stabilizers). It should be noted that the concept "polyfunctionality" may have different meanings. For example, in organic chemistry compounds are called polyfunctional if they contain several functional groups (identical or different). In polymer stabilization polyfunctionality is understood, first of all, as the ability of a stabilizer to perform various functions in the inhibition of free radical chain oxidative process, e.g. to terminate kinetic oxidation chains (accepting peroxide radicals) and nonradically decompose hydroperoxides. These polyfunctional stabilizers are potentially able to display intramolecular synergic effect. It is natural that this ability is conditioned by the presence of various functional groups in a stabilizer molecule, e.g. the presence of a sterically hindered phenol fragment and bivalent sulfur.

The functions of phosphorous acid esters are of special interest for polymer stabilization. Possessing dual reactivity due to their chemical nature they, depending on the structure and conditions, are capable of action polyfunctionality, they can terminate chains of free radical oxidation (interaction with peroxide radicals), inhibit degenerated chain branching (interaction with hydroperoxides) etc.

On the other hand, polyfunctionality of some stabilizers is conditioned by their ability to generate new compounds in the course of oxidative conversions, which can perform other functions in inhibition of oxidation process in comparison with initial stabilizers. Some di- and polyphenol stabilizers, the oxidation products of which (quinones and stable methylenequinone) are strong acceptors of alkyl radicals, are among polyfunctional stabilizers of the abovementioned type. Thus, the high inhibiting activity of polyphenols can be explained by synergic properties of the mixture of phenols (peroxide radicals acceptors) with quinones (alkyl radical acceptors).

In whole, it is the ability to inhibit polymer ageing processes according to various mechanisms that is great potential of increasing stabilizer efficiency.

When this monograph was written mainly the results of long-term investigations in the field of phosphorus- and other elements-containing compounds carried out by the chemists of the Kazan chemical School under supervision of Professor Kirpichnikov P.A., by his followers and disciples, including the authors of this study were used.

The aim of the monograph is systematization and summarizing of this theoretical and experimental material on the development of a promising class of polyfunctional stabilizers of polymers. In the book there are submitted the main results of investigations in the field of synthesis and study of a large number of compounds as polymer stabilizers, these compounds include phosphites, phosphorus dithioacids, thiocarbamide derivatives and their metal complexes, sulfur-, nitrogen- and phosphorus-containing sterically hindered phenols and some di- and polyphenols.

Some issues were developed in the process of mutual work and discussions with such outstanding scientists as Professors Minsker K.S. and Berlinn A.A., as well as with such authorities and experts in the field of polymer chemistry as Kovarskaya B.M., Piotrovskiy K.B., Matveeva E.N. and others.

The authors express their thanks to Pobedimskiy D.G., Samuilov Y.D., Cherezova E.N., Agadzhanyan S.I., Akhmadullina A.G., Kadyrova V.Kh., Okhotina N.A., Popova L.M., Sabirova L.Kh., Cherkasova O.A., Yamalieva L.N., Chebotareva E.G., Verizhnikov L.V., Kolubakina N.S., Gren G.P., Zharkova V.M., Fazlieva L.K., Krivenko L.V. and others for their participation in the investigations which have been considered in this book.

INTRODUCTION

The fast growth of production of polymeric materials requires the development of new and more radical methods of their protection. The efficiency of existing antioxidants belonging to those classes of organic compounds (phenols, aromatic amines, sulfides, etc.) which are traditionally used for inhibition of oxidation process is likely to approach limits and new groups of antioxidants and new principles of stabilization should be searched for. It is especially urgent as the usage scope of polymeric materials will be extended and new tasks of increasing their stability will arise under new conditions of working.

The idea of obtaining stabilizers, molecules of which contain the combination of various reaction centres, promoting efficient inhibition of polymer ageing (so called polyfunctional stabilizers) is developed in this monograph. Functionalized element(N,S,P)-containing compounds are of great scientific and practical interest. The concept of the monograph structure reflects approach logic to the development of polyfunctional stabilizers.

The first chapter is devoted to the purposeful synthesis of a number of classes of polyfunctional compounds. Much attention is paid to high reactivity of organophosphorus compounds capable of participating variously in inhibition processes of polymer oxidation. A retrospective review is given where the principal data on synthesis of potential and practically used phosphorus-containing inhibitors of organic substrate oxidation are summarized and systematized.

In the same chapter experimental methods of synthesis of other N,P,S-containing compounds also are submitted. The usage of high reactivity benzylating agents (3,5-di-*tert*-butyl-4-hydroxybenzylacetate, N,N-dimethyl-3,5-di-*tert*-butyl-4-hydroxybenzylamine etc.) permits to introduce sterically hindered phenol groups into molecules of various organic compounds, it conditions the possibility of obtaining a wide range of polyfunctional stabilizers.

In Chapter 2 the results of chemical modeling processes of inhibited oxidation of organic substrates reflecting kinetic regularities of reactions of polyfunctional compounds with modeling hydroperoxides and free radicals are described. The obtained kinetic parameters permit to conduct correlation analysis "kinetic parameter – structure" using a mathematical model of frontal steric effect of any substitute at any reaction center. The suggested method enables to estimate the efficiency of antioxidant action of stabilizer proceeding from the knowledge of the compound chemical structure.

The large part of this monograph (Chapter 3) is devoted to the stabilization of color of polymer, a very significant field of the total problem of polymer stabilization. Depending on account the variety of mechanisms of coloration of polymer and polymer composites in the course of their ageing, there were suggested some organophosphorus compounds, mainly organic phosphites, to prevent such coloration. It was determined that the complex process of stabilization of the original color of polymers is caused by a diversity of hetero- and homolytic reactions of organic phosphites with the products of polymer degradation, initiating and determinating the coloration of polymers.

In Chapter 4 variants of practical realization of synergism of antioxidant action (intramolecular and intermolecular) based on the knowledge stabilizer action mechanism and principles of forming synergetic effects under thermal oxidation of polymers are discussed. The effects of nonadditive strengthening of polyfunctional stabilizers by transition metal derivatives are described.

In whole, it is shown in the monograph that polyfunctional compounds are promising oxidation inhibitors and color stabilizers owing to their high reactivity and polyfunctionality of action in polymers.

This monograph is destined for researchers and specialists connected with organic chemistry and chemistry of high molecular compounds, with processing and exploitation of p

Chapter 1

SYNTHESIS OF POLYFUNCTIONAL STABILIZERS

The review of synthesis methods of a number of groups of polyfunctional compounds (namely, tri-alkyl(aryl)- and dialkyl(aryl) phosphites, dialkyl(aryl)dithiophosphates and phosphonates, thiocarbamide derivatives, P,N,S-containing sterically hindered phenols and polyphenols) being of interest as polymer stabilizers is given in the present chapter.

1.1. ORGANOPHOSPHOROUS COMPOUNDS

Theoretical basis of stabilizing action of organophosphorous compounds is the specific ability of a phosphorus atom to start electronic interactions of various types. The presence of vacant d-orbitals and unshared electronic pair, high polarizability and polarity of phosphorus bonds provide possibilities of synthesis of organophosphorous compounds of various structures [1-2].

Among compounds of this group phosphorus acid esters are of great interest owing to the technological availability, high reactivity and stabilizing efficiency if we consider polymer stabilization.

The first report about the possibility of using triphenylphosphite for rubber stabilization appeared in 1954 [3]. The summarizing of the basic results of the subsequent investigations in this field (mainly of searching nature) up to 1965 is given in the monography of J.Foigt [4].

The regular studying of phosphorus acid esters as stabilizers of polymers begun in the late sixties of the last century. The scientists of the Kazan chemical school contributed greatly to these investigations including the research of synthesis of organophosphorous stabilizers [5]. The retrospective review of these investigations is given below.

Depending on the coordination of a phosphorus atom in phosphorus acid esters and their structures the materials are given in the following order. At first the derivatives of 3-coordinated phosphorus, namely, full acyclic and cyclic alkyl(aryl) esters of phosphorus and diphosphorus acids, oxyranyl esters of phosphorus acid and polyphosphates are considered. Then the derivatives of 4-coordinated phosphorus, namely, acid acyclic and cyclic esters of phosphorus acids are discussed.

1.1.1. Trialkyl(Aryl)Phosphites

$$(RO)_3P,\quad (RO)_2POR',\quad ROP(OR')_2$$

R=aryl, optionally substituted by $(C_1\text{-}C_{10})$-alkyl group

R' =$(C_1\text{-}C_{10})$-alkyl group

Synthesis of trialkyl(aryl)phosphites is a simple one-stage process based on the classical reaction of phosphorus trichloride (or chlorphosphite) with alcohols in the presence of tertiary amines which are acceptors of hydrogen chloride [6-8]:

$$PCl_3 + 3\ ROH \xrightarrow{R'_3N} P(OR)_3 + 3\ NR'_3 \cdot HCl$$

Unlike alcohols, phenols react with phosphorus trichloride without acceptors of hydrogen chloride that is defined by durability of the bond ArO-P. Triarylphosphites are usually obtained giving high yield, but heating of the reactionary mix to 433-453 K is required for this purpose. Using catalysts, such as Lewis's acids and, especially, pyridine, allows to perform these reactions under less strict conditions.

The efficient methods of synthesis of trialkyl(aryl)phosphites are reactions of interesterification:

$$(RO)_3P + 3\ R'OH \rightarrow (R'O)_3P + 3\ ROH.$$

Triarylphosphites and phosphites of functionally substituted alcohols are exposed to alcoholysis under mild conditions, and, gradual distillation of reaction products for balance shift is not obligatory in this case.

To obtain full phosphites the reactions of alcoholysis and phenolysis of phosphorus acid amides are also used:

$$P(NR_3)_3 + 3\ ArOH \rightarrow P(OAr)_3 + 3\ NR_3\ .$$

This reaction is of special interest for phosphorylation of functionally substituted alcohols, for example, carbohydrates, as well as for hydroxyl containing polymers.

Acyclic phosphites are efficient antioxidants and color stabilizers for a large number of polymers (polyolefines, rubbers, vulcanizates, polyvinylchloride, polyethylene terephthalate, polyamides etc.) [9-18].

Sterically hindered triarylphosphites as antioxidants are interesting from practical point of view. This group of phosphites includes the following not coloring stabilizers: Irgafos TNPP, Irgafos TPP, Irgafos DDPP, Irgafos 168 which are produced by various companies (Ciba Specialty Chemical (Switzerland); Great Lakes (Germany); Degussa AG (USA) etc.)

Tris-(2,4-di-tert-butyl)phenylphosphite known, in particular, under trade mark Irgafos 168 is widely used at present time. This stabilizer is used for many commercial polymers, including mixtures with phenol antioxidants, in this case it is known under trade mark Blends.

1.1.2. Alkyl(Aryl)Alkylene(O,O'-Arylene)Phosphites

The synthesis of cyclic esters of phosphorus acids is carried out, basically, according to the two-stage scheme by the interaction of haloid derivatives of phosphorus with polyatomic alcohols or bisphenols resulting in the formation of corresponding acid chlorides and their etherification by alcohols or phenols, for example [8]:

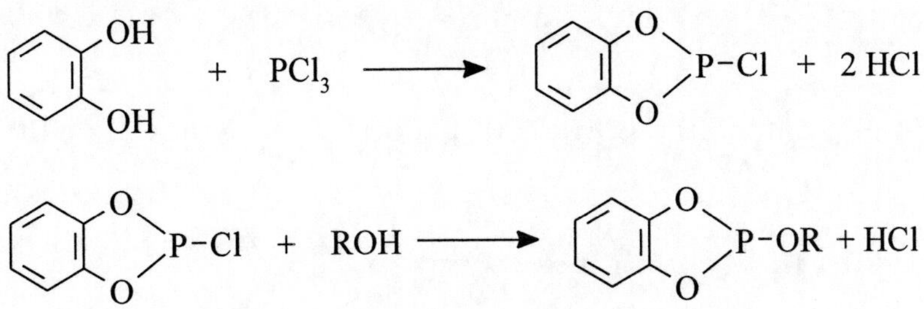

Various alkyl(aryl)alkylene(arylene)phosphites can be synthesized depending on the nature of initial hydroxyl containing components.

The basic classes of cyclic phosphites which are used or can be used for stabilization of polymers are given below. The choice of the synthesis methods depends on the nature of substitutes at a phosphorus atom. Cyclic esters of three-coordinated phosphorus acids are obtained, basically, by interaction of haloid derivatives of phosphorus with mono-and polyatomic alcohols, phenols, bisphenols, aliphatic and aromatic amines and also by phosphite interesterification.

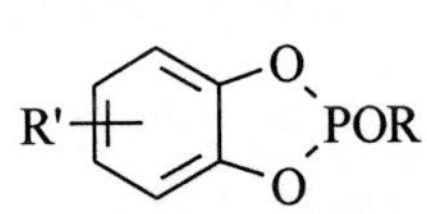

$R = C_1$-C_{18} alkyl, C_6H_5, p-$CH_3C_6H_4$, p-$C_8H_{17}C_6H_4$, p-$C_{12}H_{25}C_6H_4$, p-ClC_6H_4, α-$C_{10}H_7$, β-$C_{10}H_7$, p-$C_6H_5C_6H_4$, o-$C_6H_5C_6H_4$, C_6H_5S, Cl_4C_6H, $C_6H_5CH_2$, C_6H_5-$C(CH_3)_2$-C_6H_4,

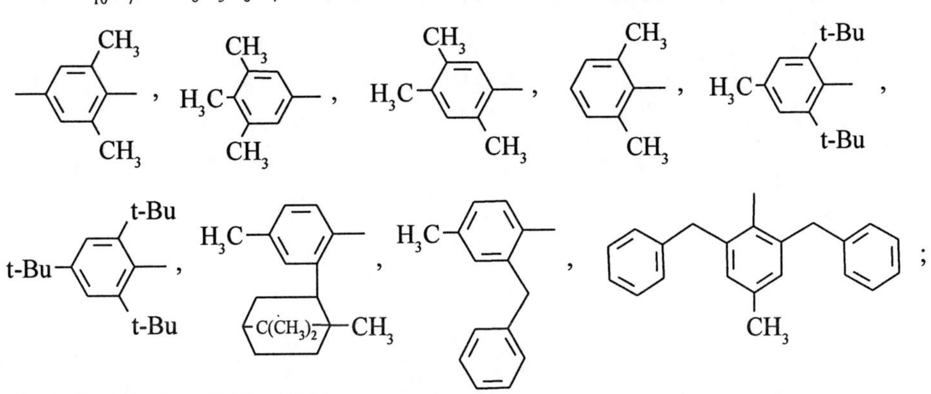

$R' = CH_3$, i-C_3H_7, t-C_4H_9, t-C_5H_{11}

o,o'-Phenylenephosphorus acid esters [19-24].

o,o′-Phenylenephosphorus acid esters are efficient stabilizers for a large number of polymers: polyolefines, polyamides, polyethylene terephthalate, polyvinylchloride etc. [25-28].

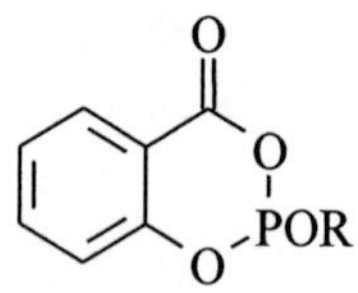

R = i-C_5H_{11}, C_6H_{13}, C_8H_{17}, $C_{12}H_{25}$, C_6H_5, p-$(CH_3)_3C$-C_6H_4, p-$CH_3C_6H_4$, o-$CH_3C_6H_4$,

m-$CH_3C_6H_4$, b- $C_{10}H_7$, 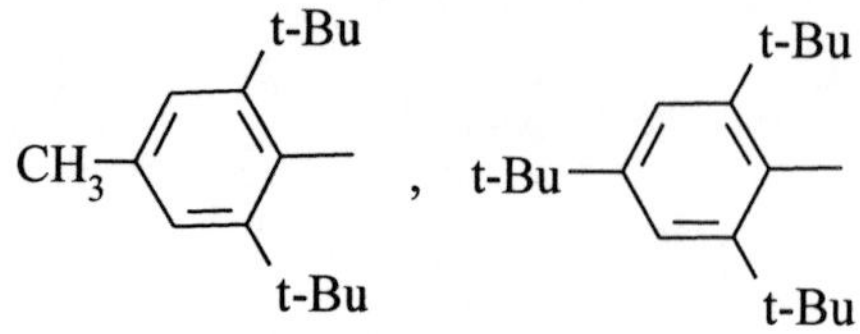

Salicylphosphorus acid esters [8 p.634, 29].

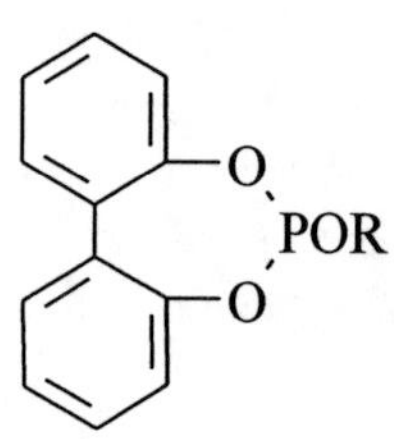

R = CH_3, C_2H_5, i-C_3H_7, n-C_4H_9, n-C_5H_{11}, n-C_6H_{13}, n-C_7H_{15}, n-C_8H_{17}, n-C_9H_{19}, n-$C_{10}H_{21}$, p-$CH_3C_6H_4$, p-$(CH_3)_3C$-C_6H_4, $\alpha-C_{10}H_7$, $\beta-C_{10}H_7$

o,o′-Diphenylphosphorus acid esters [30].

Diphenylphosphorus acid esters are investigated as stabilizers of polyvinylchloride and plasticizers [31-33].

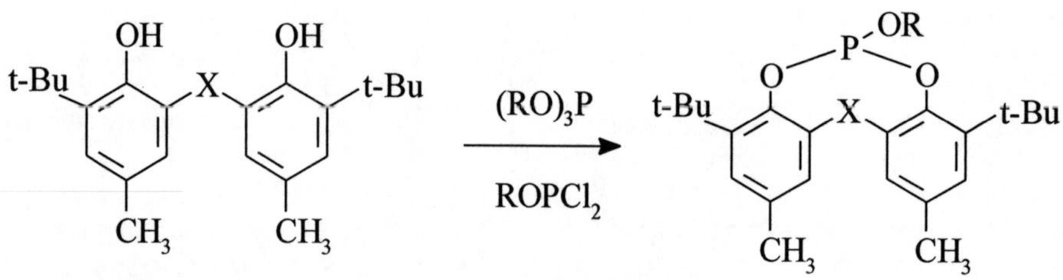

X = CH_2, S; R = C_2H_5, n-C_3H_7, n-C_4H_9, C_6H_5, p-$(CH_3)_3$-C_6H_4, $\alpha-C_{10}H_7$

Esters of methylenebis- and thiobisphenylphosphorus acids [34-36].

Higher arylenephosphites with 7-8-member cycles, being extremely stable structures (destruction temperature is above 473 K) are promising for heat-resistant heterochain polymers, primarily, as stabilizers of polymer color during high-temperature stabilization [37-39].

1.1.3. Tetraalkyl(Tetraaryl)-Alkylene(Arylene)Diphosphites

Tetraalkyl and tetraaryl esters of diphosphorus acids have, unlike trialkyl(aryl)phosphites, increased hydrolytic, thermal stability and compatibility with polymers, these characteristics increase the possibilities for using with wider range of polymers.

$$(R'O)_2POROP(OR')_2$$

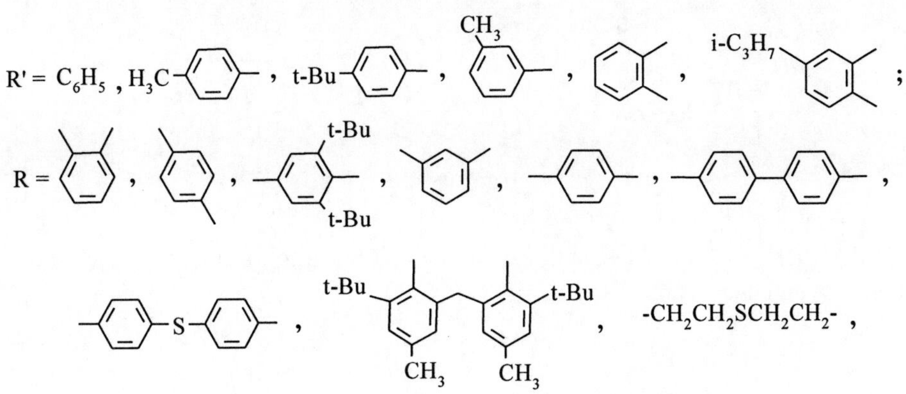

$-CH_2CH_2-$, $-CH_2CH_2OCH_2CH_2-$, $-CH_2CH_2SCH_2CH_2-$

To synthesize tetraalkyl(aryl) esters of diphosphorus acids the interaction of chloranhydrides of diarylphosphorus acids with dihydroxy compounds (pyrocatechin, hydroquinone, *p,p'*-dihydroxydiphenyl, etc.) is used. This method is the most suitable, especially, when carrying out reactions in the presence of the bases in the solution of ester or dioxane [40, 41]:

$$2\ (R'O)_2PCl + HOROH \rightarrow (R'O)_2POROP(OR')_2 + 2\ HCl.$$

The drawback of the reaction of interesterification, in particular, of triphenylphosphite by dihydroxy compounds is the time length of the process and the necessity of heating, as well as low yield of the target product.

Diphosphites of spiran and bridged structures can be obtained on the basis of pentaerythritol, depending on synthesis conditions and the nature of cocomponents [42]:

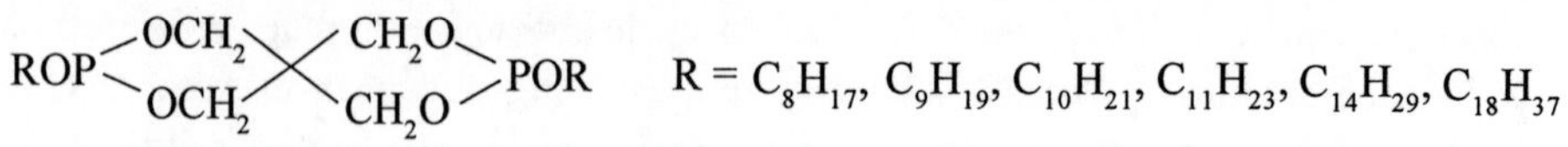

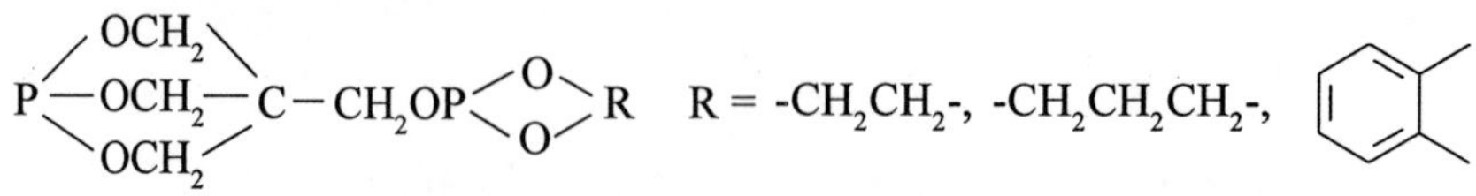

Diphosphites are efficient stabilizers of PVC and other polymers [43]. The first stabilizers on the basis of pentaerythritolphosphorus acid esters have been developed by the English company *Pure Chemicals* under trade mark Fosclere. At the present time diphosphite Irgafos 126 is widely used to stabilize various polymers (polycarbonate, ABC-plastics, polyamides, etc.).

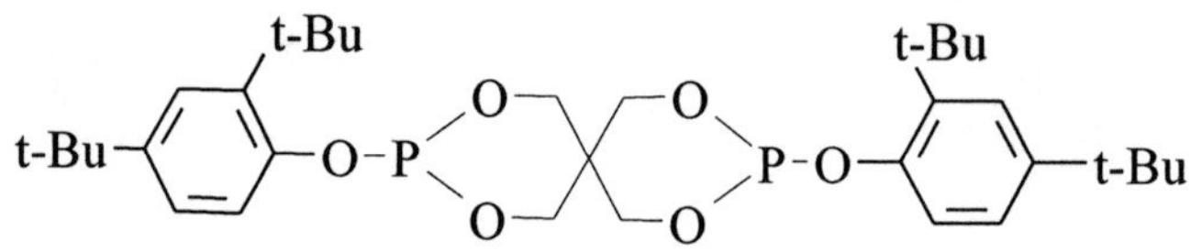

Irgafos 126

Cyclic esters on the basis of multiatom ethers (D-mannitol, D-sorbitol, dulcitol etc.) are of interest as potential stabilizers of polymers [44-49].

1.1.4. Alkyl(Aryl)-Polyalkylene(Arylene)Phosphites

Polyphosphites are synthesized by polycondensation of alkyl- or aryldichlorophosphites with dihydroxycompounds. Reactions can proceed in melt, in inert solvent in the presence of base, at boundary of phases. The interesterification of phosphorus acid esters by dihydroxycompounds is a more suitable method of obtaining polyphosphites [50-53]:

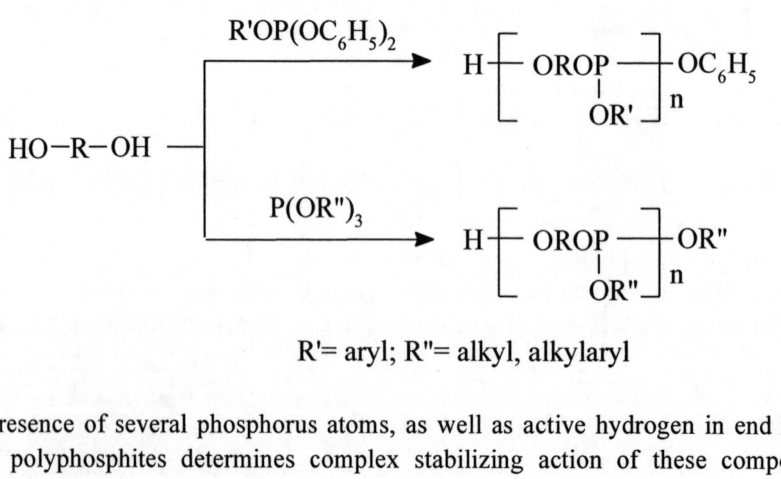

R'= aryl; R"= alkyl, alkylaryl

The presence of several phosphorus atoms, as well as active hydrogen in end hydroxyl groups of polyphosphites determines complex stabilizing action of these compounds in polyamides, rubbers, polycarbonate, polyethylene, polyvinylchloride, etc. [54, 55].

The reactions of etherification of high-molecular phenols and aminophenols by triphenylphosphites are conducted to synthesize high-molecular nontoxical nonvolatile chemicals-additives of polyfunctional action, these reactions proceed under mild conditions (363-373 K, from 2 to 3 hours), without a catalyst and with nearly total-lot yield. Monophosphite is formed at equimolar ratio of initial high-molecular one-atom phenols or aminophenols with triphenylphosphites, for example:

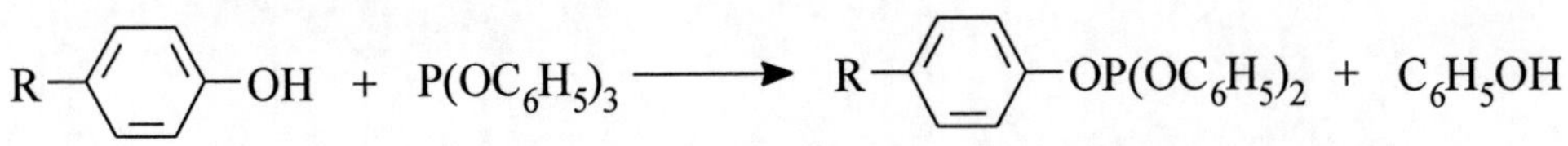

R = oligoisobutylenes, oligopropylene, oligoethylene, oligopiperylene

In case of two-atom phenols the formation both of mono- and disubstituted compounds, as well as mixtures of phosphorylated compounds is possible, the content of phosphorus depends on the ratio *high-molecular phenol : triphenylphosphite* [56].

The obtained compounds hinder the process of prepolymerization of diene monomers at the stage of their rectification, protect rubbers from thermal-oxidative destruction, providing alongside with the large period of induction the preservation of necessary plasto-elastic properties [56].

1.1.5. Alkyl(Aryl)Oxyranylphosphites

Oxyranylphosphites are synthesized by interaction of glycidol with derivatives of three-coordinated phosphorus (chloranhydrides and phosphorus acid esters) [8 p.611, 57, 58]:

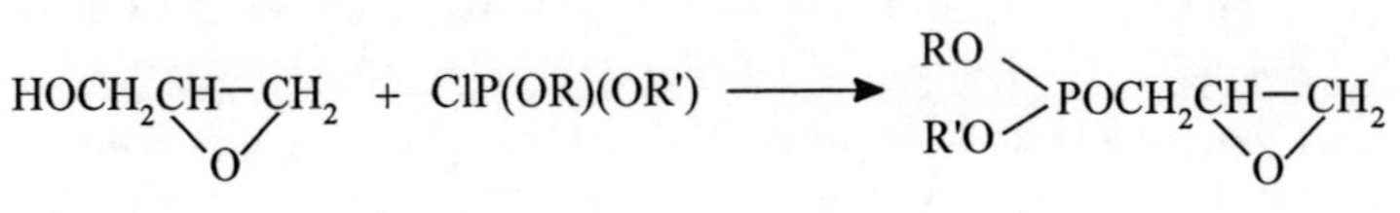

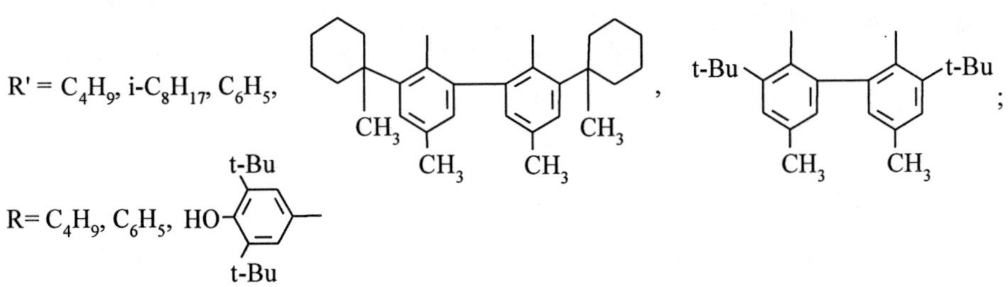

Oxyranyloctadecyl esters of phosphorus acid are synthesized by the condensation of 9,10-epoxyoctadecanol with dibutyl-and diphenylchloranhydrides of phosphorus acid in the presence of triethylamine:

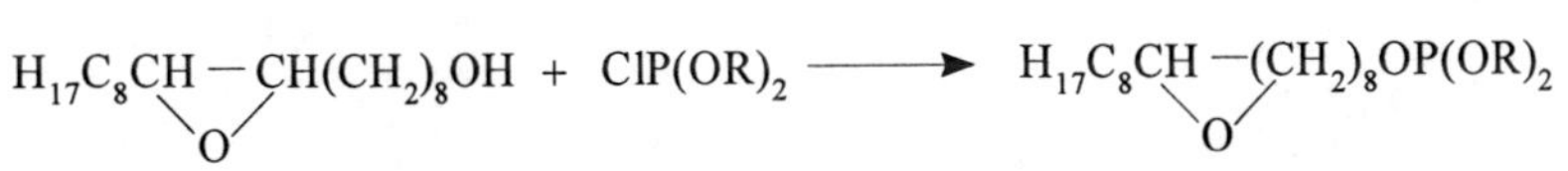

$R = C_4H_9, C_6H_5$

Oxyranylpolyphosphites are obtained by polyesterification of diphenyl-2-ethylhexyl-phosphite by diphenylolpropane with the subsequent substitution of the epoxide group for the end phenyl one:

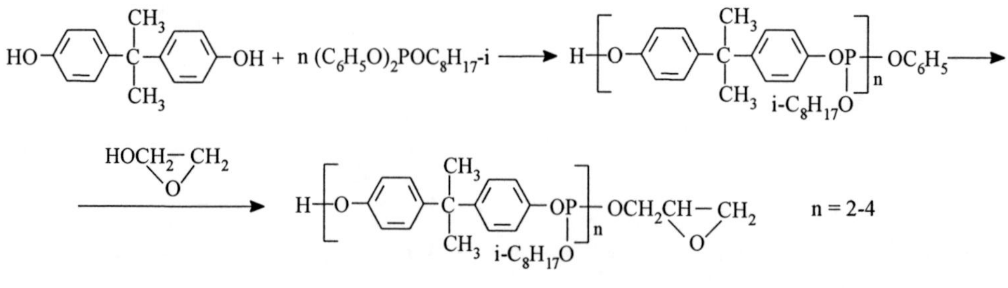

Oxyranylphosphites are able to inhibit efficiently both processes of dehydrochlorination and thermal-oxidative destruction of polyvinyl chloride and ester plasticizers [59, 60].

1.1.6. Dialkyl(Aryl)Phosphites

$(RO)_2P(O)H$

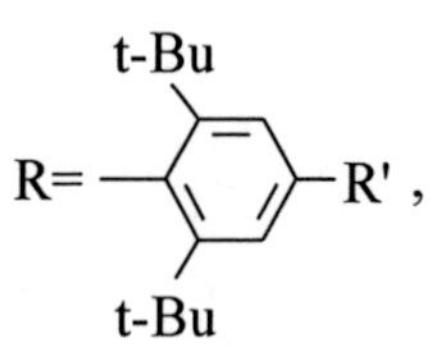

, alkyl, aryl, alkylaryl; R'= H, CH_3, $C(CH_3)_3$

All methods of obtaining dialkyl(aryl)phosphites (acidic phosphites) can be divided into two groups. The first group includes the methods where derivatives of 3-coordinated phosphorus are assumed to be used as initial substances, the second group includes the methods dealing with the substances containing fragment P(O)H, i.e. derivatives of 4-coordinated phosphorus [7, 61-63].

Acid phosphites are mostly synthesized by dealkylating trialkyl- and alkylarylphosphites by hydrogen chloride and other acid reagents:

$(RO)_3P + HCl \rightarrow (RO)_2P(O)H + RCl.$

The rate of changing of full phosphites into acid ones depends essentially on the chosen reagent, mainly, on the anion nucleophilicity of the used acid. Highly efficient dealkylating reagents are phosphorus and dialkylphosphorus acids, water in the presence of mineral and carboxylic acids. Carboxylic acids are less efficient, however they, as phosphorus acids, change triarylphosphites into diarylphosphorus acids.

Hydrolysis of relevant dialkylamides is another method of obtaining acid phosphites:

$(RO)_2PNR'_2 + H_2O \rightarrow (RO)_2P(O)H + HNR'_2.$

Etherification of phosphorus acid with the distillation of water is likely to be the most important among the second group of methods of synthesizing acid phosphites:

$$H_3PO_3 + 2\,ROH \rightarrow (RO)_2P(O)H + 2\,H_2O.$$

The reaction of etherification is particularly important when polyhydroxycompounds are phosphorylated, it gave good results at the synthesis of p,p'-arylenediphosphites:

$$2\,(RO)_2P(O)H + HO\text{-}Ar\text{-}OH \longrightarrow H(O)P(OR)\text{–}O\text{-}Ar\text{-}O\text{–}P(O)H(OR)$$

Substituted in the benzene ring diphenylenediphosphorus acids are obtained by hydrolysis of relevant p,p'-diphenylenetetrachlordiphosphites with nearly total-lot yield [63]:

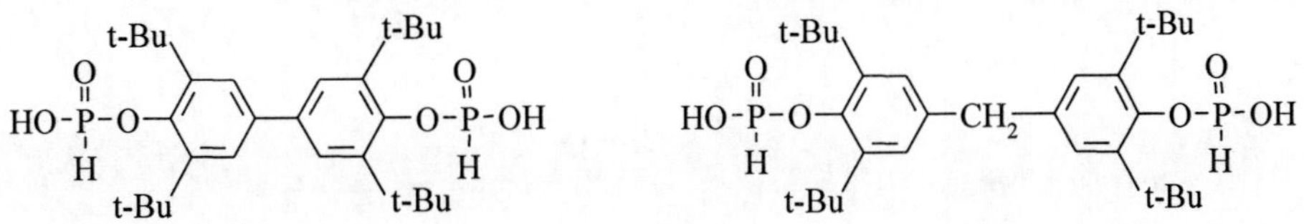

Poly(propylidenediphenylenehosphorus) acid is synthesized by the reaction of polycondensation of diphenylolpropane with diphenylphosphorus acid [64]:

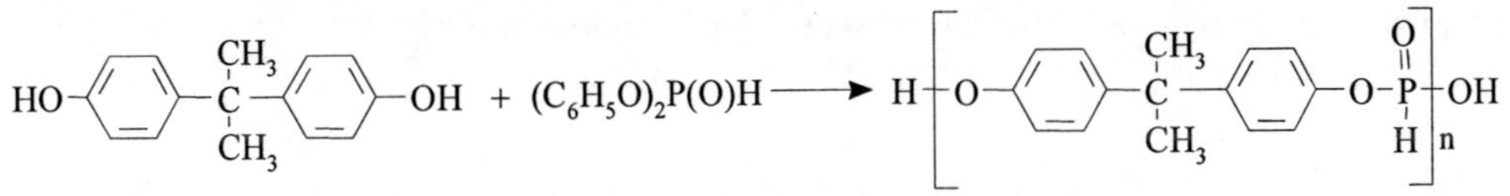

Poly(arylenephosphorus) acids are synthesized in the similar way:

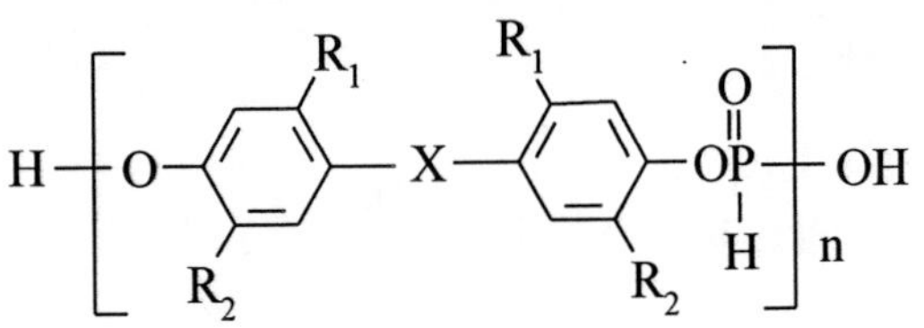

X=S, CH_2; R_1=R_2=H, CH_3; n=2-4

The regularities of inhibiting action are determined for di(alkyl)arylphosphites during the oxidation of styrene, polypropylene, oils [65, 66].

1.1.7. Alkylene(O,O'-Arylene)Phosphites

Cyclophosphorus acids with aromatic substitutes in the ring are obtained by dealkylation of full esters of phosphorus acids according to the procedure [8 p.639, 67]:

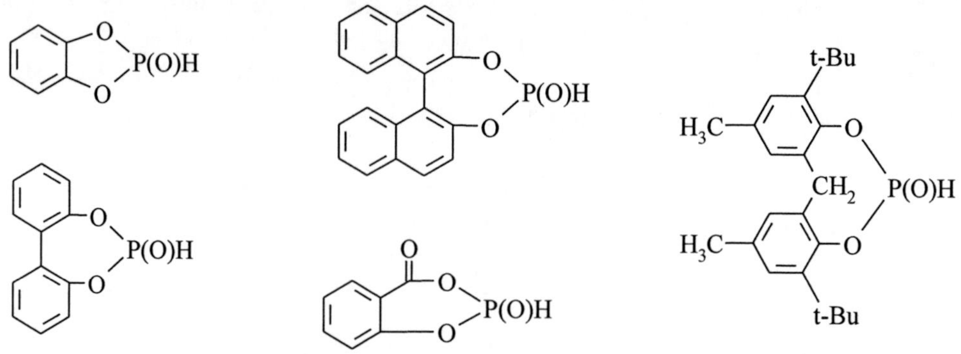

Reactions proceed when relevant acid chlorides and tert-butanol interact in the presence of pyridine with the following thermolysis of the formed tert-butyl ester of cyclophosphorus acid at 473 K:

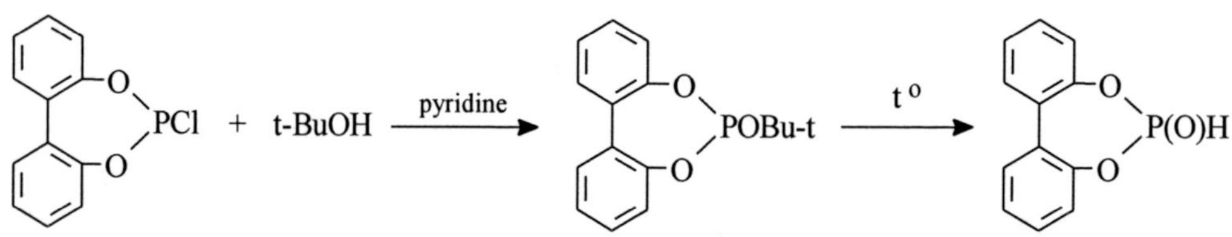

Synthesis of cyclophosphorus acids with alkylene substitutes is conducted under the interaction of phosphorus trichloride with the corresponding glycols:

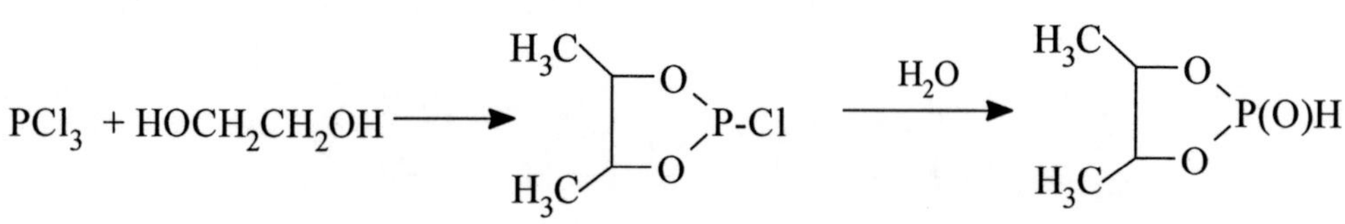

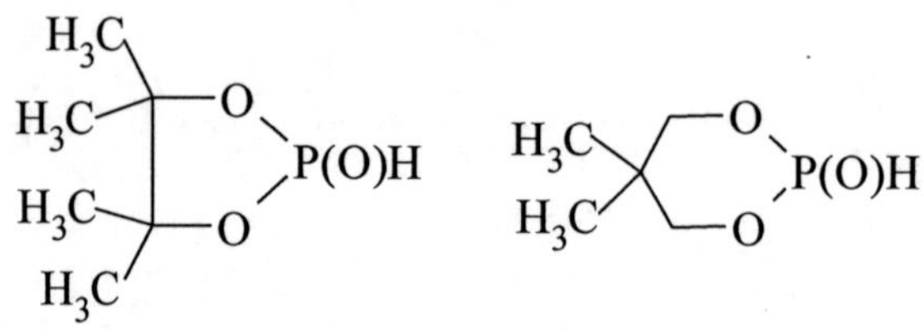

Cyclic chlorphosphites formed at the first stage are changed into the corresponding acids by hydrolysis in aqueous medium.

Cyclic monomer and polymer acids on the basis of sterically hindered bisphenols are efficient stabilizers for a number of polymers taken separately or in mixtures with antioxidants of other groups [68].

1.2. Nitrogen-, Sulfur-, Phosphorus Containing Compounds

Compounds containing simultaneously several heteroatoms in the molecule structure are of considerable interest from the standpoint of polymer stabilization that creates prerequisites for their displaying effects of intermolecular synergism.

The synthesis of P,S-containing compounds – alkyl(aryl) esters of thiophosphorus and dithiophosphorus acids, N,P-containing compounds – amides and amidoesters of phosphorus acids having various coordination, N,S-containing compounds – derivatives of thiocarbamides and their complexes with 3d-metals is described in this section.

1.2.1. Dialkyl(Aryl)Dithiophosphates and Dialkylthiophosphites

o,o'-Dialkyl(aryl)dithiophosphates (1) and alkylene(o,o'-arylene)dithiophosphates (2) are synthesized by the most popular method, that is the interaction of corresponding alcohols, phenols or glycols with phosphorus pentasulfide [69-71]:

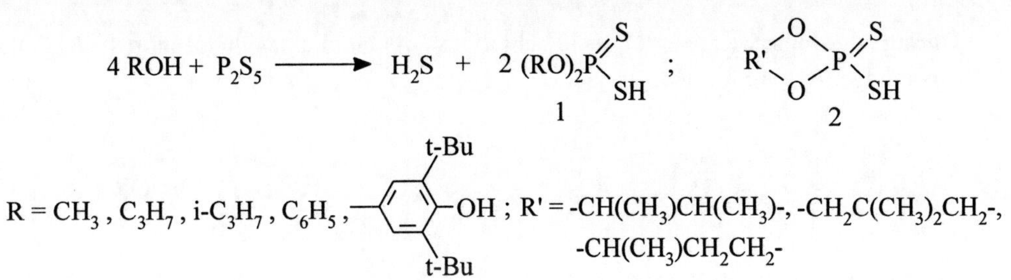

Dialkyldithiophosphates are widely used as additives to oils having simultaneously antioxidant, rust-preventive, antiwelding properties.

As a result of the reaction of hydrogen sulfide with acid chlorides of octatomic cyclic o,o'-diphenylphosphorus acids cyclic thiophosphorus acids are obtained capable to act as antioxidants [72]:

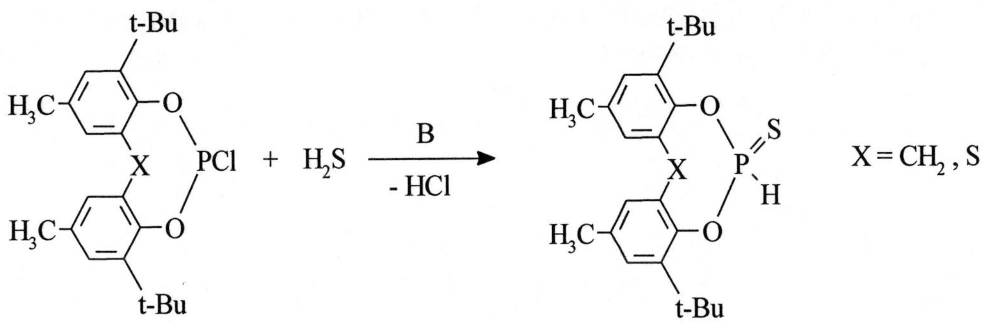

1.2.2. Amides and Amidoesters of Phosphorus Acids

Amidoesters of acids of 3- and 4-coordinated phosphorus are of interest among nitrogen- and phosphorus containing stabilizers.

Amides of *o,o'*-phenylenephosphorus and *o,o'*-phenylenephosphorus acids are obtained by the interaction of their acid chlorides with aromatic amines and diamines [73-75]:

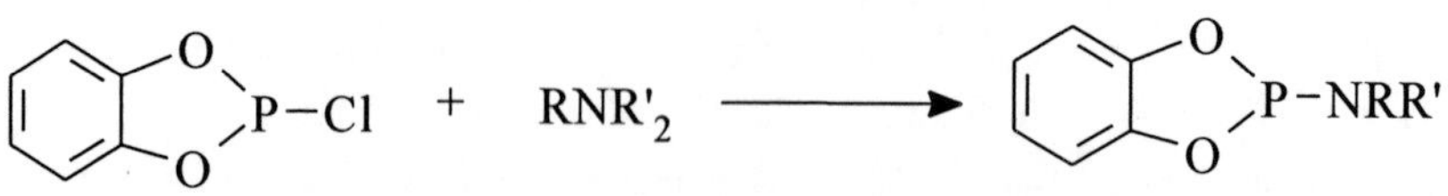

R' = H, alkyl, aryl; R' = alkyl, aryl

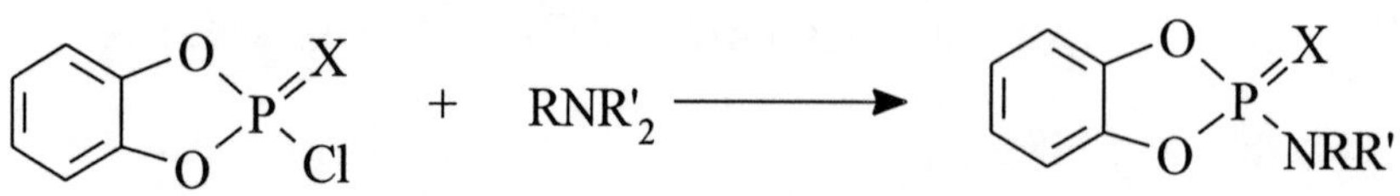

R' = phenyl, o- and p-xylenes, α- and β-naphtols; R = phenyl; X = O, S.

During the etherification of acid chlorides of amidophenylphosphorus acid the corresponding esters are formed:

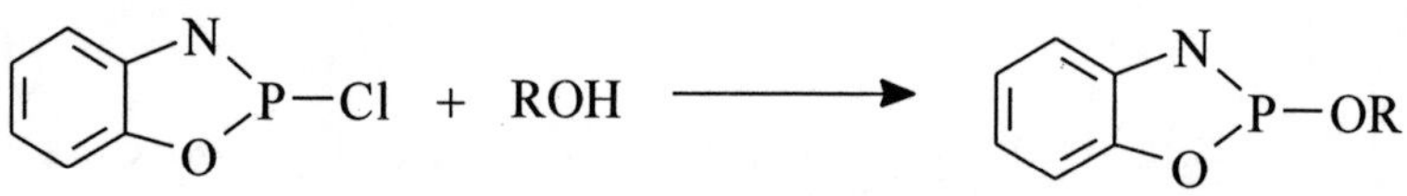

Amidophosphites bind easily sulfur, changing into corresponding amidothiophosphates, having inhibiting properties as well:

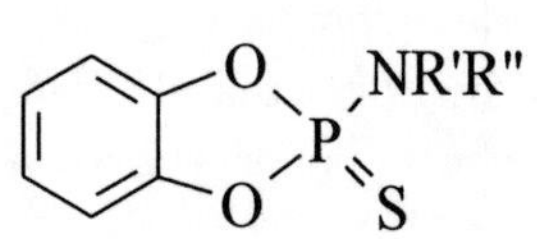

Diamidodiphosphites are synthesized when chloranhydrides of corresponding phosphorus acids interact with aromatic diamines:

$$2\ (RO)_2PCl + H_2N\text{-}R'\text{-}NH_2 \longrightarrow (RO)_2P\text{-}HN-R'-NH\text{-}P(OR)_2$$

R' =

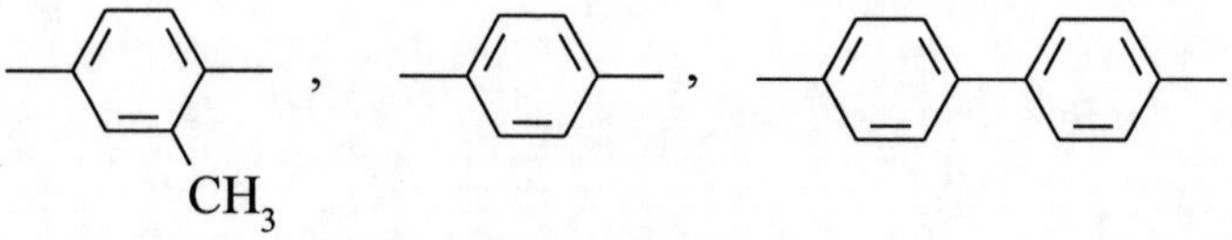

Phosphorus acid amides are efficient stabilizers for isoprene, bivinyl-styrene and other rubbers, vulcanizates, polyvinylchloride [76].

1.2.3. Thiocarbamide Derivatives and Their Complexes With 3d-Metals

The ability of isothiocyanates to react easily with compounds containing an active hydrogen atom, in particular, with amines is used for synthesizing N,N-substituted diarylthiocarbamides.

As initial amines primary alkyl- and arylamines, secondary amines including heterocyclic ones are used. The initial isothiocyanates are phenyl- and benzoyl isothiocyanates. The reactions are performed in sufficiently diluted solutions of polar solvents (acetone, dioxane) at room temperature.

The interaction of phenylisocyanate with aromatic and aliphatic diamines proceeds depending on the ratio of reagents [77].

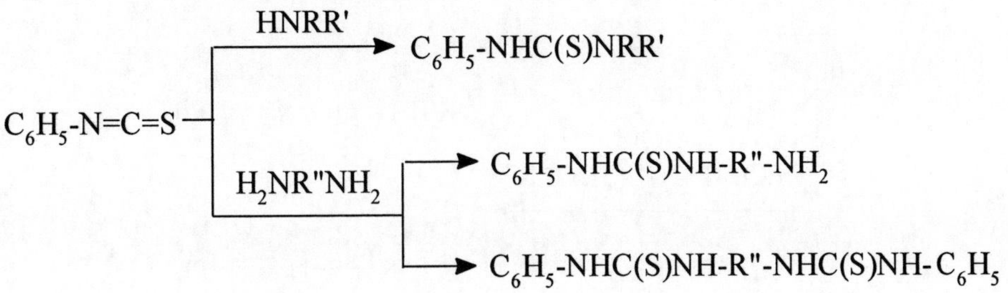

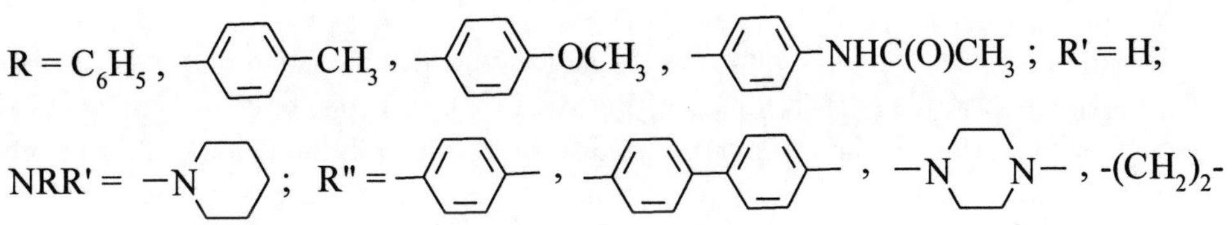

The forming of the product of binding two molecules of isothiocyanate is detected only when the reaction is conducted with the excess of phenylisothiocyanate, that is likely to be connected with the decelerating of the reaction rate of the product of the composition (1:1) with the second molecule of phenylisothiocyanate owing to the substitution of the phenylthiocarbamide fragment for the strong electron-donor NH_2-group.

The reactions of benzoylisothiocyanate with amines proceed in nonpolar solvents with molar ratio of reagents (1:1) [78]:

$$C_6H_5C(O)NCS + HNR'R'' \longrightarrow C_6H_5C(O)NHC(S)NR'R''$$

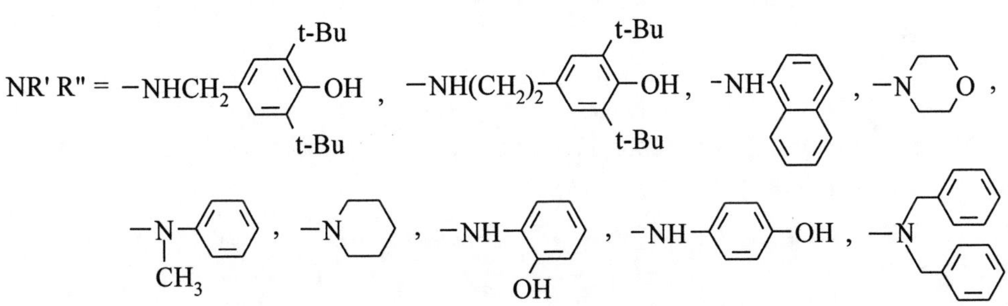

Thiocarbamides derivatives are studied as antioxidants of low pressure polyethylene, benzoylthiocarbamides demonstrate higher activity in inhibiting thermal-oxidative destruction of polymer in comparison with arylthiocarbamides [79, 80].

Metal complexes of arylthiocarbamides are synthesized when they are interacted with acetates of 3d-metals (M=Cu(II), Ni(II)) at temperature 273-323 K, in polar solvents, in neutral or alkalescent medium, at molar ratio *metal ion – ligand* (1:1) or (1:2). The results of elemental analysis of obtained compounds comply with metal complexes ML_2.

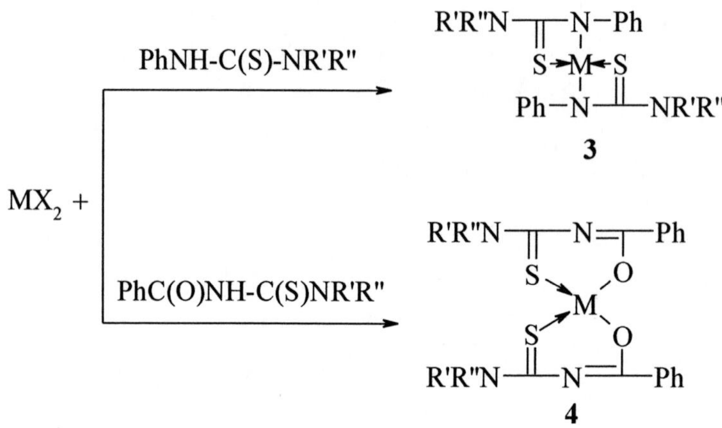

Data of infra red spectra of complexes (**3**) indicate the deprotonation of one of available amide groupings, and the connectedness of ligands in internal sphere of formed complexes with central atom through atoms of nitrogen and sulfur resulting in formation of four-member chelate cycles $M(NS)_2$. In the complex (**4**) a metal atom is coordinated with oxygen and sulfur atoms of ligand [81].

Reactions of complexation of vanadium ions are studied under conditions similar to the above described, the interaction of arylthiocarbamides with vanadium oxotrichloride and trichloride taken as an example. The composition of coordination compounds corresponds to gross formulas according to the elemental analysis $[VO(H_2L)Cl_3]$, $[V(HL)_2]Cl$.

Unlike arylthiocarbamides in the structure of benzoylthiocarbamide molecules there is an additional centre of coordination, that is an oxygen atom of a benzoyl group, owing to it these compounds can act as O,S- or N,O,S-electron-donor ligands. Benzoylthiocarbamides in most cases are bidentate oxygen-sulfur-containing ligands, and only in infrequent cases they are tridentate ones. During the complexation of Co(II), Ni(II) and Cu(II) with 1,3-disubstituted benzoylthiocarbamides the coordination compounds are formed in solutions both with deprotonated (HL^-, L_2-) and unionized (H_2L) forms with the ratio *metal ion – ligand* (1:1), (1:2), (2:2).

On the whole, thiocarbamide derivatives form stable metal complexes with ions Ni(II), Cu(II), Co(II), V(III), V(IV) by means of coordination with atoms of nitrogen and sulfur (arylthiocarbamides) and with sulfur and oxygen atoms (benzoylthiocarbamides), which are of interest as stabilizers of polymers [82, 83].

1.3. Compounds Containing Sterically Hindered Phenol Fragments

One of the approaches to obtaining phenols stabilizers is an introduction of sterically hindered phenol fragments into a molecule of an organic compound. Various 3,5-di-*tert*-butyl-4-hydroxyphenyl derivatives can be used for this purpose.

Methods of synthesis of compounds containing sterically hindered phenol fragments, using 2,4-bis(3',5'-di-*tert*-butyl-4'-hydroxyphenyl)-1,3-dithia-2,4-dithioxo-2,4-diphosphetane, N,N – dimethyl - 3,5 – di - *tert* - buty l - 4- hydroxybenzylamine and 3,5 – di – *tert* – butyl - 4- hydroxybenzylacetate are considered in this section.

1.3.1. Synthesis on the Basis of 2,4-Bis(3',5'-Di-Tert-Butyl-4'-Hydroxyphenyl)-1,3-Dithia-2,4-Dithioxo-2,4-Diphosphetane

To obtain polyfunctional phosphorus-, sulfur -containing stabilizers with sterically hindered phenol fragments the derivatives of 1,3-dithio-2,4-dithioxo-2,4-diphosphetane (high reactionary compounds containing four-memebered cycle) can be used. The motive force of the processes of interaction of these compounds with various reagents is the changing of strained four-memebered cycles into energetically more stable acyclic compounds [84].

To synthesize 2,4-bis(3',5'-di-*tert*-butyl-4'-hydroxyphenyl)-1,3-dithio-2,4-dithioxo-2,4-diphosphetane (**5**) 2,6-di-*tert*-butylphenol is used as an initial reagent:

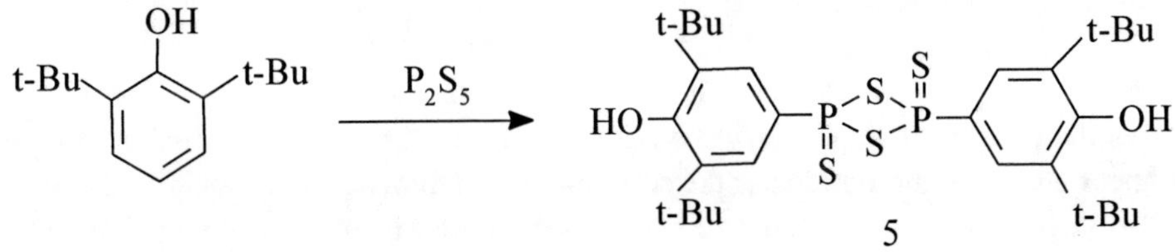

5

It is known that the presence of bulky *tert*-butyl groups in 2,6-substituted phenols makes hydroxyl group inaccessible to the attack by phosphorus pentasulfide. At the same time, sterically hindered phenols, not containing substitutes in *p*-position regarding hydroxyl groups, enter easily the reactions of electrophilic substitution and form the corresponding *p*-substituted derivatives. The total action of the mesomeric and inductive effects displayed by hydroxyl and alkyl substitutes of 2,6-di-*tert*-butylphenols promotes the increase of nucleophilicity of carbon in *p*-position of an aromatic nucleus and, thereby, facilitates the course of reaction with phosphorus pentasulfide, having distinct electrophilic properties.

Dithioxodiphosphetane (**5**) is used as a reagent for obtaining phosphorus containing compounds with a sterically hindered phenol fragment.

The interaction of dithioxodiphosphetane (**5**) with monoatomic alcohols (benzene, 293-313 K) results in obtaining O-alkyl-(3,5-di-*tert*-butyl-4-hydroxyphenyl)dithiophosphonic acids (6) (Scheme 1.1), they display antioxidant activity regarding polyethylene and raw rubbers [85, 86].

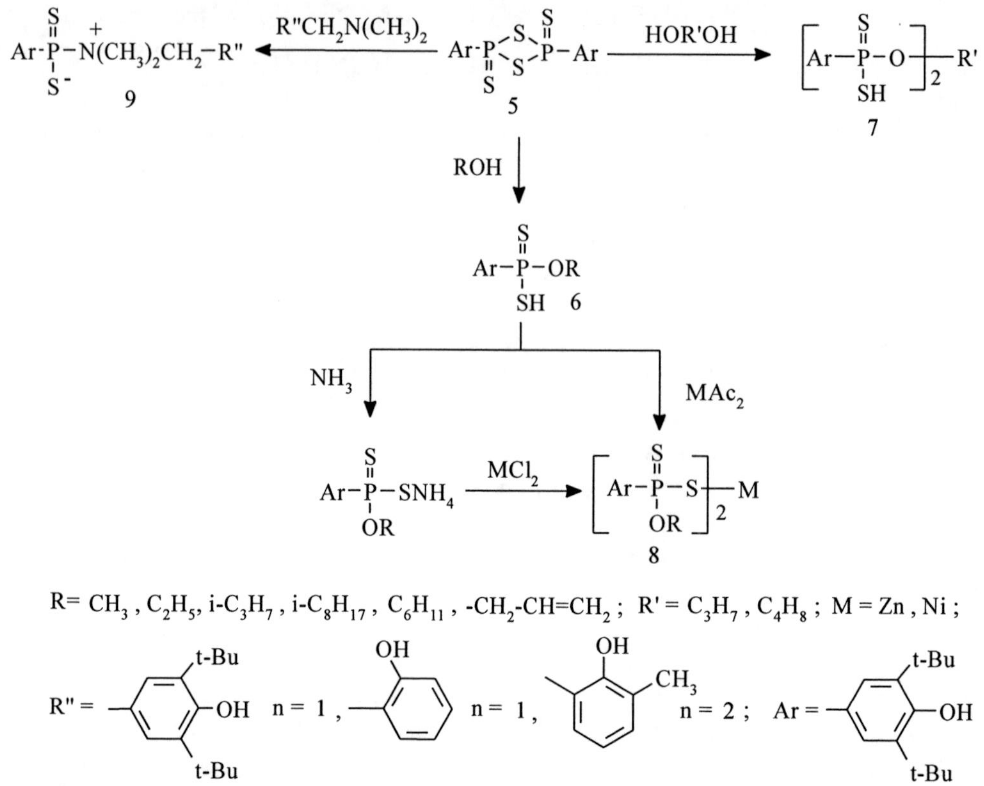

Scheme 1.1.

The reaction proceeds through the attack of electron-deficient atom of phosphorus by nucleophilic oxygen of alcohol hydroxyl group with the subsequent opening of tetramerous phoshpetane cycle and formation of the corresponding dithiophosphonic acid.

The interaction of compounds (5) with diols results in the total-lot formation of bis-dithiophosphonic acids (7) (Scheme 1.1).

Dithiophosphonic acids (6) react easily with compounds of 3d-metal with the formation of salts of dithiophosphonates (8), which are of interest from the point of their ability to efficient hydroperoxide decomposition [87].

O-alkyl-(3,5-di-*tert*-butyl-4-hydroxyphenyl)dithiophosphonates of metals (8) are synthesized by two means, i.e. by the interaction of dithiophosphonic acids (6) with ammonia and later with zinc or nickel chlorides (Scheme 1.1), and by the reaction of dithiophosphonic acids (6) with zinc or nickel acetates proceeding as one stage. In both cases the reaction proceeds easily in aqueous-alcoholic media at room temperature.

The observed (in infra red spectra) balancing of phosphorus– sulfur bonds in the triad (S-P=S) indicates the chelate structure of complex NiL_2. The diamagnetism of complexes NiL_2 confirms flat-square coordination of a nickel atom by four sulfur atoms of two chelate ligands, when the study is performed by the method of ESR (electron spin resonance).

When dithioxophosphetane (5) interacts with a number of phenolic Mannich bases N,N-dimethyl-(3,5-di-*tert*-butyl-4-hydroxybenzyl)amine, 2-dimethylaminomethylphenol, 2,6-di(dimethylaminomethyl)phenols the betaines (9) are obtained with the total-lot yield

(Scheme 1.1), betaines display high inhibiting activity during the oxidation of synthetic and mineral oils [88, 89].

1.3.2. Synthesis on the Basis of N,N-Dimethyl-(3',5'-Di-Tert-Butyl-4'-Hydroxybenzyl)Amine

Application of 3,5-di-tert-butyl-4-hydroxybenzyl derivatives, such as N,N-dimethyl-(3,5-di-tert-butyl-4-hydroxybenzyl)amines (10) and 3,5-di-tert-butyl-4-hydroxybenzylacetate (11), in the synthesis of sterically hindered phenols stabilizers is based in most cases on their ability to act as precursors of benzyl carbcation (12) and 2,6-di-tert-butylmethylenequinone (13) [90]. Formed during the acid catalysis benzyl carbcation (12) is able to interact with weak nucleophiles forming target products. As by-process the cation (12) can change into methylenequinone (13) with the following formation of products of its dimerization and disproportionation (14) and (15) [91]. In the presence of bases, as a rule, the removal of a proton of hydroxyl group of benzyl derivatives (10) and (11), the forming of methylenequinone (13) and its subsequent reaction with any nucleophilic reagent takes place (Scheme 1.2).

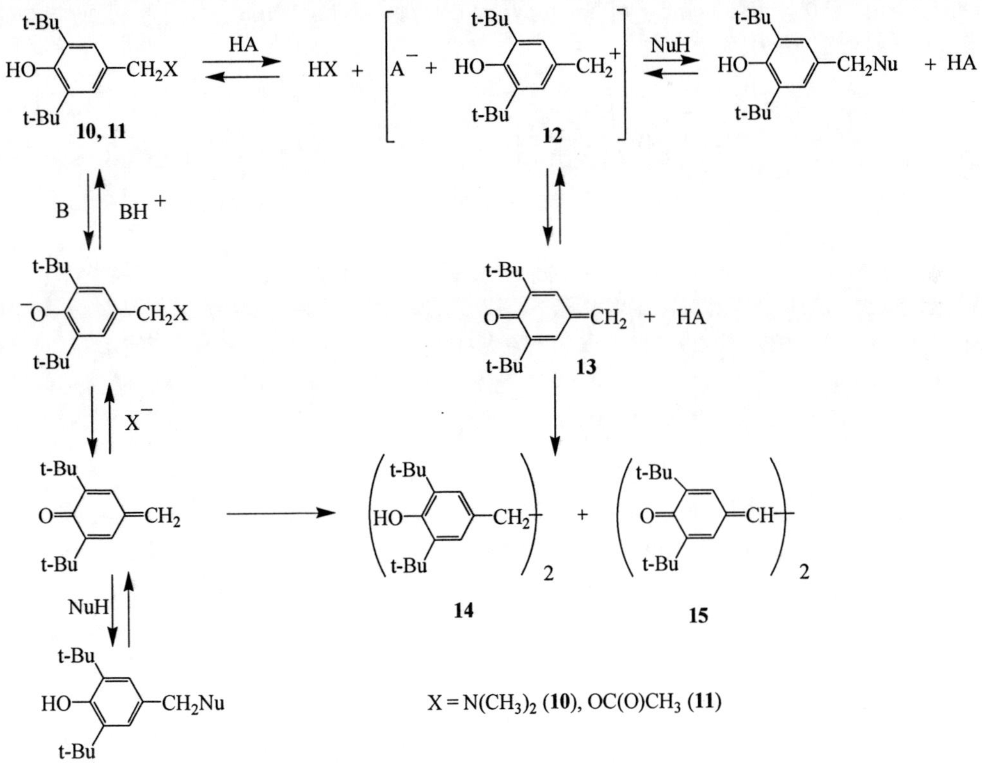

Scheme 1.2.

The interaction of N,N-dimethyl-(3,5-di-tert-butyl-4-hydroxybenzyl)amines (10) with a number of primary amines is studied to obtain compounds containing fragments of sterically hindered phenols and aromatic amine [92-95].

The reactions of transamination of a compound (10) by aliphatic and aromatic amines result in, depending on the conditions of the process, the obtaining of mono- or disubstituted compounds (Scheme 1.3).

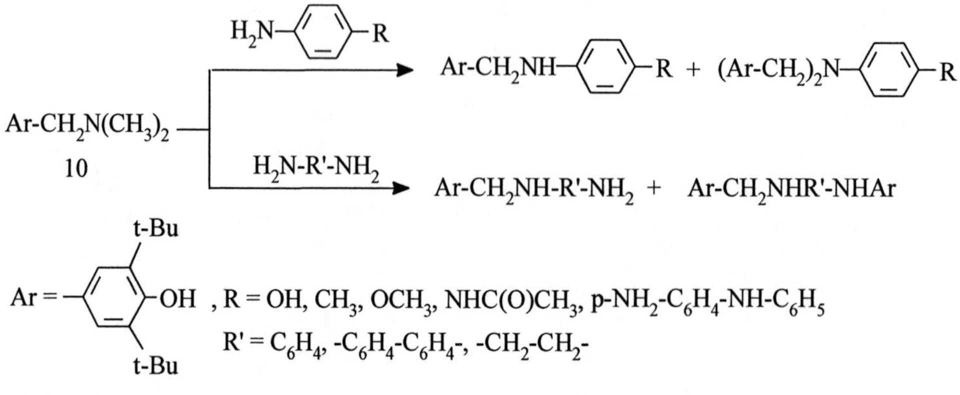

Scheme 1.3.

Efficient polyfunctional stabilizers (16), (17) and (18) are obtained by the interaction of benzylamine (10) with a number of heterocyclic compounds containing primary or secondary nitrogen atom [96-99].

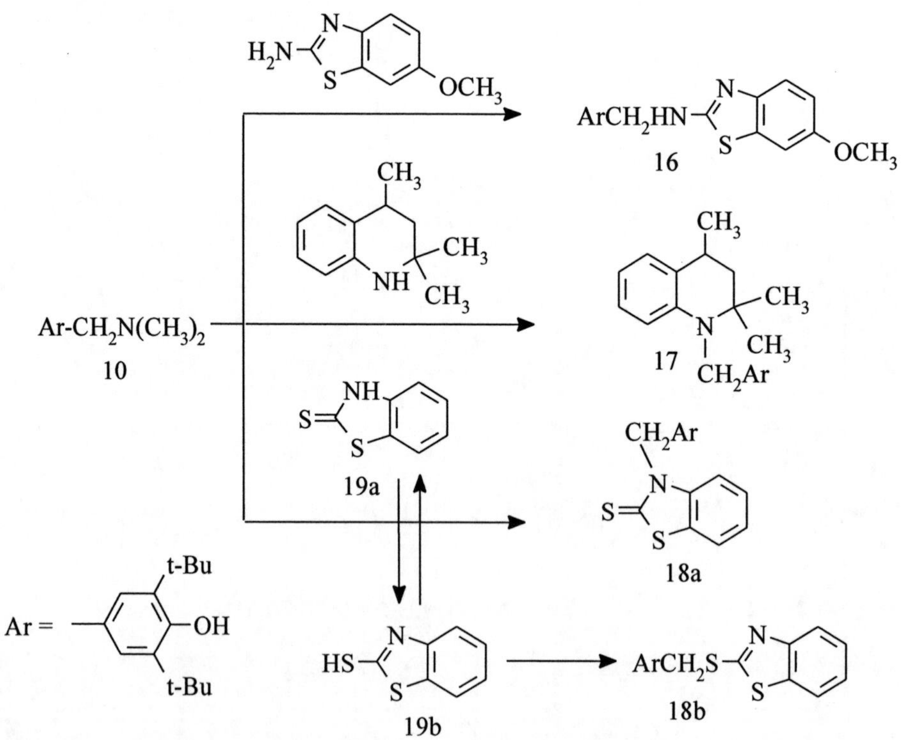

Scheme 1.4.

As 2-mercaptobenzothiazole is available in two tautomeric forms, namely thione (19a) and thiol ones (19b), the interaction of reagents may proceed by two alternative routes resulting in the formation of N-(3,5-di-tert-butyl-4-hydroxybenzyl)benzothiazolthione-2 (18a) and S-(3,5-di-tert-butyl-4-hydroxybenzyl)benzthiozole (18b) (Scheme 1.4).

According to the investigations of the structure of 2-mercaptobenzothiazole the balance is shifted into the thione group in the solid phase and in most of inert solvents, as well as in DMF and acetonitrile,, as for the alkalescent solvent the balance is shifted into the thiol group.

When the reaction of benzylamine (10) with 2-mercaptobenzothiazole in the melt is conducted the compound (18a) is a product of the reaction. In the NMR ^{13}C spectrum of these compounds there is a signal of thiocarbonyl carbon (S=C, 190.5 ppm), which is absent in the NMR spectrum of the compound obtained by counter synthesis (18b). The interaction of amine (10) with 2-mercaptobenzothiazole in the medium of monoatomic alcohols (n-PrOH, i-PrOH, n-BuOH) results in the formation of the same product (18a). In nonpolar solvents (benzene, toluene, ethylbenzene, cumene) the mixture of products (18a) and (18b) is formed.

The presence in the structure of the molecule of benzothiazolthione (18a) of nitrogen and sulfur atoms makes it possible to form complexes with compounds of 3d-metal. The interaction of compounds (18a) with $CuCl_2$ in tetrahydrofuran resulted in the formation of complexes of the composition central atom – ligand (1:2). Calculated by the ESR method g-coefficients (g =2.267 и g =2.077) confirm the formation of the paramagnetic compound with electron configuration d9.

N-phenyl-N-(3,5-di-tert-butyl-4-hydroxybenzyl)thiocarbamide (20) is synthesized by the interaction of benzylamine (10) with phenylthiourea [98].

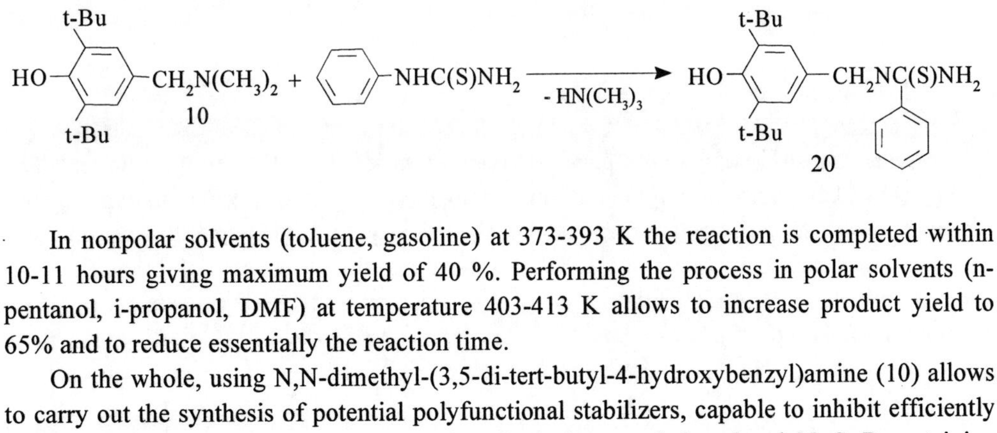

In nonpolar solvents (toluene, gasoline) at 373-393 K the reaction is completed within 10-11 hours giving maximum yield of 40 %. Performing the process in polar solvents (n-pentanol, i-propanol, DMF) at temperature 403-413 K allows to increase product yield to 65% and to reduce essentially the reaction time.

On the whole, using N,N-dimethyl-(3,5-di-tert-butyl-4-hydroxybenzyl)amine (10) allows to carry out the synthesis of potential polyfunctional stabilizers, capable to inhibit efficiently oxidation processes of polymers. The described above polyfunctional N-,S-,P-containing additives have shown considerable stabilizing efficiency in the suppression of oxidation of polyethylene, polypropylene, poly-4-methylpentene-1, butyl rubber, polyisoprene, ethylene-propylene rubber [96-99].

1.3.3. Syntheses on the Basis of 3,5-Di-Tert-Butyl-4-Hydroxybenzylacetate

3,5-Di-*tert*-butyl-4-hydroxybenzylacetate (11) can be easily obtained by the interaction of N,N-dimethyl-(3,5-di-*tert*-butyl-4-hydroxybenzyl)amine (10) with acetic anhydride [100-102]. Acetate group in benzylacetate (11) is a better outgoing group than dimethylamine one, that's why according to Scheme 1.2 (Section 1.3.2), benzylacetate (11) is a more active precursor of benzyl carbcation (12) and methylenequinone (13). Besides, acetic acid educing in the process of its application should shift the balance *benzylcarbcation* (12) – *methylenequinone* (13) towards the former, decreasing the yield of by-products of dimerization and disproportionation of methylenequinone – compounds(14) and (15) (Scheme 1.2).

1.3.3.1. Conversions of 3,5-Di-Tert-Butyl-4-Hydroxybenzylacetate in Various Media

3,5-Di-*tert*-butyl-4-hydroxybenzylacetate (11) is a highly reactive reagent undergoing conversions in various media. These transformations should be considered when methods of synthesis of sterically hindered phenol stabilizers on the basis of benzylacetate (11) are worked out.

The reactions including the stage of alkyl splitting [103 p.114] and proceeding with intermediate formation of 3,5-di-*tert*-butyl-4-hydroxybenzyl carbcation (12) are characteristic of benzylacetate (11) being ester of acetic acid and substituted benzyl alcohol.

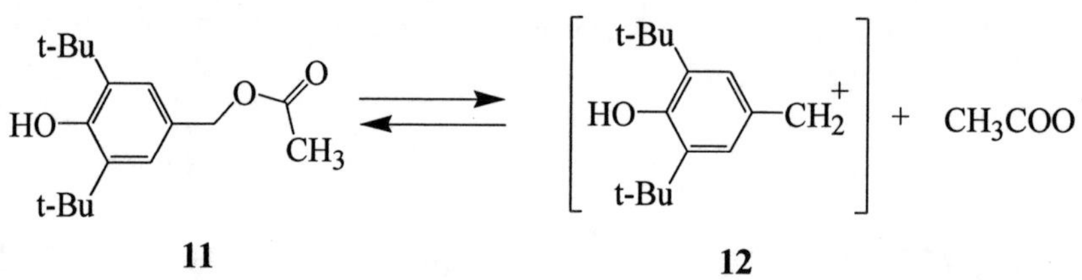

Similar transformations of benzyl derivatives proceed easily during acid catalysis, reducing nucleophilicity of outgoing group. Finally they result in the formation of products of interaction of benzyl carbcation (12) and (or) methylenequinone (13) emerging as a result of its deprotonation [91] with any nucleophilic reagent, and they result as well in the formation of the products of dimerization and disproportionation of methylenequinone (13) – compounds (14) and (15).

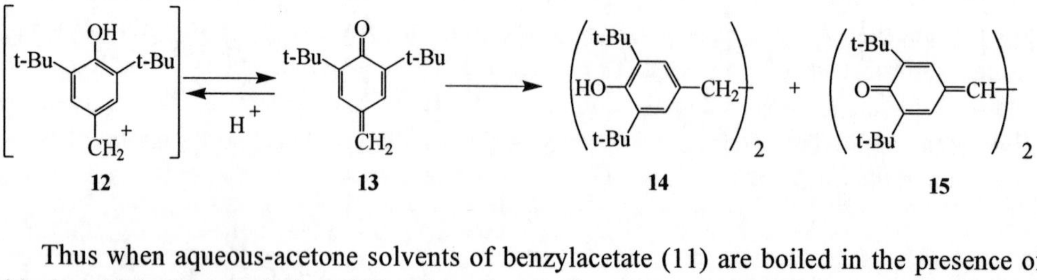

Thus when aqueous-acetone solvents of benzylacetate (11) are boiled in the presence of chloric acid the benzyl alcohol (21) and products of its subsequent transformations are formed [101]:

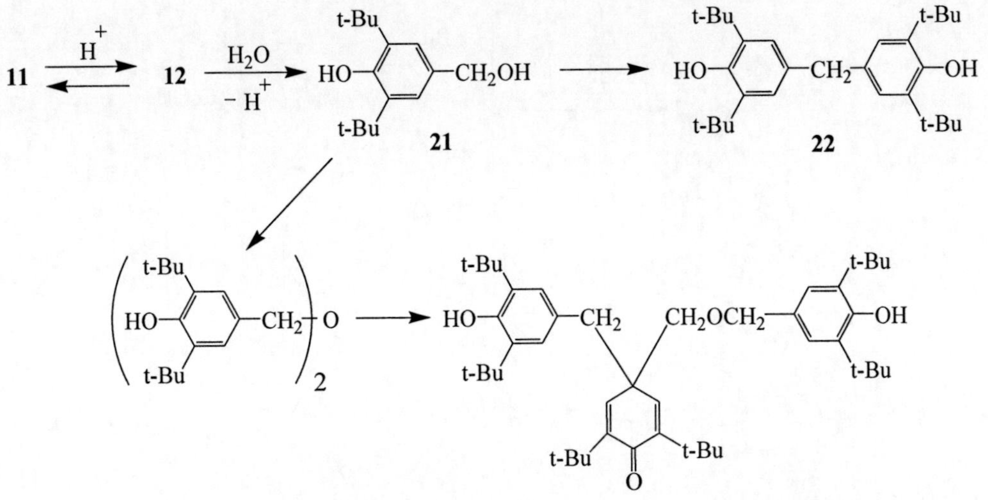

Compounds (21) and (22) are also formed during long-term keeping or boiling solutions of benzylacetate (11) in aqueous acetone in the presence of hydrochloric, phosphorus or acetic acids. It should be noted, that benzylacetate (11) stands two-hour boiling without changes in anhydrous acetic acid.

In methanol solution when acid and alkaline catalysts are absent at room temperature there is replacement of acetoxyl fragment of benzylacetate (11) by methoxyl one [101, 104].

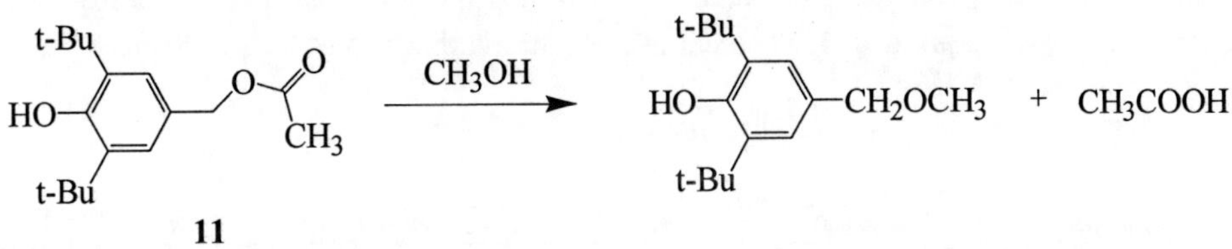

The transformation of benzylacetate (11) in solutions of deutero-methanol and deutero-ethanol, investigated by the method of NMR ^{1}H spectroscopy [105] proceeds through quickly established balance benzylacetate (11) – methylenequinone (13).

It is shown, that methylenequinone (13) in alcohol solutions of benzylacetate (11) is obtained in the process of benzyl cation (12) deprotonization which, in turn, arises in the form of ionic pair (23) as a result of monomolecular alkyl splittings (solvolysis) of benzylacetate (11) in the alcohol medium.

According to Scheme 1.5, solvolysis of benzylacetate (11) with the fast reversible stage is a reaction of the pseudo-first order. At the same time the balance position *benzylacetate* (11) – *methylenequinone* (13) influences the kinetics of this process, depending on the initial concentration of benzylacetate (11). As it is seen from Scheme 1.5, it changes according to the concentration reduction of compound (11) during the reaction.

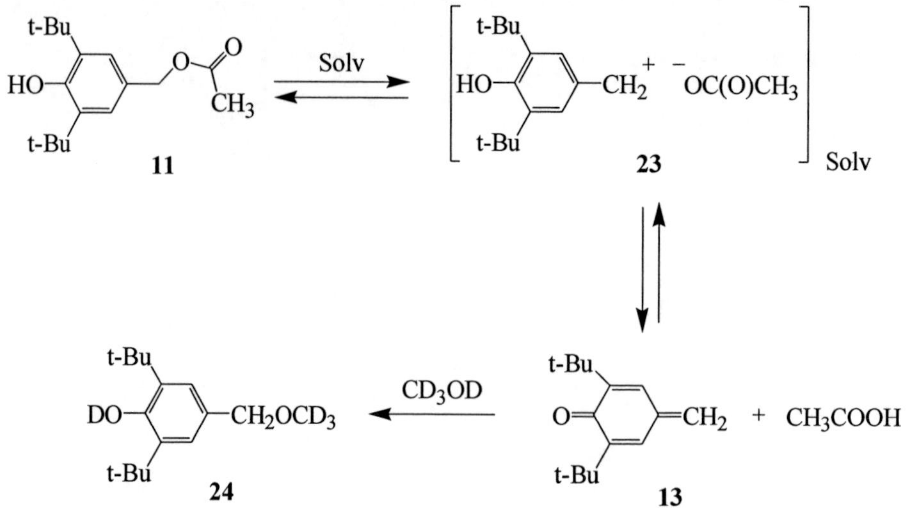

Scheme 1.5.

Table 1.1 determines the yield of ether (24) within 4 hours of the reaction and first order effective rate constants of benzylacetate (11) solvolysis in solutions with different ratio *benzylacetate* (11) – *methylenequinone* (13), depending on the concentration of an initial reagent and solvent. From the given data it is seen, that the solvolysis rate increases with the increase of methylenequinone (13) content in the solution that agrees with Scheme 1.5.

Table 1.1. Parameters of benzylacetate (11) solvolysis in methanol-d$_4$

$C_{(11)}$, mole/l	C_{D2O}, mole/l	$C_{(13)}/C_{(11)}$	Yield of ether[a] (24), %	k_{ef}^{20}, c^{-1}
0.0730	-	0.22	12	1.6×10^{-5}
0.0173	-	0.62	26	3.1×10^{-5}
0.0180	0.5	0.75	33	4.1×10^{-5}

[a] The process duration is 4 hours.

Thus, alkyl breakage of benzylacetate (11) occurs in the presence of acids and in alcohols. In the latter case such breakage results in the formation of alkoxybenzyl derivatives. In the acid media when any nucleophilic agents (water, alcohols, etc.) are absent benzylacetate (11) is relatively stable, that is likely to be caused by the shifting of balance *methylenequinone (13) – benzyl carbcation (12)* towards the latter.

In the publication [91] the products of benzylacetate (11) conversion, emerging when water alkali is added into its solution are investigated by methods of chromatomass- and electronic spectroscopy.

In the reaction mixture bis(3,5-di-*tert*-butyl-4-hydroxyphenyl)methane (22), hydrogalvinoxyl (25), 1,2-di-(3′,5′-di-*tert*-butyl-4′-hydroxyphenyl)ethylene (26), α-acetoxy-2,6-di-*tert*-butylmethylenequinone (27) were found in addition to the expected 1,2-di-(3′,5′-di-*tert*-butyl-4′-hydroxyphenyl)ethane (14) and 3,3′,5,5′-tetra-*tert*-butyltoluylene-4,4′-quinone (15), a product having weight of 494 was also found, being, obviously, a product of formal addition of acetic acid, which splits, when benzylacetate (11) is being decomposed, to toluylenequinone (15), namely, compound (28).

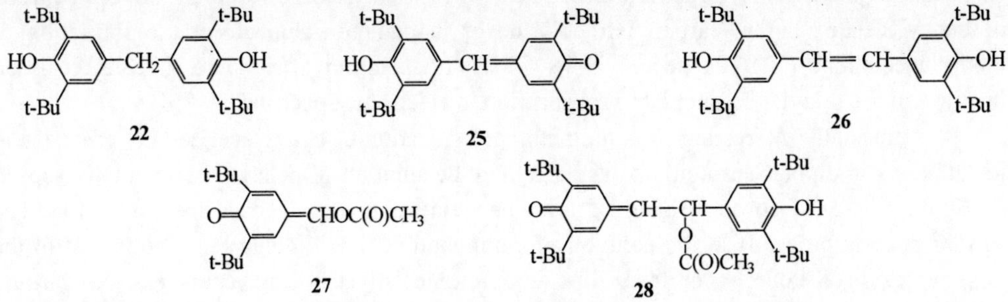

The formation of hydro-galvinoxyl (25) can explain the lilac coloring of benzylacetate (11) solutions when alkali is added to them, as compound (25) forms easily phenolates of lilac color ($\lambda_{макс.}$ 580 nanometers, ε 80000). Hydro-galvinoxyl (25) occurs in the reaction mixture owing to the process of compound (22) dehydrogenation by toluylenequinone (15). The formation of diphenylmethane (22) may be a result of conversions of 3,5-di-*tert*-butyl-4-hydroxybenzyl alcohol(21) [106-108] which, in its turn, is a product of water addition to methylenequinone (13).

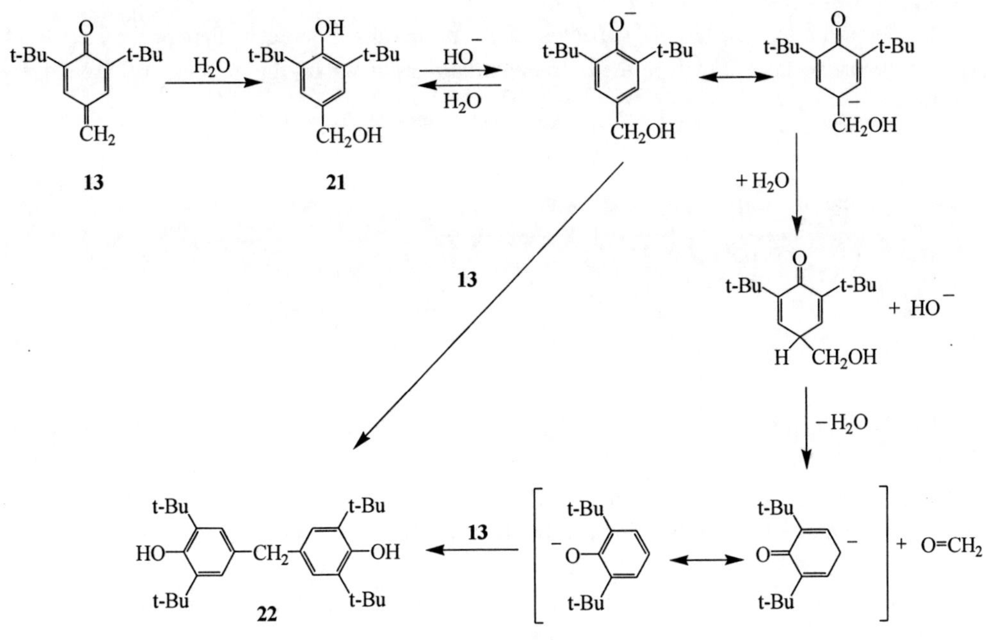

Thus, the conversions of 3,5-di-*tert*-butyl-4-hydroxybenzylacetate (11) in the alkaline medium are not limited to the formation of only 1,2-di-(3′,5′-di-*tert*-butyl-4′-hydroxyphenyl)ethane (14) and 3,3′,5,5′-tetra-*tert*-butyltoluylene-4,4′-quinone (15) from unstable 2,6-di-*tert*-butylmethylenequinone (13). This process is accompanied by dehydrogenation of phenol compounds under the action of toluylenequinone (15) resulting in the formation of methylenequinone products.

The deprotonization of benzylacetate (11), resulting in the formation of methylenequinone (13) and the products of its dimerization and disproportionation, proceeds easily under the action of organic bases. Thus, when triethylamine is added into the solution of benzylacetate (11) in carbon tetrachloride or in deuteron-chloroform the formation of methylenequinone (13) is detected in its NMR ^{1}H spectrum after some time, there is an increase of the toluylenequinone (15) absorption in electronic spectrum.

It is especially interesting to note, that deprotonization of benzylacetate (11) proceeds in its solutions in dipolar aprotonic solvents. When the solution of benzylacetate (11) is kept in DMF there is an absorption band of toluylenequinone in its electron spectrum (15). The similar process proceeds in the solution of compound (11) in dimethyl sulfoxide and in the less degree in acetone. Therefore, dipolar aprotonic solvents can act as bases regarding benzylacetate (11) and initiate the forming of methylenequinone (13) and the products of its subsequent transformations.

Thus, 3,5-di-*tert*-butyl-4-hydroxybenzylacetate (11) is an extremely labile compound, eliminating easily acetate fragment under the action of acids and bases, and a number of solvents with the formation of reactive intermediate structures and the products of their further transformations. This fact should be taken into consideration when the conditions of conducting processes with benzylacetate (11) participation and development of efficient methods of obtaining sterically hindered phenols stabilizers on its basis are defined.

1.3.3.2. Syntheses in the Presence of Bases

Nucleophilic agents, having sufficient basic properties, interact with 3,5-di-*tert*-butyl-4-hydroxybenzylacetate (11) forming corresponding products of the process of breakage – addition.

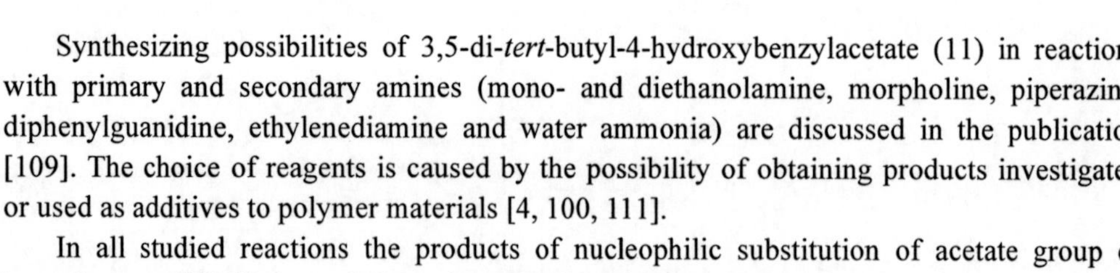

Synthesizing possibilities of 3,5-di-*tert*-butyl-4-hydroxybenzylacetate (11) in reactions with primary and secondary amines (mono- and diethanolamine, morpholine, piperazine, diphenylguanidine, ethylenediamine and water ammonia) are discussed in the publication [109]. The choice of reagents is caused by the possibility of obtaining products investigated or used as additives to polymer materials [4, 100, 111].

In all studied reactions the products of nucleophilic substitution of acetate group of benzylacetate (11) (Scheme 1.6) are formed giving high yield.

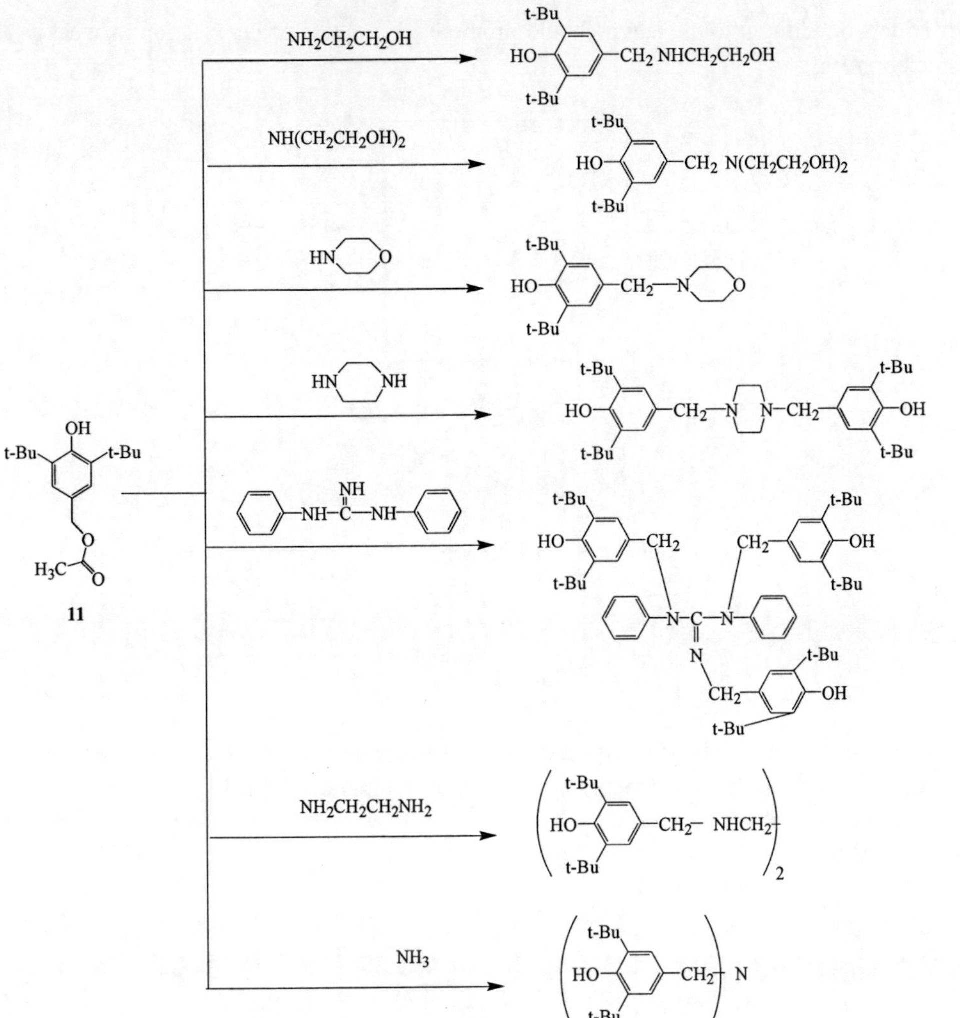

Scheme 1.6.

The reactions proceed easily at room temperature or when heated to 318 K, without the catalyst in acetone, acetonitrile or in the solution of the corresponding amine. Depending on the ratio of reagents the products of different degree of benzylation can be obtained. In nonpolar solvents the reactions under discussion go more slowly.

It is known, that N-aryl-substituted 3,5-di-*tert*-butyl-4-hydroxybenzylamines according to their antioxidizing action surpass considerably N-alkyl-substituted derivatives. Thus aniline, benzidine, *p*-phenylendiamines, and phenylhydrazine do not react with benzylacetate (11) in inert, nonpolar solvents, such as chloroform, carbon tetrachloride, benzene. In acetone and acetonitrile the reactions of aromatic amines with benzyl acetate (11) proceed slowly.

According to Scheme 1.2 (Section 1.3.2) the inertness of aromatic amines and phenylhydrazine in relation to benzylacetate (11) can be connected with their insufficient basicity. In that case the addition (in a reaction mixture) of stronger basis, capable to provide transformation of benzylacetate (11) into methylenequinone (13), but not forming the products of addition to the latter, should promote the process of benzylation of weakly basic nucleophiles.

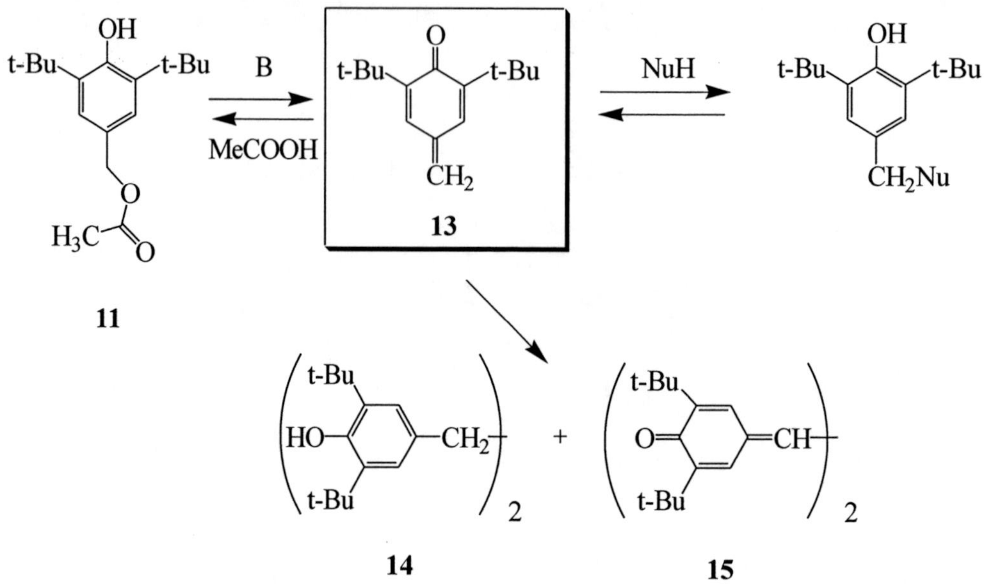

In fact the synthesis of benzyl derivatives of aniline (29) and (30) proceeds easily and selectively in the presence of triethylamine at room temperature [112].

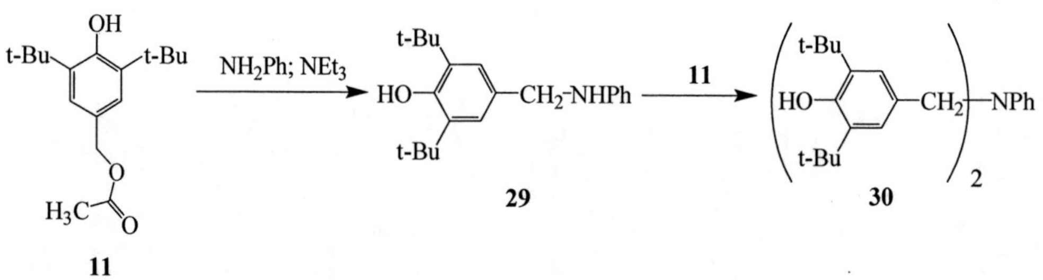

In DMF the interaction of benzylacetate (11) with aniline, benzidine and phenylhydrazine proceeds under the same mild conditions, as in the presence of triethylamine, thus depending on the ratio of reagents it is possible to obtain products of different degree of benzylation [112]. Also it is easy to obtain the product of exhaustive benzylation of hydrazine in dipolar aprotonic solvents. Mono- and di-3,5-di-*tert*-butyl-4-hydroxybenzyl derivative of hydrazine are described as stabilizers, implanting to nitrile rubbers during vulcanization [113].

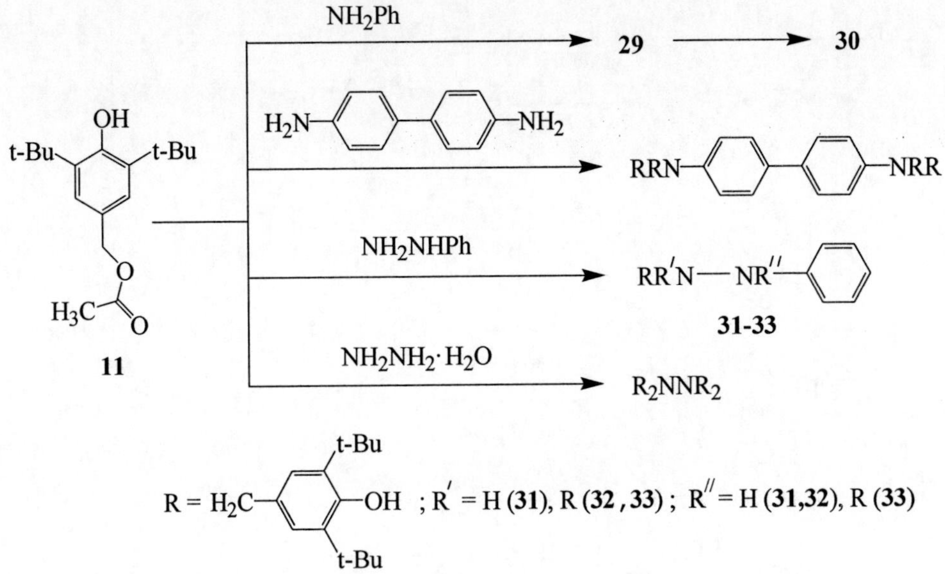

In acetone benzylation of aniline proceeds much more slowly, than in DMF, that corresponds to intensity of the methylenequinone (13) formation from benzylacetate (11) in these solvents.

The application of dipolar aprotonic solvents allows obtaining easily products of benzylation of thioamides, thioureas and hydrazides:

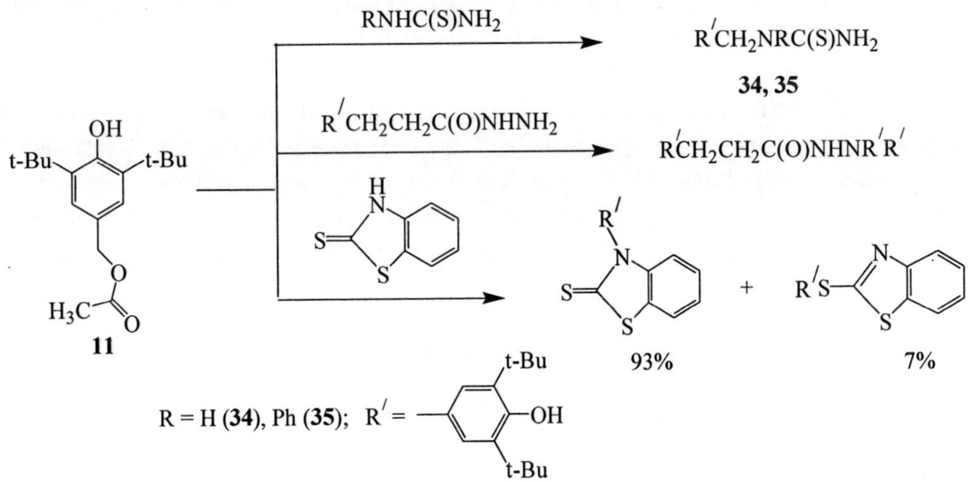

As it is shown above, the quickly established balance of *benzylacetate (11) – methylenequinone (13)* is observed in elementary alcohols. Owing to this circumstance, and also to the fact that alcohols are weaker nucleophiles in comparison with aromatic amines, benzylation of the latter is managed to be conducted selectively in methanol and ethanol medium [105]. So, for example, mono- and di-(3,5-di-*tert*-butyl-4-hydroxybenzyl)aniline can be obtained depending on the ratio of reagents.

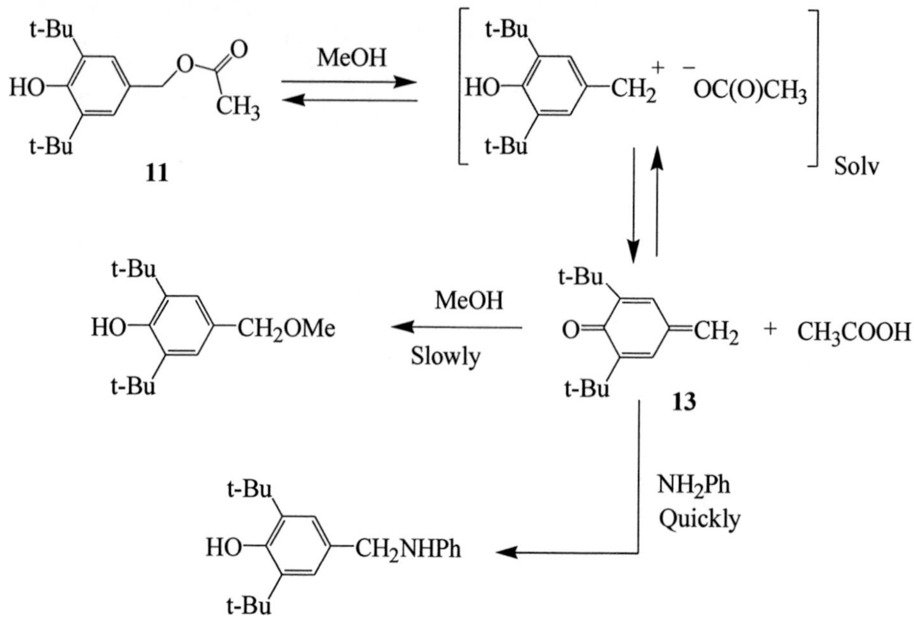

Less nucleophilic thiocarbamides also form products of benzylation under the action of benzylacetate (11) in methanol medium. N-phenyl-N-(3,5-di-*tert*-butyl-4-hydroxybenzyl) thiourea (35) is formed under these conditions giving high yield.

Thus, reactivity of benzylacetate (11) relative to weakly basic nucleophile can be increased by the addition of bases or by using some solvents. In this case reactivity is defined by the activity of used reagents in the reaction with methylenequinone (13).

Among weak nucleophiles the indole derivatives are of great interest from the point of obtaining polyfunctional stabilizers on their basis. Indole derivatives are known as stabilizers of polyvinyl chloride and its copolymers [114], as well as of polymers and lubricating oils used at high temperatures [115]. Functionalisation of indole derivatives by introducing sterically hindered phenol fragments into molecules permits to obtain polyfunctional stabilizers, containing combinations of several reactive centers capable to inhibit free radical chain oxidizing processes.

The tendency of indole to polymerization process in acid medium [116] prevents from obtaining the product of its interaction with benzylacetate (11) through benzyl carbcation (12). All attempts to obtain the product of the condensation of indole with benzylacetate (11) in the presence of bases and in solutions of dipolar aprotonic solvents failed. Obviously, indole has no sufficient reactivity towards methylenequinone (13) owing to low basicity (pK_a –2.4 [117, vol.2, p.232]) and therefore, nucleophilicity. The main components of reaction mixtures formed under the given conditions are diphenylethane (14) and toluylenequinone (15) (the products of side conversions of methylenequinone (13) and not reacted indole.

Only when the reaction is conducted in methanol 4-[3,3-bis-(3,5-di-*tert*-butyl-4-hydroxybenzyl)-3H-indole-2-yl-methylene]-2,6-di-*tert*-butylcyclohexa-2,5-dienon (36) with 15%-yield, as well as methyl ether of 3,5-di-*tert*-butyl-4-hydroxybenzyl alcohol with 20%-yield are isolated from the reaction mixture by fractional crystallization.

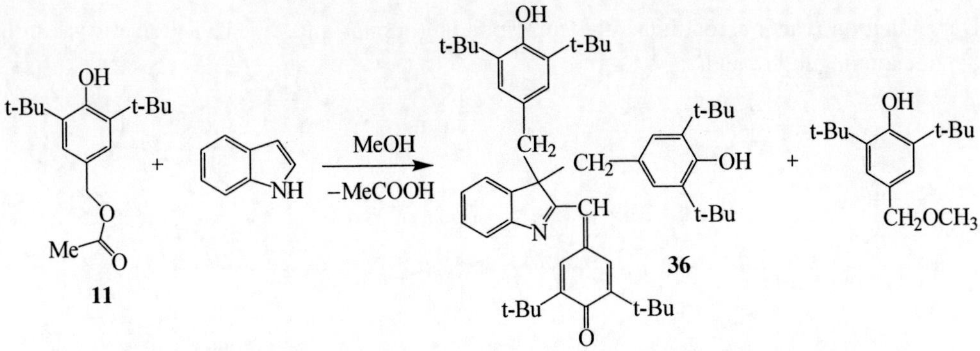

Tryptophan having more distinct nucleophilic properties {pK_a (NH_2) 2.38 [117, vol.5, p.5]} in comparison with indole reacts with benzylacetate (11) in the solution of dimethyl sulfoxide under mild conditions forming mono-benzyl derivatives (37):

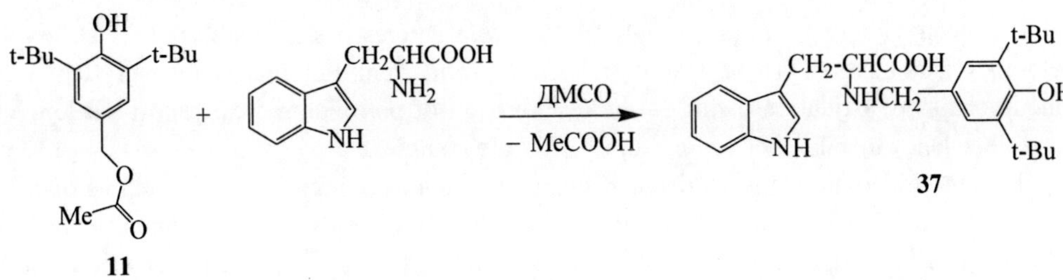

The reaction of benzylacetate (11) with isatin resulting in the formation of compounds (38) proceeds rather easily in dipolar aprotonic solvents (dimethyl sulfoxide).

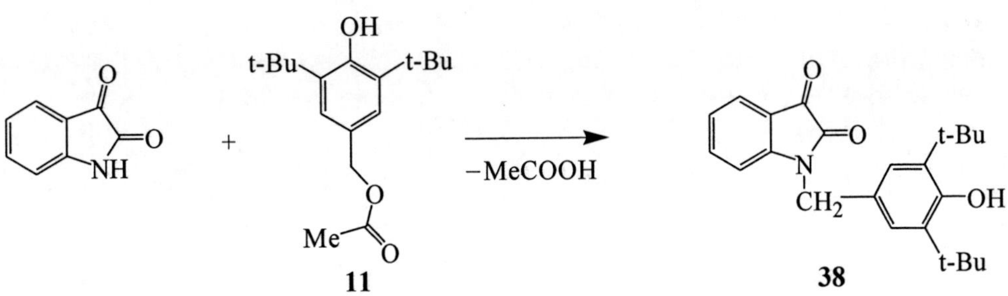

As nucleophilic properties of amine group in isatin are less distinct than in indole it might be assumed that its reaction with benzylacetate (11) starts with the protonation of carbonyl group of methylenequinone (13), formed in the dimethyl sulfoxide solution, by acid NH-proton. The reaction of benzyl acetate (11) with hydrogen sulfide in the solution of dimethyl sulfoxide or in the presence of bases proceeds in similar way.

3,5-di-*tert*-butyl-4-hydroxybenzyl derivatives of acetyl acetone, malonic and acetoacetic ethers are described as efficient stabilizers of polyethylene, polyamides, ABC–plastic, polystyrene [118].

The interaction of the mentioned C–nucleophiles (39-41) and benzoyl acetone (42) with benzylacetate (11) in the presence of triethylamine proceeds at room temperature [112, 119].

Depending on reaction conditions the corresponding mono- and dibenzyl derivatives can be obtained giving high yield.

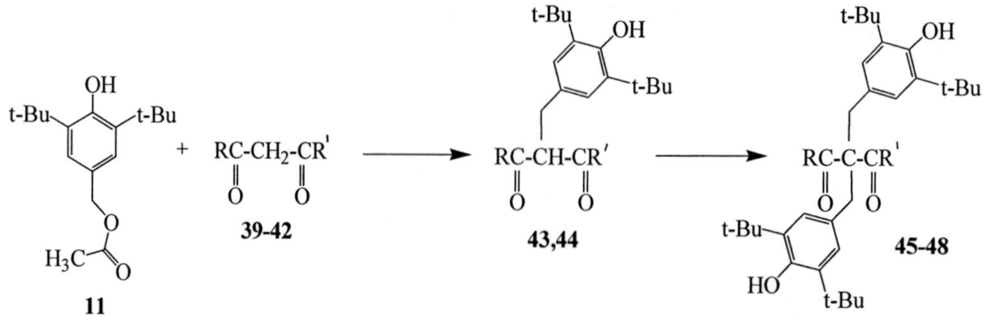

R = R′ = Me (**39,43,45**); R = R′ = OEt (**40,46**); R = Me, R′ = OEt (**41,47**); R = Ph, R′ = Me (**42,44,48**)

Monobenzyl derivatives (43) и (44) are of special interest, as they preserve the ability to keto-enol tautomerism. These stabilizers can perform dual function: they can terminate kinetic chains of oxidation owing to the deactivation of peroxide radicals and bind ions of variable valency metals, such as Fe, Cu, Co, Ni, into complexes.

The interest to sterically hindered phenols compounds, containing sulfur atoms and, in particularly, sulfide fragments is determined by the ability of such compounds to display intramolecular synergism, that is to combine functions of a peroxide radical acceptor and a nonradical hydroperoxide decomposer [120].

One of the most known specimens of sulfur containing phenol stabilizers is bis-(3,5-di-*tert*-butyl-4-hydroxybenzyl) sulfide (49). This product is obtained under mild conditions when benzyl acetate (11) and sodium sulfide interact.

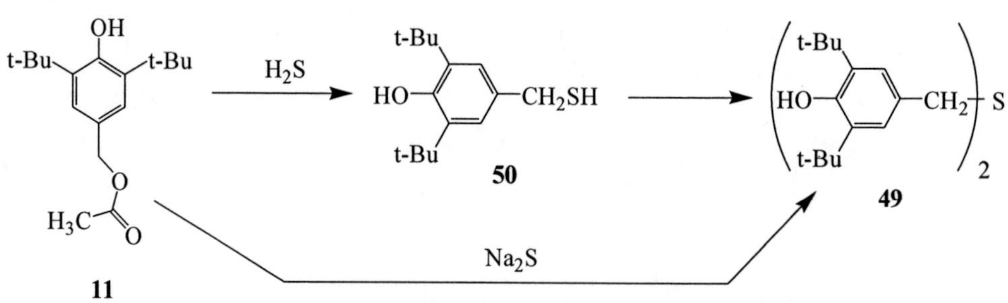

The interaction of benzyl acetate (11) with hydrogen sulfide is of practical interest, it proceeds at room temperature in aqueous DMF and in aqueous acetone in the presence of triethylamine. Depending on the conditions of a reaction both sulfide (49) and mercaptan (50) can be obtained giving high yield. In ethanol the reaction proceeds considerably slower, and in acetone it does not take place at all without additional base. Compound (50) is a stabilizer implanting into rubbers and ABC-plastic.

When base and acid catalysts are absent 3,5-di-*tert*-butyl-4-hydroxybenzylacetate (11) interact only with simple alcohols (methanol, ethanol). At the same time all attempts to use alkaline catalysis in the reaction of benzyl acetate (11) with polyatomic alcohols aiming to get products of their exhaustive benzylation didn't give positive results. Side transformations of methylenequinone (13), formed from benzyl acetate (11) under the action of alkali compete effectively with the target reaction of benzyl acetate (11) alcoholysis.

Thus, it can be stated, that owing to small nucleophilicity alcohols interact insufficiently with 2,6-di-*tert*-butylmethylenequinone (13), it results in the accumulation of products of its side transformations and in incomplete benzylation of polyatomic alcohols.

The reactions of 3,5-di-*tert*-butyl-4-hydroxybenzylacetate (11) with phenols are more successful, they are ionized in greater degree in comparison with alcohols and, therefore, more nucleophilic.

The reaction of benzyl acetate (11) with benzophenone (51) was conducted with the equimolar ratio of reagents at 323 K, in acetone medium, in the presence of triethylamine (2 mole of triethylamine for 1 mole of benzylacetate (11). 2,4-Dihydroxy-3-(3′,5′-di-*tert*-butyl-4′-hydroxybenzyl)benzophenone (52) is obtained from the reaction mixture giving 50% yield.

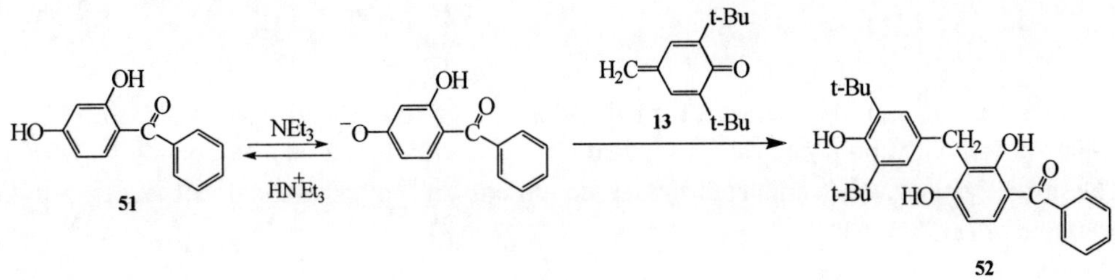

1.3.3.3. Syntheses in the Presence of Acid Catalysts

The selectivity of acid catalyzed reactions of 3,5-di-*tert*-butyl-4-hydroxybenzylacetate (11) with alcohols is defined by the competition between the process of interaction of the latter with intermediate formed benzyl cation (12) and its side transformations. The reversibility of acid catalyzed reactions of benzyl acetate (11) with alcohols may influence negatively the selectivity of compounds (53) forming.

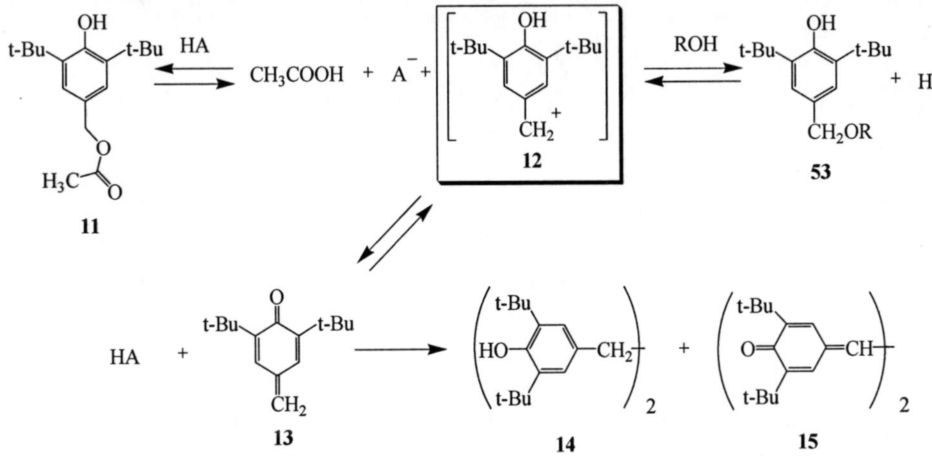

To get individual 3,5-di-*tert*-butyl-4-hydroxybenzyl derivatives of alcohols the reactions of their benzylation should be conducted under conditions of acid catalysis in the surplus of alcohols, that permits to inhibit the forming of methylenequinone (13) and the products of its side transformations. Thus, when benzyl acetate (11) and the surplus of diethyleneglycol interact under conditions of acid catalysis mono-3,5-di-*tert*-butyl-4-hydroxybenzyl ether of diethyleneglycol is formed as the only product.

At the same time conducting comprehensive benzylation of polyatomic alcohols is not possible due to side transformations of benzyl acetate (11). Thus in reactions of compound (11) with 2,2-dimethyl-1,3-propanediol at the ratio of reagents 2:1 respectively the mixture of mono- (54) and diethers (55), as well as products of side transformations of benzyl acetate (11) are formed.

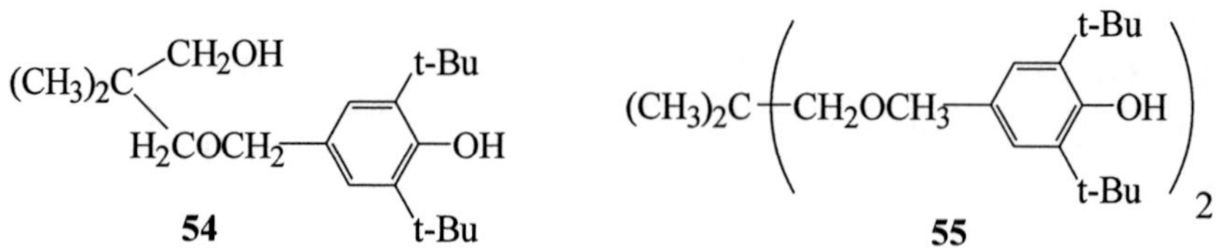

The process of modification of polyethylene glycol by benzyl acetate (11) in order to impart it antioxidation properties is worked out by Bochkov A.F. and Stroganov N.S. [121]. The modification was conducted under conditions of polyethyleneglycol excess in the presence of acid catalyst.

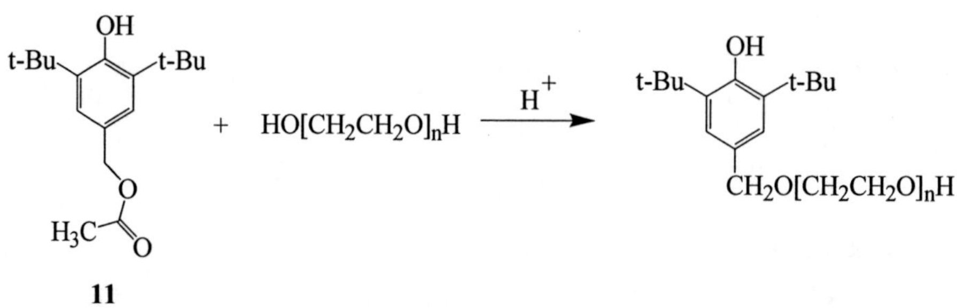

Acid catalyzed benzylation of aromatic compounds by 3,5-di-*tert*-butyl-4-hydroxybenzylacetate (11) gives large possibilities for synthesizing new and advanced technologies of producing existing polyphenol stabilizers [122, 123] (Scheme 1.7).

Compounds (56)-(58) and (60) have considerable inhibiting activity in polypropylene and in rubber vulcanizates [124], the product (59) as a stabilizer of lubricating oils is not inferior to commercial antioxidants, namely, zinc dialkylphosphate and antioxidant BHT [125]. 2,4,6-Tri-(3′,5′-di-*tert*-butyl-4′-hydroxybenzyl)mesitylene (63) is a polymer stabilizer which is known under the trade mark *Ethyl Antioxydant 330* (Ethyl Corporation) and *Ionox 330* (Ciba).

The interest in 2,4,6-tri(3′,5′-di-*tert*-butyl-4′-hydroxybenzyl)resorcinol (61) is conditioned by the fact that the efficiency of antioxidant action of phenol stabilizers in a number of cases is determined by its spatial structure [126]. Thus, the high antioxidant activity of 2,4,6-tri(3′,5′-di-*tert*-butyl-4′-hydroxybenzyl)mesitylene (63) is connected with the features of spatial arrangement of sterically hindered hydroxybenzyl fragments in a molecule of this compound. 2,4,6-Tri-(3′,5′-di-*tert*-butyl-4′-hydroxybenzyl)resorcinol (61) can be considered as a structural analogue of compound (63), having the same spatial structure [127]. A molecule of the compound (61) on the whole, has the structure of «basket» (figure

1.1), the bottom of which is a resorcin ring, and "side walls" consist of 3,5-di-*tert*-butyl-4-hydroxybenzyl fragments, situated on one side from the 'bottom". The edges of the «basket» are a trap for capturing peroxide radicals, and their deactivation by hydroxyl groups takes place here.

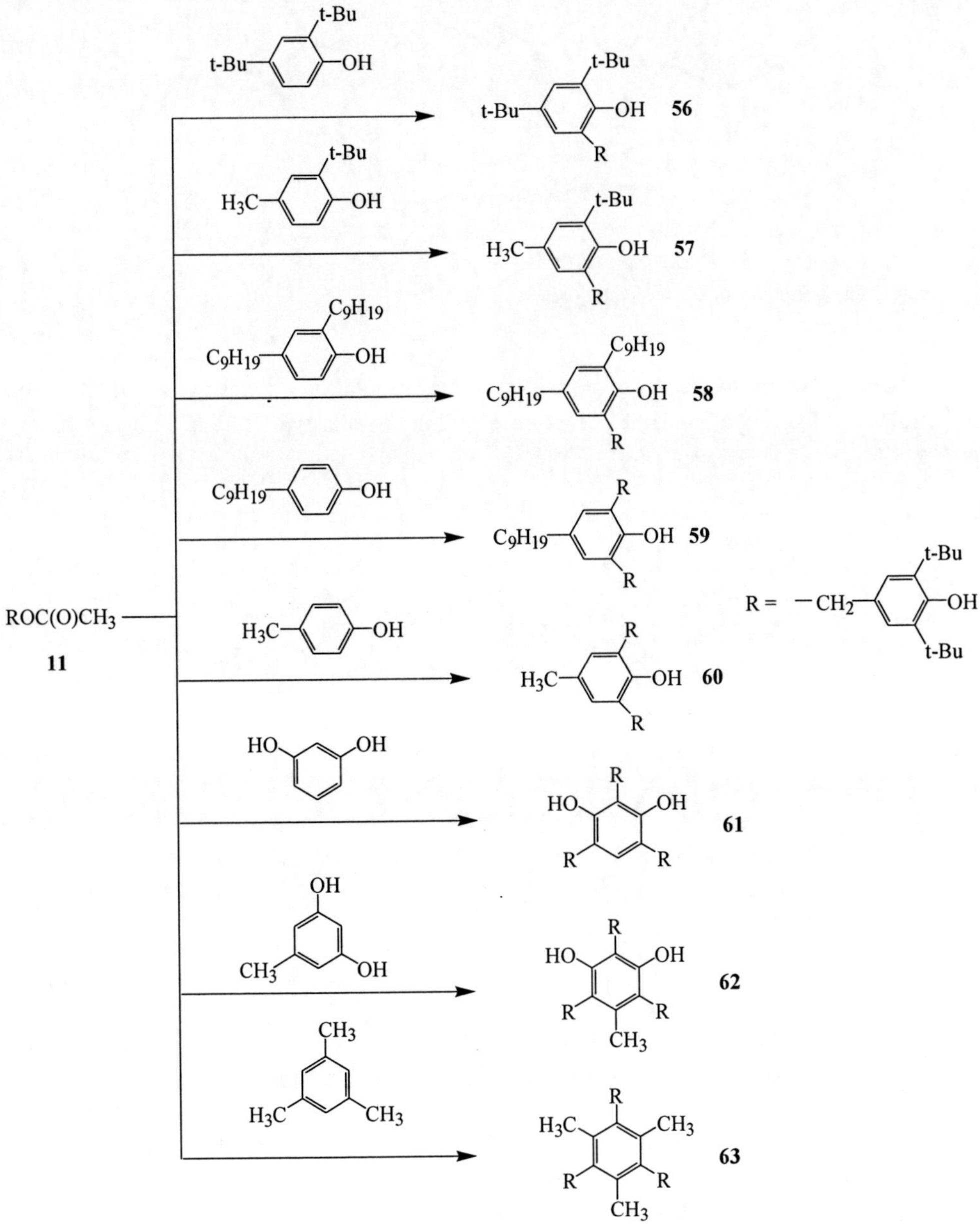

Scheme 1.7.

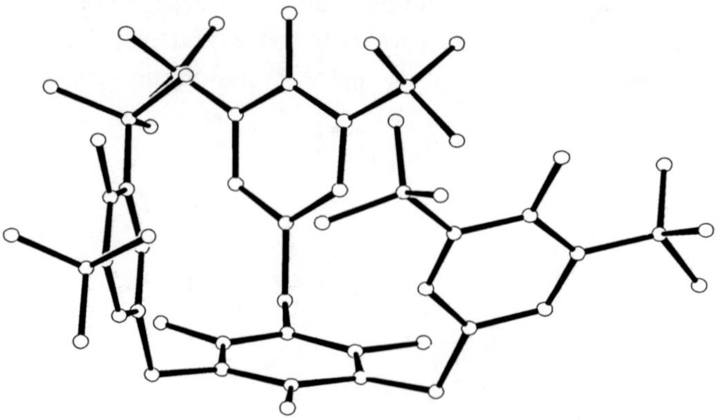

Figure 1.1. The geometry of the molecule (61) in crystal.

Retaining the conformation of a "basket" type in solutions of resorcine (61) promotes intramolecular hydrogen bonds of OH···π type having the energy of 1.48 kJ/mole for one OH group [128].

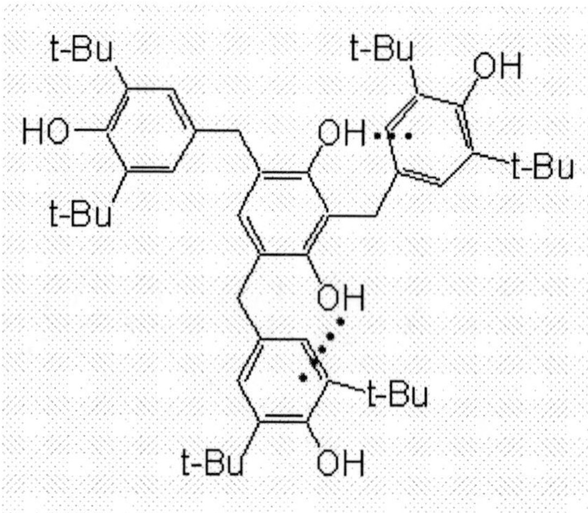

2,4-di-hydroxybenzophenone (51) and its methyl and octyl ethers (64), (65) are known as light stabilizers of polymeric materials [129]. The reactions of 3,5-di-*tert*-butyl-4-hydroxybenzylacetate (11) with these compounds and with 4-formyl resorcine (66) proceed under the same mild conditions as the reaction with alkylphenols [130]. In case of 2,4-dihydroxybenzophenone (51) and formyl resorcinol (66) the products of dibenzylation (67) и (68) are easily formed, which are present in considerable amounts in the reaction mixture even at equimolar ratio of reagents.

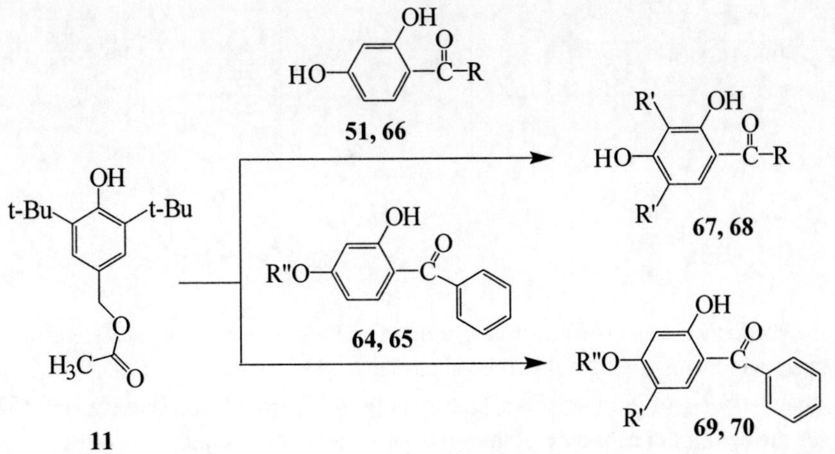

$R = C_6H_5$ (**51,67**), H (**66,68**); $R'' = CH_3$ (**64,69**), C_8H_{17} (**65,70**); $R' = H_2C$– 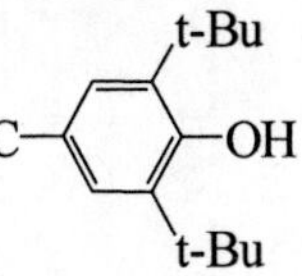

Compounds (67-70) combine a light stabilizing fragment (carbonyl group in *o*-position to hydroxyl one) and a fragment of antioxidant (sterically hindered phenol group) in one molecule. In these compounds there is the best combination of photophysical and photochemical principles of stabilization, as the fragments of sterically hindered phenols and an ultraviolet-absorber can act independently of each other [131].

Hydroquinones form also the products of C-benzylation under conditions of acid catalysis. Thus, in the reaction of hydroquinone with excess benzyl acetate (11) 2,3,5-tris(3′,5′-di-*tert*-butyl-4′-hydroxybenzyl)hydroquinone (71) is obtained:

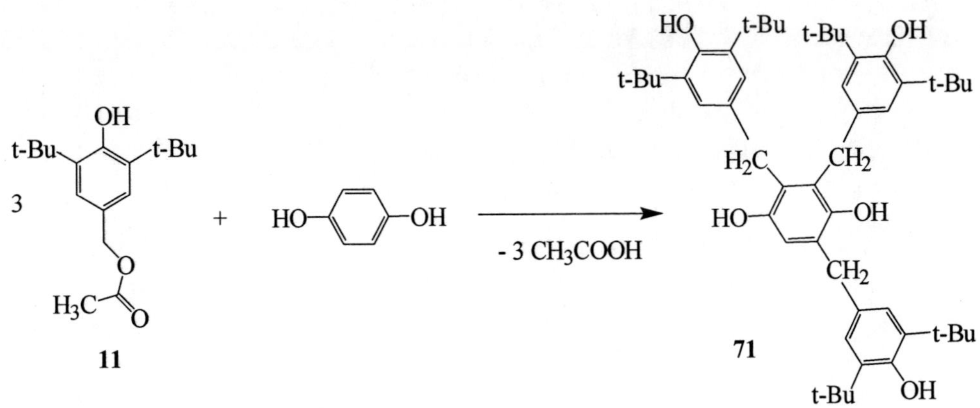

In all described above acid catalyzed reactions of benzyl acetate (11) sulfuric and chloric acids were used as an acid catalyst. Owing to oxidation-reduction processes in the presence of these acids it is impossible to conduct reactions of benzylation of aromatic compounds containing electrondonor fragments, e.g. amine groups.

In the solution of formic acid benzyl acetate (11) converts quickly into the corresponding benzyl formate (72).

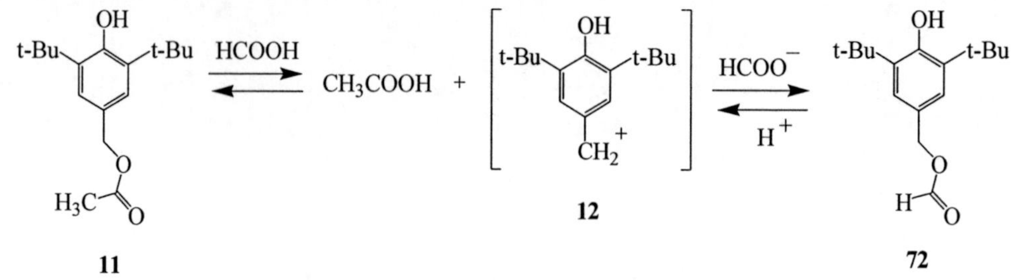

Therefore, formic acid has sufficient acid properties for protonation of acetate group and removal of acetic acid from a molecule of the compounds (11).

In fact, the synthesis of 2,4,6-tris(3′,5′-di-*tert*-butyl-4′-hydroxybenzyl)mesitylene (63) in medium of formic acid, in the absence of mineral acids can be performed.

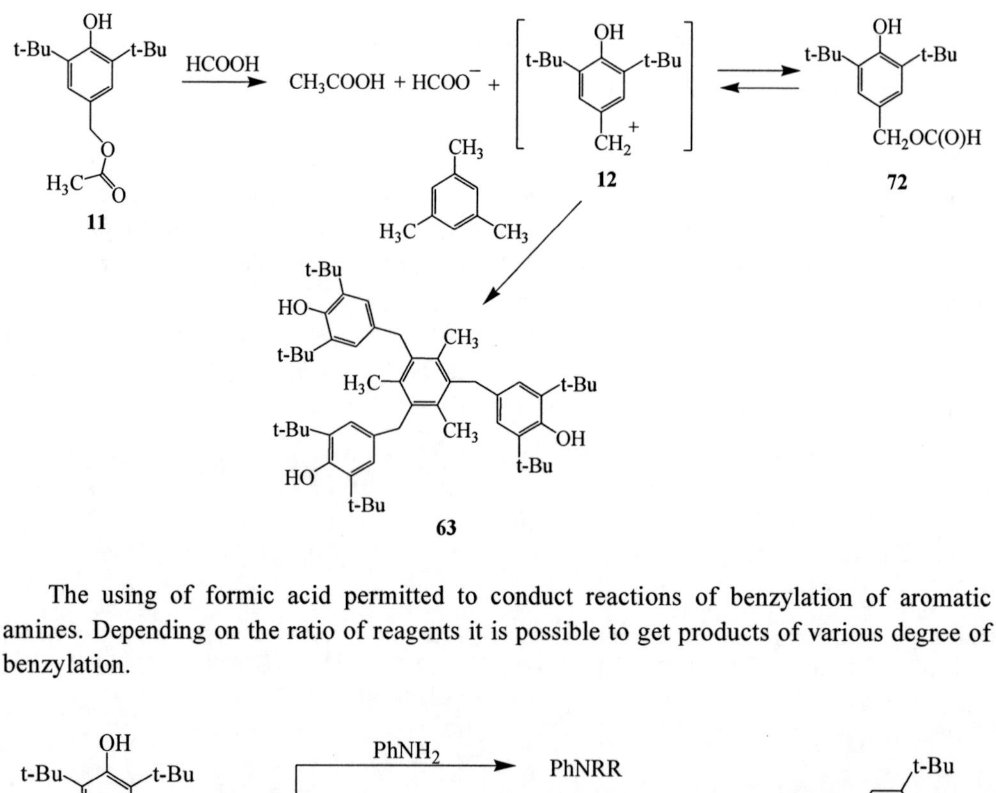

The using of formic acid permitted to conduct reactions of benzylation of aromatic amines. Depending on the ratio of reagents it is possible to get products of various degree of benzylation.

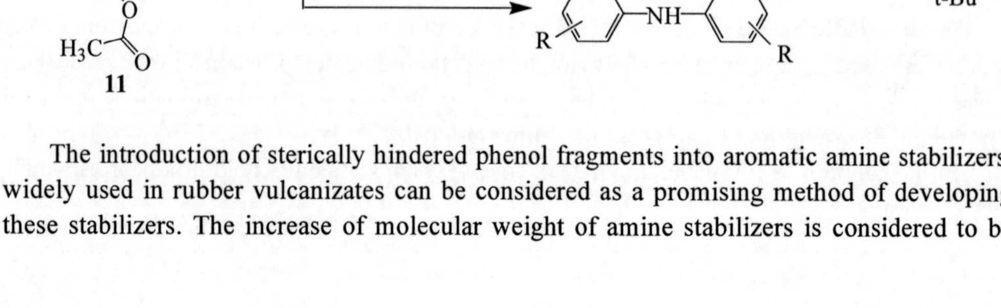

The introduction of sterically hindered phenol fragments into aromatic amine stabilizers, widely used in rubber vulcanizates can be considered as a promising method of developing these stabilizers. The increase of molecular weight of amine stabilizers is considered to be one of the ways of decreasing their diffusion capacity and, therefore, decreasing the pollution of the environment and prolongation of protective action [132]. Besides, the effects of intramolecular synergism are possible in systems containing simultaneously amine and sterically hindered phenol groups [133].

The interaction of aromatic amine stabilizers, namely, phenyl-β-naphthylamine and N-phenyl-N′-isopropyl-*p*-phenylenediamine with benzyl acetate (11) in formic acid results in forming corresponding mixtures of isomer products of benzylation:

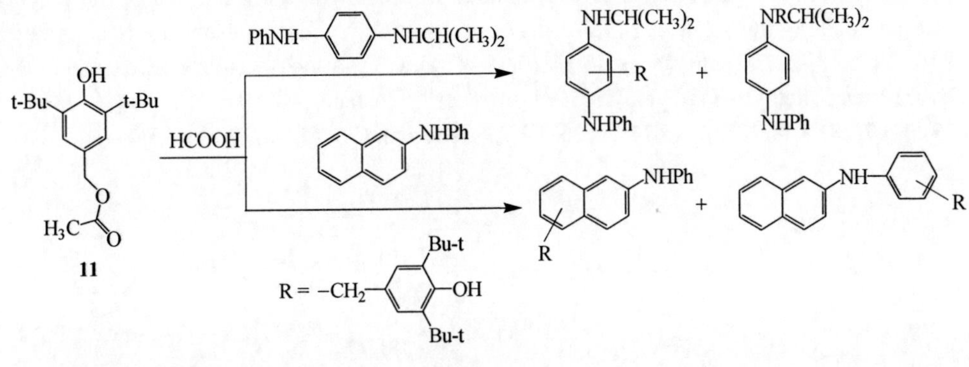

High benzylation activity of the system *benzyl acetate* (11) – *formic acid* allows to introduce 3,5-di-*tert*-butyl-4-hydroxybenzyl fragments into molecules of such weak bases, as thioamides, thiocarbamides, carbamides and even isatin.

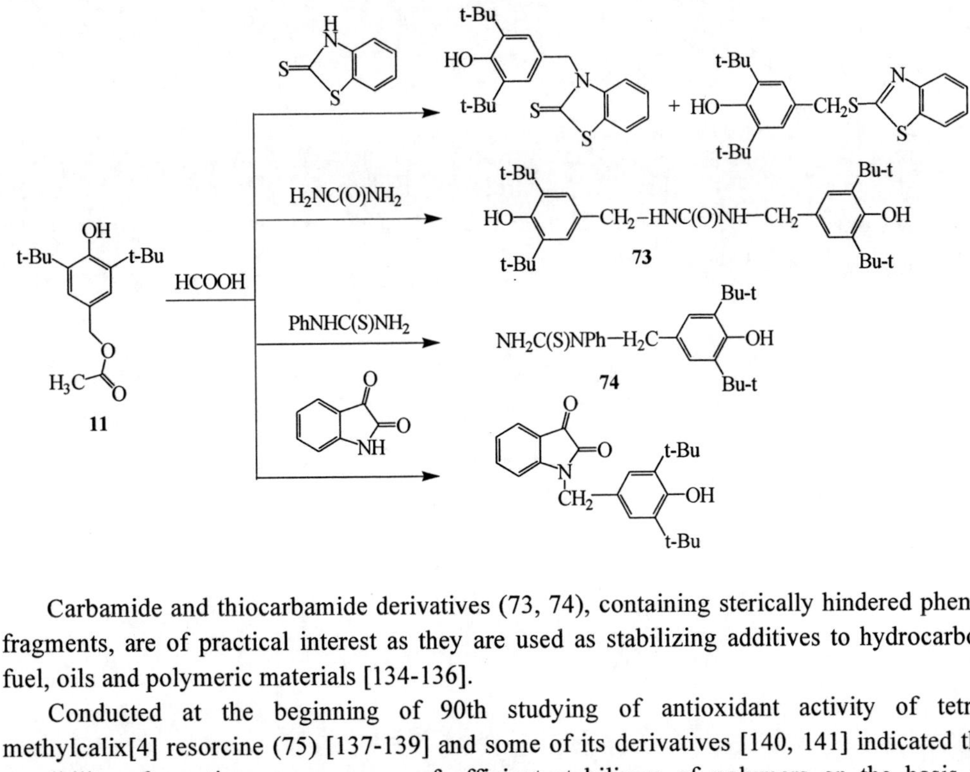

Carbamide and thiocarbamide derivatives (73, 74), containing sterically hindered phenol fragments, are of practical interest as they are used as stabilizing additives to hydrocarbon fuel, oils and polymeric materials [134-136].

Conducted at the beginning of 90th studying of antioxidant activity of tetra-methylcalix[4] resorcine (75) [137-139] and some of its derivatives [140, 141] indicated the possibility of creating a new group of efficient stabilizers of polymers on the basis of calixarenes. Calixarene matrix provides wide possibilities for obtaining polyfunctional stabilizers capable to inhibit thermal-oxidative destruction of polymer according to various mechanisms. On the basis of calixarenes it is possible to get stabilizers having various combination of polar and nonpolar structural fragments (by hydrophilic – lipophilic balance). Thus, stabilizers on the basis of calixarenes should combine the advantages of polyfunctional stabilizers and oligomerous stabilizers characterized by increased compatibility with polymer.

The reaction of methylcalix[4]resorcinol (75) with 3,5-di-*tert*-butyl-4-hydroxy-benzylacetate (11) in the presence of chloric acid proceeds with the formation of the reaction mixture containing 70% 2,4,6-tris(3,5-di-*tert*-butyl-4-hydroxybenzyl)resorcinol (61) and 30% 4,6,10,12,16,18,-22,24-octahydroxy-5,11,17,19-tetrakis(3,5-di-*tert*-butyl-4-hydroxybenzyl)-2,8,14,20-tetra-methylpenta-cyclo[19.3.1.1^{3,7}.1^{9,13}.1^{15,19}]octacosa-1(25),3,5,7(28),9,11,13(27),15,17,19(26),21,23-dodecaen (76) [142].

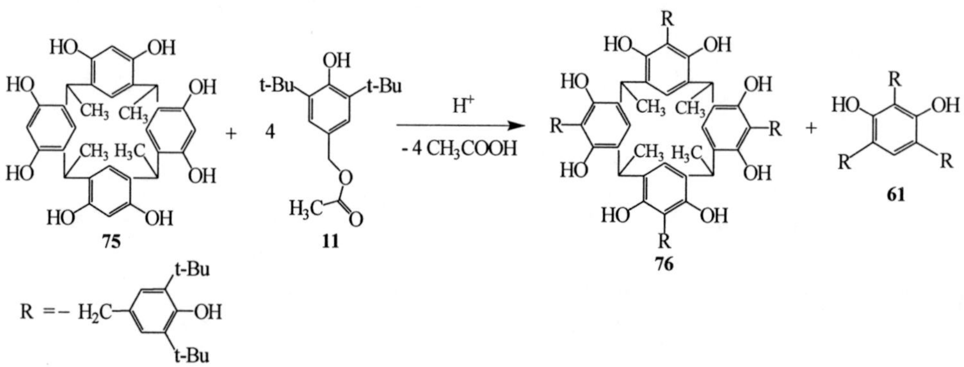

The using of formic acid allows to increase the content of calixarene (76) in the reaction mixture to 70% [143].

The investigation of supramolecular structure of the compound (76) by methods of one- and two-dimensional NMR ^{1}H and ^{13}C spectroscopy and infra-red spectroscopy indicated that it had the conformation ‘cone”. In solutions such conformation is stabilized by intramolecular hydrogen bonds OH···O and OH·of π - type.

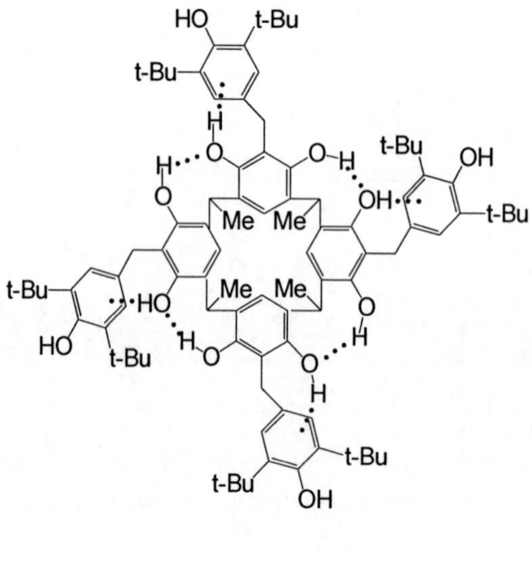

As a result of the reaction of tetraethyl-, tetrapropyl-, tetrapentylcalix[4]resorcinols (77-79) with benzyl acetate (11) new tetrabenzylized derivatives of calix[4]resorcinols (80-82) are obtained [143].

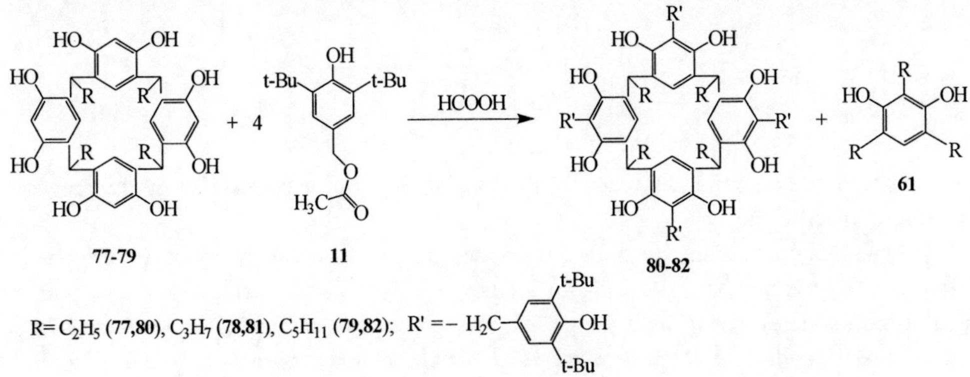

The further investigation of the reaction of calix[4]resorcinols benzylation indicated that the amount of introduced 3,5-di-*tert*-butyl-4-hydroxybenzyl group depends considerably on the nature of alkyl substitute on the bottom "band" of calixarene matrix. The interaction of calixarenes (83-85) containing heptyl, octile and nonyl fragments on the bottom "band" with 3,5-di-*tert*-butyl-4-hydroxybenzylacetate (11) in the presence of formic acid results in the formation of compounds (86-88) containing only two 3,5-di-*tert*-butyl-4-hydroxybenzyl groups.

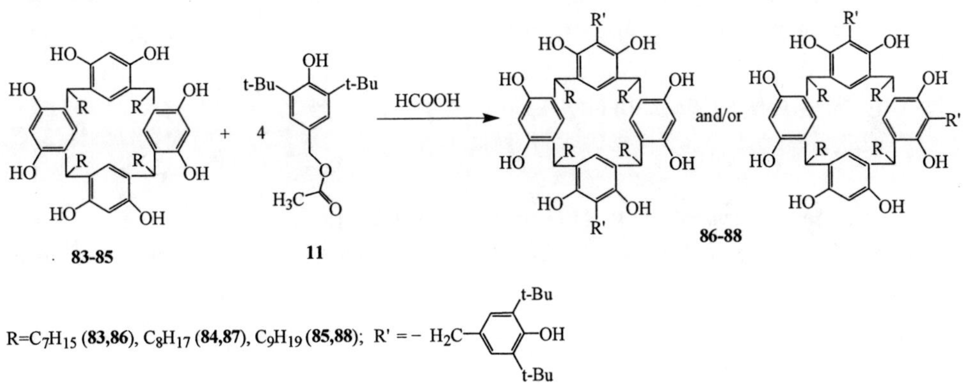

Considering calix[4]resorcinol (89), containing an undecyl fragment on the bottom "band", monobenzylized calixarene (90) is formed as the only product in the reaction with 3,5-di-*tert*-butyl-4-hydroxybenzylacetate (11).

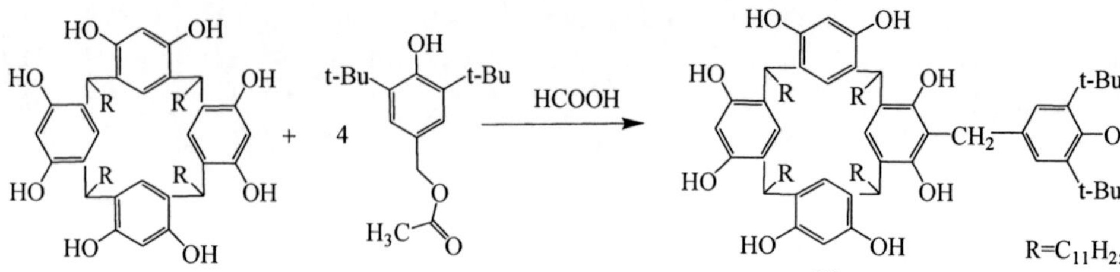

89 **11** **90**

R=$C_{11}H_{25}$

It should be noted that the reaction of benzylation is not accompanied by the process of opening of macrocycle.

Phosphorus containing cavitands (91-94) are obtained by phosphorylation of tetra-(3,5-di-*tert*-butyl-4-hydroxybenzyl)calix[4]resorcines (76) and (80-82) by methyl dichlorphosphonates giving high yield. The interaction was conducted in acetonitrile in the presence of triethylamine at temperature 343 K within 18 hours, reagent ratio being 1:4:4.

76, 80-82 **91-94**

R = CH_3 (**76,91**), C_2H_5 (**80-92**), C_3H_7 (**81,93**), C_5H_{11} (**82,94**), R' =

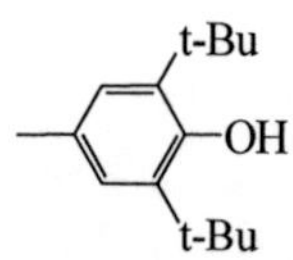

Acyclic derivatives of calix[4]resorcinol (95-97) are obtained by the reaction of calixarene (75), (77) and (79) with 3,5-di-*tert*-butyl-4-hydroxybenzyldichlorphosphonate [144].

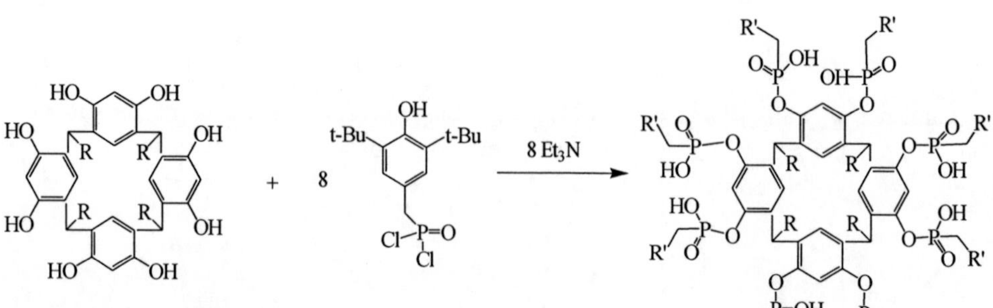

75, 77, 79

R=CH_3 (**75,95**), C_2H_5 (**77,96**), C_5H_{11} (**79,97**); R' =

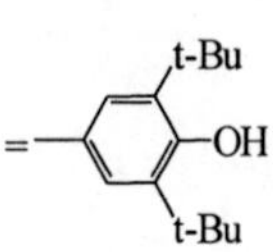

95-97

1.3.4. Synthesis on the Basis of 3,3',5,5'-Tetra-Tert-Butyl-4,4'-Diphenoquinone

3,3',5,5'-Tetra-tert-butyl-4,4'-diphenoquinone (98) is a product of oxidation of 2,6-di-tert-butylphenols which is conducted in the presence of alkaline [145-147] or metal complex [148-160] catalysts. The chemistry of diphenoquinone (98) is presented nearly solely by the reactions of its reduction to 3,3',5,5'-tetra-tert-butyl-4,4'-dihydroxybiphenyl (99) (Ethyl Antioxidant 712).

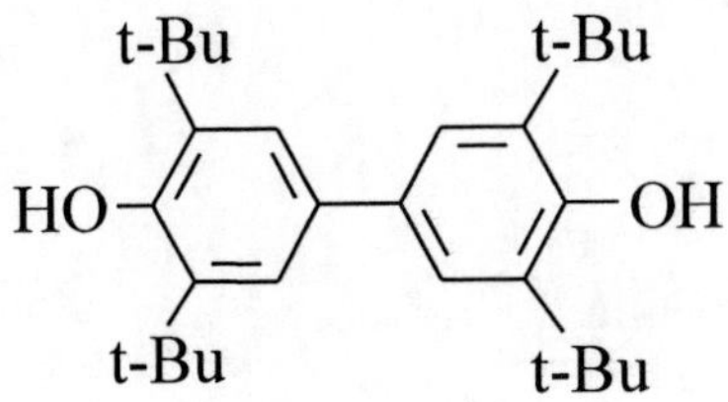

Ethyl Antioxidant 712

It should be noted that quinones are often used as mild dehydrating agents in the synthetic organic chemistry. At that such quinones (active in the reactions of dehydrogenation) as tetrachlorbenzoquinone-1,2,3,4-dichlor-5,6-dicyanobenzoquinone, tetrachlorbenzoquinone-1,2 are inclined to side reactions of nucleophylic substitution, addition and diene synthesis. Diphenoquinone (98) due to being sterically hindered and the absence of halogen atoms is deprived of these shortcomings to a great extent. The low inclination of diphenoquinone (98) to reactions of addition and substitution provides high selectivity of this compound in oxidation–reduction reactions.

The most important commercial use of 3,3',5,5'-tetra-tert-butyl-4,4'-diphenoquinone (98) is the synthesis of 3,3',5,5'-tetra-tert-butyl-4,4'-dihydroxybiphenyl (99) by the reaction of oxidative dehydrogenation of 2,6-di-tert-butylphenols. Bisphenol (99) is recommended for the stabilization of many types of polymeric materials, it is used as an intermediate product in the synthesis of 4,4′-dihydroxydiphenyl [161-164], applied for producing engineering plastics, it is patented as component of physiologically active substance [165]. Diphenoquinone (98) was used as a dehydrating agent of other sterically hindered phenols. The mixtures of stabilizers were obtained, one component of them is bisphenols (99) [166-169]. Thus, in the reaction of diphenoquinone (98) with bis(3,5-di-tert-butyl-4-hydroxyphenyl)methane (22) alongside with bisphenol (99) hydro-galvinoxyl (25) is formed capable to interact both with peroxide and alkyl radicals and to terminate repeatedly kinetic chains of oxidation [170]. Dehydrogenation of sterically hindered phenols by diphenoquinone (98) is conducted at temperature 423-473 K in an autoclave (Scheme 1.8).

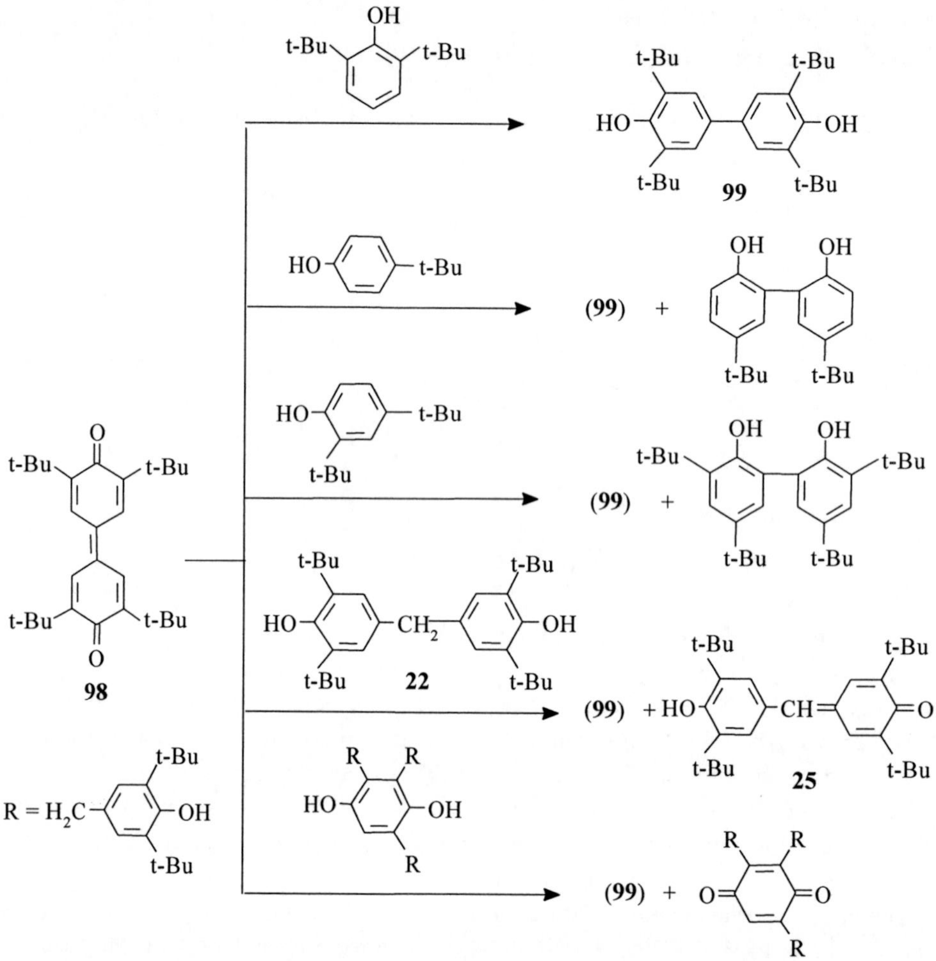

Scheme 1.8.

Diphenoquinone (98) reacts with electron-donor compounds, which have bonds hydrogen – heteroatom, under milder conditions. Thus, when mercaptophenol (50) was dehydrogenated bis(3,5-di-tert-butyl-4-hydoxyphenyl)disulfide (49) was obtained.

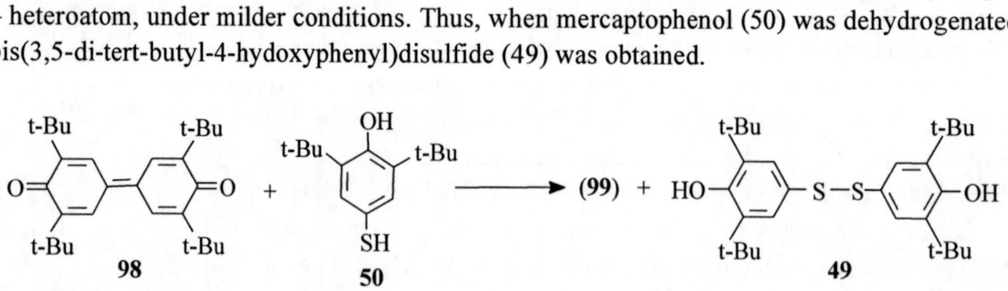

The dehydrogenation of Mannich bases (100) and (101) results in the formation of corresponding Shiff bases.

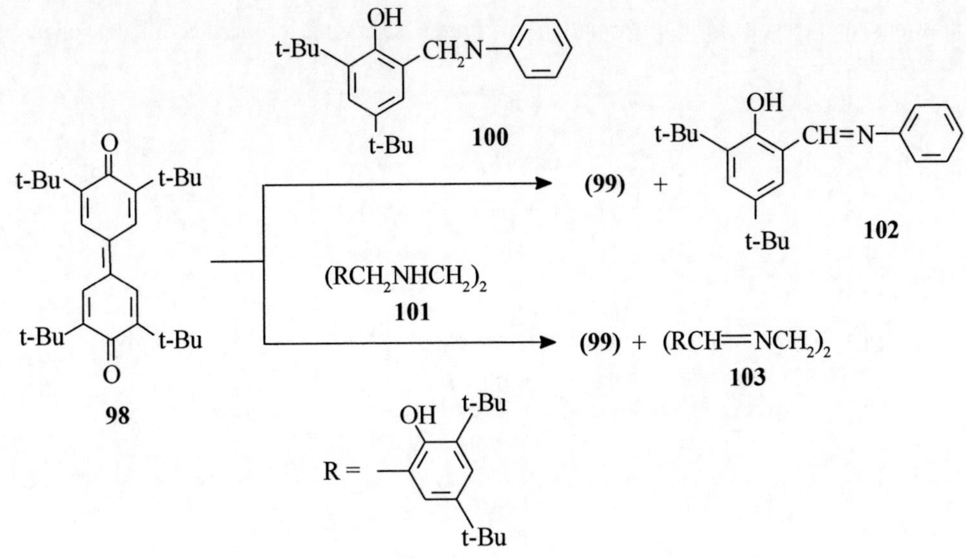

Compounds (102), (103) form complexes having high thermal stability with variable valency metals [171, 172]. Compounds containing in their structure simultaneously azomethine and sterically hindered phenols fragments can perform twofold function. On the one hand, they are able to deactivate variable valency metals due to complexation, these metals are known to be able to act as initiators of thermal-oxidative destruction of polymer, on the other hand they are able to terminate kinetic chains of oxidation. Besides, similar compounds can act as acceptors of hydrogen chloride in chlorinated polymers such as polyvinylchloride [173].

It should be taken into consideration that sterically hindered hydroxybenzaldehydes can be used not always for obtaining the corresponding azomethins. For example, 3,5-di-tert-butyl-2-hydroxybenzaldehyde (104) is difficult to be obtained, unlike benzaldehyde (105) it can't be obtained in the reaction of 2,4-di-tert-butylphenol with formylating agents. According to Reimer-Tieman and Duff reactions and formylation by o-formic ether the formation of other products from 2,4-di-tert-butylphenol takes place [174, 175]. On the other hand, the reactions of 3,5-di-tert-butyl-4-hydroxybenzaldehyde (106) [176] with amines do not always proceed smoothly. So some aromatic amines don't react with it or give insignificant yield of corresponding azomethins [177].

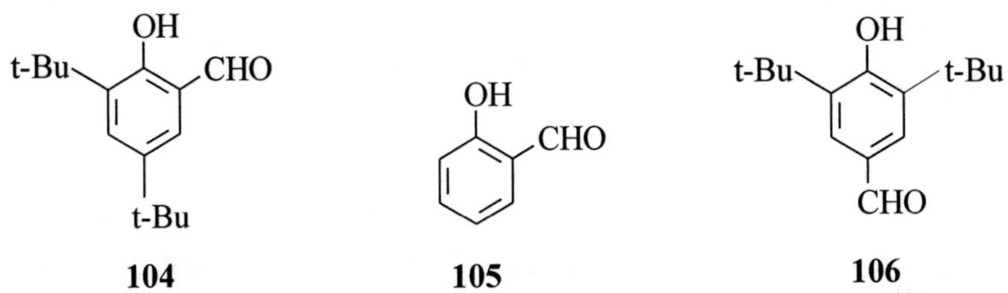

The reactions of dehydrogenation of 2-mercaptobenzothiazole and dialkyldithiophosphorus acids by diphenoquinone (98) [178, 179] proceed under rather mild conditions and result in the formation of products having properties of vulcanization accelerators.

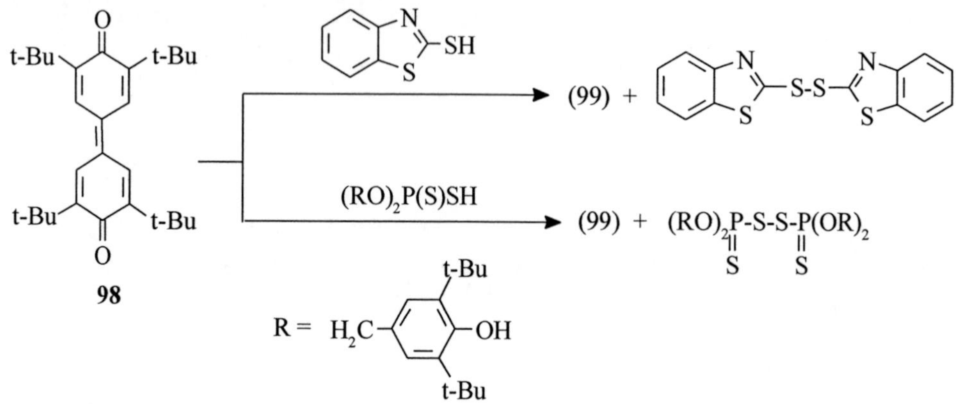

1.3.5. Syntheses of O,O'-Dialkyl(Aryl)-(3,5-Di-Tert-Butyl-4-Hydroxybenzyl)Phosphonates

Using phosphonates containing fragment of sterically hindered phenols as thermal- and color-stabilizers of polymers and lubricating oil stimulated the development of simple and manufacturable methods of their synthesis.

The methods of obtaining phosphonates of such type are known, however carrying out the reactions can be multi-stage, it requires the presence of a catalyst and repeated excess of one of the reagents or it doesn't result in sufficiently high yield of the target product.

The most reactive initial reagents for the synthesis of phosphonates are compounds, containing 3,5-di-*tert*-butyl-4-hydroxybenzyl fragment in their molecules.

When 3,5-di-*tert*-butyl-4-hydroxybenzylacetate **(11)** interacts with full esters of phosphorus acid the corresponding phosphonates are obtained in the melt (363-368 K) at equimolar ratio of reagents with the yield to 80% [180].

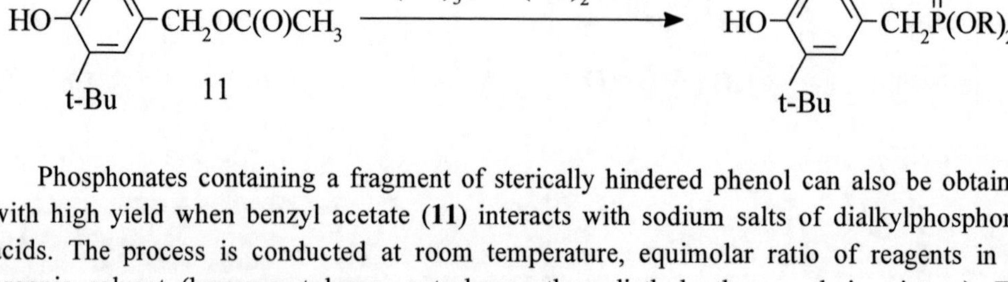

Phosphonates containing a fragment of sterically hindered phenol can also be obtained with high yield when benzyl acetate **(11)** interacts with sodium salts of dialkylphosphorus acids. The process is conducted at room temperature, equimolar ratio of reagents in an organic solvent (benzene, toluene, petroleum ether, diethyl ether or their mixture). The constants of phosphonates, obtained by different methods coincide [181-183].

The reactions of 3,5-di-*tert*-butyl-4-hydroxybenzyldimethylamine (**10**) with dialkylphosphorus acids proceeds in the melt or in the excess of the corresponding dialkylphosphites. The yield of the product makes 76-87% [184, 185].

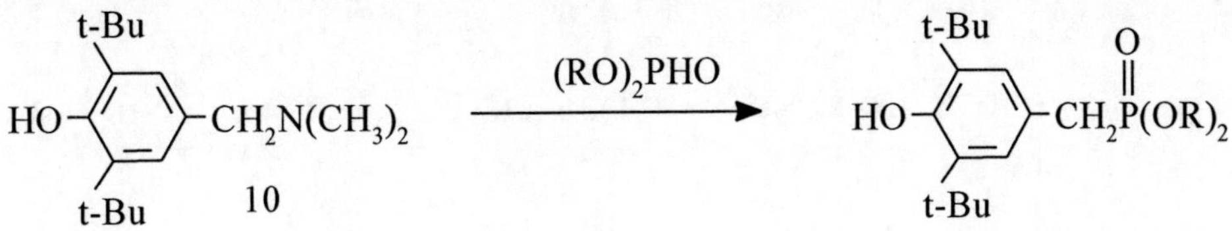

The reaction of trialkylphosphites with methyl ether of 3,5-di-*tert*-butyl-4-hydroxybenzyl alcohol (107) (an intermediate product in the synthesis of mononuclear alkylphenols) proceeds in toluene in the presence of anhydrous aluminium chloride at 383 K within 10 hours giving yield of 77-94% [186].

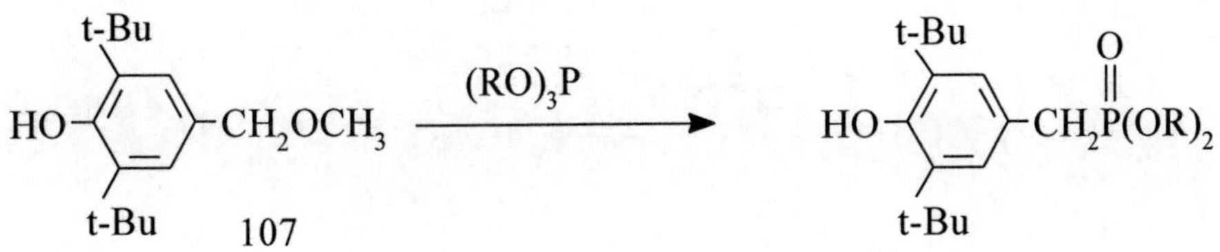

The suitable methods of synthesis of new functionalized phosphorus containing derivatives of 2,6-di-*tert*-butyl-4-methylphenols are developed, they are of interest as efficient ligands and antioxidants. Thus, phosphoniums (**108**) and phosphonates (**109**) are obtained by addition reaction of bis(trimethylsiloxy)phosphine (**110**) and trimethylsilylphosphites (**111**) to 3,5-di-*tert*-butyl-4-hydroxybenzaldehyde (**112**) [187].

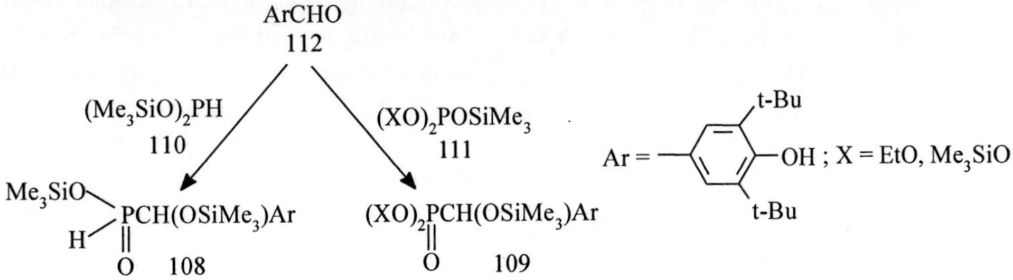

Diphosphonates (113) and (114) and functionalized diphosphonates (115) and (116) were obtained by the interaction of trimethylsilyl esters of a number of acids of trivalent phosphorus with 3,5-di-*tert*-butyl-4-hydroxybenzaldichloride (117) and 3,5-di-*tert*-butyl-4-hydroxybenzoylchloride (118) [187].

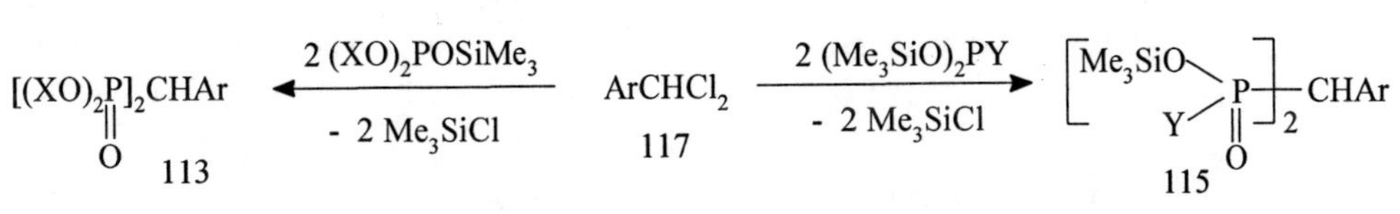

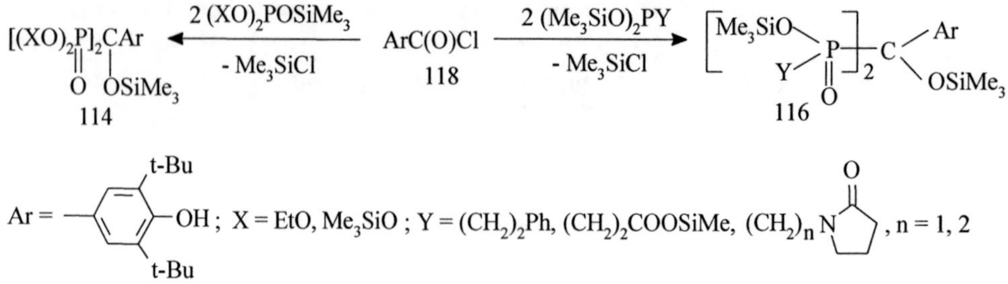

Compounds (108), (109), (113-116) react easily with methanol or sodium methylate in methanol, producing new water-soluble acids or their derivatives.

3,5-Di-*tert*-butyl-4-hydroxybenzylidenedichloride (119) reacts with esters of P(III) and P(V) acids according to scheme of Arbusov reaction resulting in the formation of diphosphorylated sterically hindered phenols with high yields [188].

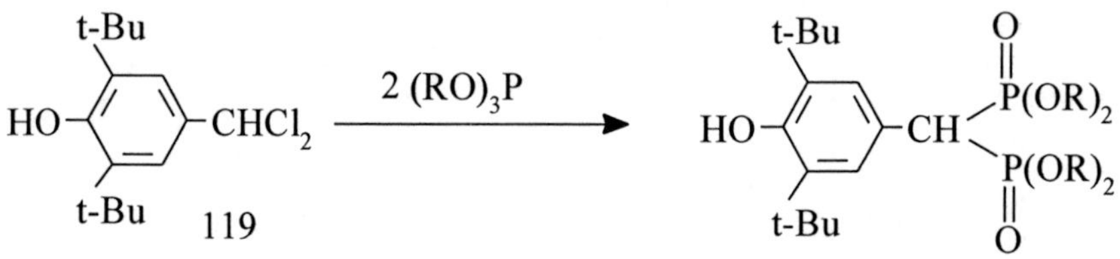

Independently of the ratio of reagents (1:1 or 1:2) the reaction results in the formation of the products of disubstitution only.

* * *

By now a large number of element (N,P,S) containing compounds of various classes, types, structures and complexity (e.g. trialkyl- and dialkyl(aryl)phosphites, di-, tri- and polyphosphites, derivatives of dithiophosphoric and dithiophophonic acids, aryl- and benzoylthiocarbamides and their metal complexes, sterically hindered phenols, etc) have been obtained as potential polyfunctional stabilizers for polymeric materials with the help of main synthetic procedures and methods described in this chapter.

The presence of such typical nucleophilic heteroatoms as N, P, S (individually or in combination) in molecules of stabilizers of polymers was shown to be an important prerequisite of their high reactivity in the course of stabilization.

It is established that introduction of sterically hindered fragments into molecules of N,P,S-containing stabilizers may enhance their antioxidative activity. 3,5-Di-*tert*-butyl-4-hydroxybenzylacetate, N,N-dimethyl-3,5-di-*tert*-butyl-4-hydroxybenzylamine, 1,3-dithia-2,4-dithioxophos-phetane etc are commonly used as universal benzylation agents. This method made it possible to synthesize a wide range of sterically hindered polyfunctional stabilizers.

The development of synthesis methods of the certain compounds (a kind of prostabilizers), which are capable to generate novel efficient stabilizers under the conditions of inhibited polymer thermooxidation, assumed to be fruitful.

Some original approaches to the synthesis of new promising polyfunctional stabilizers on the basis of calixarenes and their derivatives were determined.

The abovementioned examples make it possible to suppose how many new stabilizers, including elementcontaining polyfunctional ones can be synthesized by using the information of this chapter.

REFERENCES

[1] Kirby, A.J.,Warren, S.G. *The Organic Chemistry of Phosphorus*. Amsterdam: Elsevier; 1967.

[2] Hudson, R.F. *Structure and Mechanism in Organo-phosphorus Chemistry*. London – New-York: Academic Press.; 1969.

[3] Hunter, B.A., Barnhart, R.R., Provost, R.L. – *Ind. Eng. Chem.*, 1954, vol. 46, №7, p. 1524-1530.

[4] Voigt, J. *Die Stabilisierung der Kunsfstoffe Gegen Licht und Wärme*. Berlin: Springer-Verlag; 1966.

[5] Kirpichnikov, P.A., Mukmeneva, N.A. In: *Starenie i stabilizatsiya polimerov (Ageing and Stabilization of Polymers)*, Ed. by Kuzminskii A.S. Moscow: *Nauka;* 1966; p. 165-195.

[6] Arbuzov, A.E. *Izbrannye trudy po khimii fosfororganicheskikh soedinenii (The Sselected Works on Chemistry of Organophosphorus Compounds)*. Moscow: *Nauka;* 1976.

[7] Nifant'ev, E.E. *Khimiya fosfororganicheskikh soedinenii (Chemistry of Organophosphorus Compounds)*. Moscow: Moscow State University Press.; 1971.

[8] Purdela, D., Vylchanu, R. *Khimiya organicheskikh soedinenii fosfora (Chemistry of Organic Compounds of Phosphorus)*. Moscow: *Khimiya*; 1972.

[9] Kirpichnikov, P.A., Angert, L.G., Kuzminskii, A.S. – *Zh. Prikl. Khim.*, 1963, vol.34, №10, p. 2270-2276.

[10] Angert, L.G., Kirpichnikov, P.A., Kuzminskii, A.S. et al. – *Zh. Prikl. Khim.*, 1964, vol.35, №3, p. 856-857.

[11] Tokareva, L.G., Mikhailov, N.V., Rosova, N.N. – *Khim. Volokna,* 1962, №3, p. 23-25.

[12] Tokareva, L.G., Kirpichnikov, P.A., Mikhailov, N.V. – *Khim. Volokna,* 1968, №3, p. 214-216.

[13] Kirpichnikov, P.A., Kolyubakina, N.C., Mukmeneva, N.A. et al. – *Vysokomol. Soed.,* 1970, vol. B12, №3, p. 189-192.

[14] Mironova, Z.I., Smirnov, L.N., Rubezhova, T.E. et al. –*Khim. Volokna,* 1975, №6, p. 32.

[15] Minsker, K.S, Kirpichnikov, P.A., Mukmeneva, N.A. et al. – *Vysokomol. Soed.*, 1968, vol. A10, №11, p. 2500-2512.

[16] Mikhailov, N.V., Tokareva, L.G., Burovchenko, K.K. et al. – *Vysokomol. Soed.*, 1962, vol.4, №8, p. 1186-1192.

[17] Kovarskaya, B.M., Tanunina, P.M., Levantovskaya, I.I. et al. – *Plast. Massy,* 1965, №2, p.15-17.

[18] Kirpichnikov, P.A., Kadyrova, V.Kh., Levantovskaya, I.I. et al. – *Plast. Massy,* 1971, №1, p. 44-46.

[19] Arbusov, A.E., Valitova, F.G. – Izv. AN SSSR, *Otdel Khim. Nauk,* 1940, p. 529-544.

[20] Gurvich, Ya.A, Kirpichnikov, P.A., Tsyryul'nikov, N.V. et al. – *Khim. Prom.,* 1965, vol.2, p. 100-101.

[21] Kirpichnikov, P.A., Kuzminskii, A.S, Popova, L.M. – *Trudy Kaz. Khim.-Tekhnol. Inst., Ser. Khim. Nauki,* 1962, vol. 30, p. 47-51.

[22] Kirpichnikov, P.A., Popova, L.M. – *Zh. Obshch. Khim.,* 1965, vol. 35, №6, p. 1026-1027.

[23] Verizhnikov, L.V., Kirpichnikov, P.A. – *Trudy Kaz. Khim.-Tekhnol. Inst., Ser. Khim. Nauki,* 1964, vol. 33, p. 287-292.

[24] Kirpichnikov, P.A., Popova, L.M., Richmond, G.Ja. – *Zh. Obshch. Khim.,* 1966, vol. 36, №6, p. 1143-1147.

[25] Matveeva, E.N., Kirpichnikov, P.A., Kremen, M.Z. et al. – *Plast. Massy,* 1964, №2, p. 37-39.

[26] Kirpichnikov, P.A., Kolyubakina, N.S., Mukmeneva, N.A. et al. – *Plast. Massy,* 1966, №12, p. 24-26.

[27] Kirpichnikov, P.A., Kolyubakina, N.S., Mukmeneva, N.A. et al. – *Plast. Massy,* 1971, №4, p. 45-47.

[28] Kirpichnikov, P.A., Popova L.M., Levin, P.I. – *Trudy Kaz. Khim.-Tekhnol. Inst., Ser. Khim. Nauki,* 1964, vol. 33, p. 269-273.

[29] Kirpichnikov, P.A., Gren, G.P., Mukmeneva, N.A. – *Zh. Obshch. Khim.,* 1965, vol. 35, №4, p. 744-745.

[30] Verizhnikov, L.V., Kirpichnikov, P.A. – *Zh. Obshch. Khim.,* 1967, vol. 37, №6, p. 1335-1358.

[31] Verizhnikov, L.V., Kirpichnikov, P.A., Kolyubakina, N.S. – *Trudy Kaz. Khim.-Tekhnol. Inst., Ser. Khim. Nauki,* 1969, vol. 40, p. 382-386.

[32] Kirpichnikov, P.A., Minsker, K.S., Kolyubakina, N.S. et al. – *Vysokomol. Soed.,* 1968, vol. A10, №11, p. 2500-2512.

[33] Verizhnikov, L.V., Kirpichnikov, P.A., Kolyubakina, N.S. et al. – *Vysokomol. Soed.,* 1971, vol. A13, №3, p. 714-718.

[34] Kadyrova, V.Kh., Kirpichnikov, P.A., Mukmeneva, N.A. et al. – *Zh. Obshch. Khim.,* 1971, vol. 41, №8, p. 1688-1691.

[35] Mukmeneva, N.A., Kadyrova, V.Kh., Zharkova, V.M. et al. – *Zh. Obshch. Khim.,* 1986, vol. 56, №10, p. 2267-2271.

[36] Vershinin, P.V. et al. The USSR Author Certificate 794016 (1981); *Bull. Izobret.* (Rus), 1981, №1.

[37] Verizhnikov, L.V., Voskresenskaya, O.V., Kadyrova, V.Kh. et al. – *Zh. Obshch. Khim.,* 1971, vol. 41, №10, p. 2162-2164.

[38] Ordukhanyan, K.A. et al. The USSR Author Certificate 897797 (1980); *Bull. Izobret.* (Rus), 1982, №2.

[39] Kirillova, E.I. and Shul'gina, E.S. *Starenie i stabilizatsiya termoplastov (Ageing and Stabilization of Thermoplasts).* Leningrad: *Khimiya;* 1988.

[40] Mukmeneva, N.A., Kirpichnikov, P.A., Pudovik, A.N. – *Zh. Obshch. Khim.,* 1962, vol. 32, №7, p. 2193-2196.

[41] Mukmeneva, N.A., Kirpichnikov, P.A., Pudovik, A.N. – *Zh. Obshch. Khim.,* 1963, vol. 33, №10, p. 3192-3196.

[42] Mukmeneva, N.A., Kadyrova, V.Kh., Zharkova, V.M. – *Zh. Obshch. Khim.,* 1987, vol. 57, №12, p. 2796-2797.

[43] Kirpichnikov, P.A., Mukmeneva, N.A., Kadyrova, V.Kh. et al. – *Izv. Vysshikh Uchebn. Zavedenii, Ser. Khim. Technol.,* 1974, №3-4., p. 59-62.

[44] Voskresenskaya, O.V., Makarova, N.A., Kirpichnikov, P.A. et al. – *Izv. AN SSSR, Ser. Khim.,* 1968, №6, p. 1393-1396.

[45] Voskresenskaya, O.V., Makarova, N.A., Kirpichnikov, P.A. et al. – *Izv. AN SSSR, Ser. Khim.*, 1969, №7, p.1624-1625.

[46] Voskresenskaya, O.V., Makarova, N.A., Kirpichnikov, P.A. et al. – *Izv. AN SSSR, Ser. Khim.,* 1969, №7, p. 1626-1627.

[47] Voskresenskaya, O.V., Makarova, N.A., Kirpichnikov, P.A. et al. – *Izv. AN SSSR, Ser. Khim.*, 1969, №8, p. 1833-1835.

[48] Voskresenskaya, O.V., Makarova, N.A., Kirpichnikov, P.A. et al. – *Izv. AN SSSR, Ser. Khim.*, 1970, №7, p. 1666-1668.

[49] Kirpichnikov, P.A., Voskresenskaya, O.V., Mukmenev, E.T. – *Izv. AN SSSR, Ser. Khim.,* 1972, №6, p. 1395-1339.

[50] Shermergorn, J.M., Kuznetsov, E.V., Belyaeva, V.A. – *Trudy Kaz. Khim.-Tekhnol. Inst., Ser. Khim. Nauki,* 1962, vol. 30, p. 70-79.

[51] Kirpichnikov, P.A., Kadyrova, V. Kh. – *Trudy Kaz. Khim.-Technol. Inst., Ser. Khim. Nauki,* 1964, №33, p. 193-197.

[52] Kadyrova, V. Kh., Kirpichnikov, P.A., Mukmeneva, N.A. et al. – *Zh. Obshch. Khim.,* 1971, vol. 41, №8, p. 1688-1691.

[53] Kadyrova, V.Kh., Kirpichnikov, P.A., Mukmeneva, N.A. et al. – *Vysokomol. Soed.,* 1972, vol. A14, №5, p. 1081-1086.

[54] Kirpichnikov, P.A., Tarasova, Z.N., Baeva, N.A. et al. – *Vysokomol. Soed.*, 1965, vol.7, №8, p. 1368-1372.

[55] Itinskaya, G.P., Kotrelev, V.N., Shkarpeikina, G.A. – *Plast. Massy,* 1973, №10, p. 68-70.

[56] Mukmeneva, N.A., Cherezova, E.N., Biglova, R.Z. et al. – *Vysokomol. Soed.,* 1997, vol. A39, №6, p. 953-959.

[57] Rizpolozhenskii, N.I., Boiko, L.V., Zvereva, M.A. - *Dokl. AN SSSR,* 1964, vol.155, №5, p. 1137-1139.

[58] Petrov, K. A., Gol'tsova, R.G. – *Usp. Khim.,* 1966, vol. 35, №8, p. 1477-1494.

[59] Minsker, K.S., Mukmeneva, N.A., Kolesov, S.V. et al. – *Dokl. AN SSSR,* 1981, vol.259, №4, p. 890-894.

[60] Yamalieva, L.N., Kolesov, S.V., Mukmeneva, N.A. et al. – *Vysokomol. Soed.*, 1983, vol. A25, №10, p. 2157-2161.

[61] Tsvetkov, E.N. and Kabachnik, M.I. – In: *Reaktsii i metody issledovaniya organicheskikh soedinenii (Reactions and Methods of Investigation of Organic Compounds)*, Ed. by Kazanskii, B.A. Leningrad: *Khimiya;* 1964, vol.13, p. 267-427.

[62] Nifant'ev, E.E. – *Usp. Khim.,* 1978, vol. 47, №9, p. 1565-1608.

[63] Mukmeneva, N.A. et al. The USSR Author Certificate 744000 (1980); *Ref. Zh. Khim.* (Rus), 1981, 6N164.

[64] Kirpichnikov, P.A. et al. The USSR Author Certificate 582257 (1977); *Ref. Zh. Khim.* (Rus), 1978, 22N222.

[65] Zaichenko, L.P., Nechaev, Yu.D., Babel, V.G. et al. – *Zh. Prikl. Khim.,* 1971, vol. 44, №9, p. 2137-2138.

[66] Mukmeneva, N.A., Akhmadullina, A.G., Pobedimskii, D.G. et al. – *Vysokomol. Soed.,* 1974, vol. B16, №8, p. 635-637.

[67] Ovchinnikov, V.V., Cherkasova, O.A., Verizhnikov, L.V. – *Zh. Obshch. Khim.,* 1982, vol. 42, №3, p. 714-717.

[68] Pudovik, A.N., Akhmadullina, A.G., Pobedimskii, D.G. et al. – *Zh. Obshch. Khim.,* 1968, vol.38, №2, p. 307-311.

[69] Pudovik, A.N. and Golitsyna, G.A. – *Zh. Obshch. Khim.,* 1964, vol. 34, №3, p. 876-881.

[70] Pudovik, A.N., Cherkasov, R.A., Kondrat'eva, R.M. – *Zh. Obshch. Khim.,* 1968, vol.38, №2, p.307-313.

[71] Mukmeneva, N.A., Kadyrova, V.Kh., Zharkova, V.M. et al. – *Zh. Obshch. Khim.,* 1986, vol.56, №10, p. 2430-2431.

[72] Pudovik, A.N., Batyeva, E.S., Sinyashin, O.G. *Tioproizvodnye kislot trekhvalentnogo fosfora (Thioderivates of Tervalent Phosphorus Acids).* Moskow: *Nauka;* 1990.

[73] Baeva, N.A., Kirpichnikov, P.A., Chebotareva, E.T. – *Trudy Kaz. Khim.-Technol. Inst., Ser. Khim. Nauki,* 1965, vol. 34, p.296-298.

[74] Kirpichnikov, P.A., Ivanova, M.V., Gurvich, Ya.A. – *Zh. Obshch. Khim.,* 1966, vol. 36, №6, p. 1147-1149.

[75] Kirpichnikov, P.A., Mukmeneva, N.A., Baeva, N.A. – *Trudy Kaz. Khim.-Technol. Inst., Ser. Khim. Nauki,* 1965, vol.34, p.375-377.

[76] Kirpichnikov, P.A., Tarasova, Z.N. Senatorskaya, L.G. et al. – *Kauch. Resina,* 1965, №7, p. 5-10.

[77] Kiselev, V.D., Malkov, V.B., Murzin, D.G. et al. – *Dokl. AN SSSR,* 1989, vol.308, №1, p. 111-113.

[78] Mukmeneva, N.A., Cherezova, E.N., Cherkasova, O.A. et al. – *Zh. Obshch. Khim.,* 1994, vol. 64, №6, p. 1025-1027.

[79] Mukmeneva, N.A., Cherezova, E.N., Tuzova, N.G. et al. – *Zh. Prikl. Khim.,* 1994, vol.67, №4, p. 633-635.

[80] Krivenko, L.V., Cherezova, E.N., Mukmeneva, N.A. – *Zh. Prikl. Khim.*, 2000, vol.73, №7, p. 1193-1197.

[81] Cherezova, E.N., Zhukova, R.S., Tatarintseva T.B. et al. – *Zh. Obshch. Khim.,* 1997, vol. 67, №8, p. 1356-1360.

[82] Tatarintseva, T.B., Krivenko, L.V., Cherezova, E.N. et al. – *Abstracts of Reports.* The Conference dedicated to A.N. Nesmeyanov's Centenary, Moscow, Russia, 1999, vol.2, p. 212.

[83] Tatarintseva, T.B., Brus'ko, V.N., Krivenko, L.V. et al. – *Abstracts of Reports.* The 37-th IUPAC Congress, Berlin, Germany, 1999, p. 105.

[84] Cherezova, E.N., Mukmeneva, N.A., Cherkasova, O.A. et al. – *Zh. Obshch. Khim.,* 1987, vol.57, №8, p. 1915-1917.

[85] Mukmeneva, N.A., Cherkasova, O.A., Cherezova, E.N. et al. – *Zh. Obshch. Khim.,* 1987, vol.57, №12, p. 2497-2507.

[86] Mukmeneva, N.A., Cherkasova, O.A., Cherezova, E.N. et al. – *Zh. Prikl. Khim.,* 1991, vol.64, №9, p. 1967-1971.

[87] Cherezova, E.N., Cherkasova, O.A., Mukmeneva, N.A. et al. – *Zh. Obshch. Khim.,* 1992, vol.62, №4, p. 781-785.

[88] Cherezova, E.N., Cherkasova, O.A., Mukmeneva, N.A. et al. – *Zh. Obshch. Khim.*, 1990, vol.60, №11, p. 2478-2480.
[89] Mukmeneva, N.A. et al. The USSR Author Certificate 1437368 (1988); *Bull. Izobret.* (Rus), 1988, №42.
[90] Ershov, V.V., Nikiforov, G.A., Volod'kin, A.A. *Prostranstvenno-zatrudnennye fenoly (Sterically Hindered Phenols).* Moscow: *Khimiya;* 1972.
[91] Volod'kin, A.A., Ershov, V.V. – *Usp. Khim.*, 1988, vol.57, №4, p. 595-624.
[92] Nugumanova, G.N., Zhukova, R.S., Cherezova, E.N. et al. – *Zh. Obshch. Khim.*, 1997, vol.67, №8, p. 993-996.
[93] Nugumanova, G.N., Zhukova, R.S., Cherezova, E.N. et al. – *Zh. Prikl. Khim.*, 1997, vol.70, №9, p. 1520-1524.
[94] Cherezova, E.N., Nugumanova, G.N., Sofronova, O.V. et al. – *Kauch. Resina,* 1998, №2, p.34-36.
[95] Mukmeneva, N.A., Cherezova, E.N., Rusina, I.F. et al. – In: *Ageing of Polymers, Polymer Blends and Polymer Composites*, Ed. By G.E. Zaikov, A.L. Buchachenko, V.B. Ivanov. New-York; 2002, vol.2, p. 197-205.
[96] Mukmeneva, N.A., Kadyrova, V.Kh., Cherezova, E.N. et al. – *Zh. Org. Khim.*, 1990, vol.26, №12, p. 2493-2497.
[97] Mukmeneva, N.A. et al. Patent 2059629, Russia (1997); *Ref. Zh. Khim.* (Rus), 1997, 7N95P.
[98] Cherezova, E.N., Mukmeneva, N.A., Kadyrova, V.Kh. et al. – *Vysokomol. Soed.*, 1988, vol.B30, №5, p. 378-380.
[99] Mukmeneva, N.A., Cherezova, E.N., Rusina, I.F. et al. – *J. Polymer. Mater.*, 2000, vol. 47, p. 317-328.
[100] Davis, B., Gash, D., Woodgate, P. – *J. Chem. Soc., Perkin Trans I,* 1982, vol.7, p. 1499-1507.
[101] Nikiforov, G.A., Bannikov, G.F., Romero Maldonado, J.K.A. et al. – *Izv. AN SSSR, Ser. Khim.*, 1989, №12, p. 2765-2770.
[102] Liakumovich, A.G. et al. Patent 2017717, Russia (1994); *Bull. Izobret.* (Rus), 1994, №6.
[103] March, G. *Organicheskaya Khimiya (Organic Chemistry).* Moscow: Mir; 1988, vol.2.
[104] Batanov, I.A., Volyeva, V.B., Nikiforov, G.A. et al. – *Izv. AN SSSR, Ser. Khim.*, 1984, №10, p. 2327-2332.
[105] Bukharov, S.V., Saykaev, V.V., Podyacher, G.N. et al. – *Zh. Obshch. Khim.*, 2002, vol.72, №6, p. 988-991.
[106] Ogorodnikov, S.K. *Formaldegid (Formaldehyde).* Leningrad: *Khimiya*; 1984.
[107] Egidis, F.M., Grebenyukova, E.K., Volkotrub, M.N. In: *Sintez I issledovanie effektivnosti khimikatov-dobavok dlya polimernykh materialov (Synthesis and Study of efficiency of chemicals-additives for polymer materials)*, Ed. by Maslova, I.P., Ignatov, V.A., Mishina, I.M. Tambov: *Tambovskaya Pravda,* 1969.
[108] Maatschappij, N.V. *GB Patent* 892261 (1962); C.A., 1963, vol.8, 6748.
[109] ukharov, S.V., Nugumanova, G.N., Mukmeneva, N.A. – *Zh. Obshch. Khim.*, 1998, vol.68, №10, p. 1678-1681.
[110] Abramov, V.I. et al. The USSR Author Certificate 883016 (1981); *Bull. Isobret.* (Rus), 1981, №43.
[111] Gurvich, Ya.A. and Kumok, S.T. – *Kauch. Resina,* 1985, №6, p. 12-14.

[112] Bukharov, S.V., Nugumanova, G.N., Mukmeneva, N.A. – *Zh. Obshch. Khim.,* 2003, vol.73, №3, p. 437-441.

[113] Sigov, O.V., Marchev, Yu.M., Yadreev, F.I. – *Kauch. Resina,* 1992, №5, p. 20-22.

[114] Glazunova, N.A. et al. *Stabilizatory polivinilkhlorida i sopolimerov polivinilkhlorida. Tematitseskii obzor (Stabilizers of Polyvinylchloride and Copolymers of Polyvinylchloride. Thematic Review).* Moscow: CNIITENeftekhim; 1968.

[115] Zhungietu, G.I. *Izatin i ego proizvodnye (Isatin and its Derivatives).* Kishinev: Shtiintsa; 1977.

[116] Kochetkov, N.K. *Obtschaya organicheskaya khimiya (General Organic Chemistry).* Moscow: *Khimiya;* 1985.

[117] *Khimicheskaya entsyklopediya (Chemical Encyclopedia).* Moscow: *Sov. Entsyklopediya;* 1990. vol.2.

[118] Volod'kin, A.A. et al. The USSR Author Certificate 825501 (1981); *Ref. Zh. Khim.* (Rus), 1982, 1N119.

[119] Bukharov, S.V., Nugumanova, G.N., Mukmeneva, N.A. – *Zh. Obshch. Khim.,* 1999, vol.69, №4, p. 699.

[120] Terakh, E.I., Pinko, P.I., Zaitseva, O.V. – *Zh. Prikl. Khim.,* 2003, vol.76, №2, p. 261-265.

[121] Bochkov, A.F. et al. The USSR Author Certificate 891702 (1981); C. A., 1982, vol.96, 163671r.

[122] Mukmeneva, N.A., Kadyrova, V.Kh., Bukharov, S.V. et al. – *Zh. Obshch. Khim.,* 1996, vol.66, №10, p. 1725-1727.

[123] Gurvich, Ya.A. et al. The USSR Author Certificate 749821 (1980); *Ref. Zh. Khim.* (Rus), 1981, 9N131P.

[124] Gurvich, Ya.A., Arzamanova, I.G., Kumok, S.T. et al. – *Kauch. Resina,* 1985, №5, p. 23-24.

[125] Seichiro, H., Michihiro, O. – *J. Jap. Petrol. Inst.,* 1982, vol. 25, №6, p. 398-400.

[126] Gurvich, Ya.A., Arzamanova, I.G., Zaikov, G.E. – *Khim. Phiz.,* 1996, vol.15, №1, p. 23-42.

[127] Bukharov, S.V., Litvinov, I.A., Gubaydullin, A.T. et al. – *Zh. Obshch. Khim.*, 2002, vol. 72, №2, p. 290-293.

[128] Chernova, A.N., Shagidullin, R.R., Bukharov, S.V. et al. – *Zh. Obshch. Khim.,* 2006, vol.76, №6, p. 1034-1036.

[129] Gurvich, Ya.A., Zolotarevskaya, L.K., Kumok, S.T. *Fenol'nye stabilizatory. Tematicheskii obzor (Phenol Sstabilizers. Tnematic Review).* Moscow: CNIITENeftekhim; 1978.

[130] Kasymova, E.M., Burilov, A.R., Vagapova, L.I. et al. – *Izv. RAN,* 2006, №7, p. 1506-1510.

[131] Mukmeneva, N.A., Bukharov, S.V., Nugumanova G.N. et al. – *Vysokomol. Soed.,* 1998, vol.B40, №9, p. 1506-1510.

[132] Legotskii, P., Kavun, S.M., Syrovyi, L. – Abstracts of Papers. The IX Conference "Destruction and Stabilization of Polymers", Moscow, Russia, 2001, p. 106-107.

[133] Emanuel, N.M. and Buchachenko, A.L. *Khimicheskaya physika stareniya i stabilizatsii polimerov (Chemical Physics of Ageing and Stabilization of Polymers).* Moscow: *Nauka;* 1988.

[134] Vishnyakova, T.P. et al. The USSR Author Certificate 534449 (1976); *Ref. Zh. Khim.* (Rus), 1978, 23P243P.

[135] Krivenko, L.V., Cherezova, E.N., Mukmeneva, N.A. – *Zh. Prikl. Khim.,* 2000, vol.73, №7, p. 1193-1197.

[136] Glebova, E.V., Vishnyakova, T.P., Lebedev, V.N. et al. – *Neftekhimiya,* 1982, vol. 22, №1, p. 82-85.

[137] Ehrhardt, D. – *Gummi, Fasern Kimststoffe,* 1992, bd.45, №5, s. 231-239.

[138] Ehrhardt, D. – *Gummi, Fasern Kimststoffe,* 1992, bd.45, №7, s. 358-361.

[139] Ehrhardt, D. Germany (DD) Patent 290427 (1991); C. A., 1991, vol.115, 258104h.

[140] Mehrer, M. Appl. 19734964, Germany (1999); *Ref. Zh. Khim.* (Rus), 2000, 19N89P.

[141] Ehrhardt, D. Germany (DD) Patent 291902 (1991); C. A., 1991, vol.115, 258105j.

[142] Bukharov, S.V., Nugumanova, G.N., Mukmeneva, N.A. et al. – *Zh. Org. Khim.,* 2003, vol.39, №5, p. 735-737.

[143] Bukharov, S.V., Nugumanova, G.N., Mukmeneva, N.A. et al. – *Zh. Org. Khim.*, 2004, vol.40, №3, p. 327-334.

[144] Kasymova, E.M., Burilov, A.R., Bukharov, S.V.et al. – Absracts of papers. XVIII Mendeleev's Congress on General and Applied Chemistry, Moscow, 2007, vol.5, p. 197.

[145] Walter, R.T. US Patent 4238627 (1980); *Ref. Zh. Khim.* (Rus), 1981, 16R122P.

[146] Mina, G.L. et al. US Patent 4847434 (1989); C.A., 1989, vol.111, 214224K.

[147] Utida, H. Jpn. Patent 58-140034 (1983); *Ref. Zh. Khim.* (Rus), 1984, 21N125P.

[148] Strom, R.M. US Patent 4410736 (1983); *Ref. Zh. Khim.* (Rus), 1984, 17N103P.

[149] Kruse, W.M. et al. US Patent 4487977 (1984); *Ref. Zh. Khim.* (Rus), 1985, 17R141P.

[150] Tolstikov, G.A., Pantukh, B.I., Vakhitova, M.Sh. – *Zh. Prikl. Khim.*, 1985, vol.58, №5, p. 863-867.

[151] Shideru, T., Tom, T., Mitsuo, M. – J. *Catal.,* 1984, vol.89, p. 511-519.

[152] Walter, K.M. et al. US Patent 4482755 (1984); C.A., 1985, vol. 102, 78556e.

[153] Gorodetskaya, T.A. et al. The USSR Author Certificate 1155580 (1985); *Bull. Izobret.* (Rus), 1985, №18.

[154] Hirano, M., Isbii, T., Morimoto, T. – *Bull. Chem. Soc. Japan,* 1991, vol.64, №4, p. 1434-1436.

[155] Bir, S., Pappy, M.S., Venkateshwar, R.T. et al. – *Tetr. Lett.,* 1994 №28, p. 5083-5084.

[156] Kothary, V.M., Tazunna, Z.Z. – *J. Catalysis,* 1976, vol.41, p. 180-189.

[157] Tada, M., Katsu, T. – *Bull. Chem. Soc.Japan,* 1972, vol.45, p. 2558-2559.

[158] Fulleron, T.Z., Ahern, S.P. – *Tetr. Lett.*, 1976, №2, p. 139-142.

[159] Schmidt, T., Harting, W., Wasgestion, F. – *Inorg. Chem. Acta.,* 1998, vol.274, №1, p. 126-129.

[160] Turk, H., Ford, W.T. – *J. Org. Chem.,* 1988, vol. 53, p. 460-462.

[161] Zaku, A., et al. Jpn. Patent 61200935 (1986); C.A., 1987, vol.106, 32557k.

[162] Walter, K. M., Stephen, F.I. US Patent 4487978 (1984); C.A., 1985, vol.102, 113028h.

[163] Larry, D., et al. US Patent 4447656 (1984); C.A., 1984, vol.101, 72421q.

[164] Adigamov,E.R., Vakhitova, M.Sh., Logutov, I.I. USSR Patent 1011626 (1983); C.A., 1983, vol.99, 87808k.

[165] Lerner, S. US Patent 4115590 (1974); C.A., 1979, vol.90, 43811g.

[166] Mukmeneva, N.A., Bukharov, S.V., Kadyrova, V.Kh. et al. – *Zh. Obshch. Khim.,* 1996, vol.66, №9, p. 1526-1529.

[167] Fazlieva, L.K., Mukmeneva, N.A., Bukharov, S.V. et al. –*Zh. Prikl. Khim.,* 1996, vol.69, №8, p. 1404-1406.

[168] Liakumovich, A.G. et al. The USSR Author Certificate 1766903 (1992); *Bull. Izobret.* (Rus), 1992, №37.

[169] Mukmeneva, N.A., Bukharov, S.V., Kadyrova, V.Kh. et al. – *Zh. Obshch. Khim.* 1994, vol.64, №9, p. 1507-1509.

[170] Grassi, N. and Scott, G. *Destruktsiya i stabilizatsiya polimerov. (Destruction and Stabilization of Polymers).* Moscow: Mir; 1988.

[171] Pozdneeva, N.N., Aleksandrov, A.L., Yakushenko, I.K. – Abstracts of Papers. The II International Conference on Kinetics of Radical Liquid-phase Reactions, Kazan, Russia, 1995, p. 31.

[172] *Khimiya koordinatsionnykh soedinenii. Sbornik. (Chemistry of Co-ordination Compounds. Collection).* Ed. by Baylar, J. Moscow: *Izdatinlit*; 1950.

[173] Kolesnikov, V.A., Klep, V.Z., Novikov, E.E. et al. – *Zh. Obshch. Khim.,* 1990, vol.26, №2, p. 370-373.

[174] Dokukina, M.A., Vol'eva, V.B., Belostotskaya, I.S. – *Izv. AN SSSR, Ser. Khim.*, 1992, №10, p. 2380-2384.

[175] Komissarova, N.L., Belostotskaya, I.S., Shubina, O.V. et al. – *Zh. Obshch. Khim.* 1992, vol. 62, №1, p. 188-191.

[176] Coppinger, G. and Campbell, T. – *J. Amer. Chem. Soc.,* 1953, vol.75, №3, p. 734-736.

[177] Bruk, Yu.A., Rachinskii, F.Yu., Zolotova, L.V. et al. – *Zh. Obshch. Khim.,* 1972, vol.42, №7, p. 1603-1608.

[178] Liakumovich, A.G., Kadyrova, V.Kh., Mukmeneva, N.A. et al. – *Zh. Obshch. Khim.* 1991, vol.61, №1, p. 260-261.

[179] Mukmeneva, N.A., Kadyrova, V.Kh., Liakumovich, A.G. – *Zh. Prikl. Khim.,* 1987, vol.60, №3, p. 658-661.

[180] Mukmeneva, N.A., Cherezova, E.N., Zhukova, R.S. – *Zh. Obshch. Khim.,* 1994, vol.64, №6, p. 1049-1051.

[181] Feldman, E., Soenchen, E. FRG Patent 2312920 (1973); C.A., 1974, vol.80, 133624e.

[182] Gross, H., Seibt, H., Keitl, I. – *J.fur hrakt.Cliemie,* 1975, vol.317, №6, p. 890-896.

[183] Jositamu Sajaku, K.K. Jpn. Patent 56-19335 (1981); *Ref. Zh. Khim.* (Rus), 1982, 15N99P.

[184] Parekh, M. US Patent 42633232 (1986); *Ref. Zh. Khim.* (Rus), 1987, 5N15OP.

[185] Kirpichnikov, P.A. et al. The USSR Author Certificate 619486 (1978); *Bull. Izobret.* (Rus), 1978, №30.

[186] Mukmeneva, N.A., Cherkasova, O.A., Zhukova, R.S. – *Zh. Obshch. Khim.,* 1992, vol.62, №2, p. 468-469.

[187] Pritschenko, A.A., Livantsov, M.V., Novikova O.P. et al. – Abstracts of Papers. The XVIII Mendeleev's Congress on General and Applied Chemistry, Moscow, 2007, p. 2104.

[188] Gazizov, M.B., Ismagilov, R.K., Shamsutdinova, L.P. et al. – Abstracts of Papers. The XVIII Mendeleev's Congress on General and Applied Chemistry, Moscow, 2007, p. 394.

Chapter 2

Modeling of Chemical Processes of Polymers Oxidation Inhibition by Polyfunctional Stabilizers

2.1. The Mechanism of Thermal-Oxidative Degradation of Polymers and Actions of Antioxidants

Practically any ageing of polymers occurs while contacting with oxygen, as a result oxidation of polymers is the predominant process in practice. For this reason the main and basic studies on the ageing mechanism of carbochain polymers are devoted to the problems of their oxidation [1-13].

The theory of the chain branched and degenerated branched reactions in the liquid phase [1-5] are the basis of the concept of thermal-oxidative degradation of polymers.

According to the accepted conception a chain reaction of oxidation of organic substrata including polymers consists of a large number of elementary stages which, depending on their role and place in cumulative process, are divided into stages of initiation, development, branching and termination of chains.

The initiation of kinetic chains of oxidation (reactions 0.1 – 0.2):

$$RH + O_2 \xrightarrow{k_0} R^\bullet + HO_2^\bullet \qquad (0.1)$$

$$2\,RH + O_2 \xrightarrow{k_0} R^\bullet + H_2O_2 + R^\bullet \qquad (0.2)$$

The development of kinetic chains is performed by alteration of reactions (1) and (2).

$$R^\bullet + O_2^\bullet \xrightarrow{k_1} ROO^\bullet \qquad (1)$$

$$ROO^\bullet + RH \xrightarrow{k_2} ROOH + R^\bullet \qquad (2)$$

In the course of reaction (2) there is accumulation of hydroperoxide ROOH and its further decomposition leading to the formation of free radicals capable to generate new chains of oxidation. Hydroperoxide is the main branching agent. It is the course of the reactions of generating from neutral compounds of several high reactivity particles that causes autocatalytic effect of oxidation.

The main mechanisms of hydroperoxide decomposition (reactions 3–3.2) are:

$$ROOH \longrightarrow RO^{\bullet} + HO^{\bullet} \quad (3)$$

$$2\,ROOH \longrightarrow ROO^{\bullet} + H_2O + RO^{\bullet} \quad (3.1)$$

$$ROOH + RH \longrightarrow ROO + H_2O + R^{\bullet} \quad (3.2)$$

Secondary radicals react again quickly.

$$RO^{\bullet} + RH \longrightarrow ROH + R^{\bullet} \quad (3.3)$$

$$HO^{\bullet} + RH \longrightarrow H_2O + R^{\bullet} \quad (3.4)$$

The destruction of kinetic chains occurs, as a rule, by means of square-law termination (reactions 4–6):

$$R^{\bullet} + R^{\bullet} \longrightarrow \text{R-R or RH + olefin} \quad (4)$$

$$R^{\bullet} + ROO^{\bullet} \longrightarrow ROOR \quad (5)$$

$$ROO^{\bullet} + ROO^{\bullet} \longrightarrow ROOR + O_2 \text{ or } ROH + O_2 + R'R''C{=}O \quad (6)$$

Oxidation of carbochain polymers in the solid phase compared with liquid-phase oxidation of hydrocarbons has a number of the important features (the go-ahead mechanism of transferring valency owing to segmental diffusion, dependence of oxidation rate on the polymer rigidity, the increased stability of alkyl macroradicals etc.) [7, 11]. Nevertheless, these processes proceed according to the common mechanism of the chain autoinitiated reaction. The similarity of kinetic laws of processes of oxidation in liquid and solid phases is the basis for describing the mechanism of solid-phase oxidation by the same scheme which is accepted in liquid-phase oxidation.

The conclusion about the chain degenerated-branched character of process is common for all polymers containing C-H bonds, capable to oxidation. It makes the quantitative basis for forecasting the lifetime of polymer as material, and also allows to understand and to justify scientifically the methods of their stabilization.

To stop the chain degenerate-branched process is possible either by increasing the rate of termination of kinetic chains, i.e. reducing their length, or by reducing the rate of initiation and branching of chains by destruction of initiators and branching products [14].

Stabilizers deactivate active products of chain process, or destroy the intermediate products which are the source of active radicals.

The following basic approaches to the stabilization of polymers are known:

- accepting alkyl radicals (decrease of initiation rate of oxidation) – reactions (7.1) and (8.1);
- accepting peroxide radicals (decrease of the rate of developing the oxidation chain) – reactions (7), (8), (8.2) and (8.3);
- hydroperoxide decomposition without generating free radicals (suppression of degenerated branching of the oxidation chains) – reactions (11) and (11.1).

$$ROO^{\bullet} + InH \xrightarrow{k_7} ROOH + In^{\bullet} \qquad (7)$$

$$R^{\bullet} + InH \longrightarrow RH + In^{\bullet} \qquad (7.1)$$

$$ROO^{\bullet} + In^{\bullet} \longrightarrow \text{Molecular products} \qquad (8)$$

$$R^{\bullet} + In^{\bullet} \longrightarrow RIn \qquad (8.1)$$

$$ROO^{\bullet} + In^{\bullet} \longrightarrow ROOIn \qquad (8.2)$$

$$ROO^{\bullet} + In^{\bullet} \longrightarrow InH + \text{Molecular products} \qquad (8.3)$$

$$In^{\bullet} + In^{\bullet} \longrightarrow In\text{-}In \qquad (9)$$

$$In^{\bullet} + RH \xrightarrow{k_{10}} InH + R^{\bullet} \qquad (10)$$

$$InH + ROOH \longrightarrow \text{Molecular products} \qquad (11)$$

$$InH + ROOH \longrightarrow \text{Free radicals} \qquad (11.1)$$

Formed according to reaction (7) radical $In^{\cdot}$ can react with other free radicals producing molecular products. At the same time this radical should be of low-activity and should not enter into the reaction (10) so that new chains of oxidation couldn't arise. The ratio of rate constants k_{10}/k_2 of reactions (10) and (2) should be small, and the ratio k_7/k_2 should be large for the effective stabilization.

Guided by the kinetic scheme of inhibition it is possible to state that an inhibitor is an obstacle to the chain reaction of oxidation, mainly, due to either terminating chains or decomposing hydroperoxides. Duration of the inhibitory action of an inhibitor depends on the mechanism of its action, a set of reactions leading to the inhibition of side reactions which result in unproductive consumption of inhibitor. According to the above-stated scheme, in oxidized substrata (polymers) there is an accumulation of various (depending on the nature of polymers) products of radical and molecular nature which promote the ageing of polymers. The efficiency of stabilizers is mostly defined by their ability to react with these products and also by the course of reactions.

2.2. Modeling of Reactions of Polyfunctional Stabilizers with Products of Polymer Oxidation

For getting the scientific basics of stabilization it is of great importance to use methods of modeling chemical processes which take place when thermal-oxidative degradation of polymers occurs.

However kinetic investigations of inhibiting activity of additives during oxidation of the polymeric environment in most cases do not make it possible to define unambiguously stabilizer function at inhibiting oxidation. To identify the mechanism of the antioxidant activity, as a rule, liquid-phase processes of the antioxidants' interaction with the compounds modeling separate products, arising at oxidation of substrata are studied, these processes are easily checked (from the experimental point of view).

According to the classical notion the oxidation of stabilizers by hydroperoxides and free radicals can serve as an adequate model for understanding the mechanisms of influence of polyfunctional inhibitors at thermal oxidation of polymers.

Compounds of various classes, viz. phosphorous acid esters, phosphorus dithioacids, thiocarbamides derivatives and their metal complexes, arylamines are considered in the given section.

2.2.1. Reactions with Hydroperoxides

The main branching agent in the processes of polymer oxidation is hydroperoxide therefore the investigation of kinetics of its decomposition in the presence of stabilizers is an important aspect of the estimation of the efficiency of their antioxidizing effect. Particularly, as the regularities of changing the inhibiting ability of compounds in oxidation reactions in a liquid phase hold true for the similar reactions with polymeric hydroperoxides [14].

Methods of polarography, iodometry, chemiluminescence and NMR (nuclear magnetic resonance) spectroscopy are used for studying the reactions of polyfunctional compounds with hydroperoxide [15-17]. Reactions with hydroperoxides were carried out in the temperature range of 293-373 K. Rate constants, the orders of reactions are calculated according to the initial rate of the process [18]. Activation energy was defined by the temperature dependence of the reaction rate constant according to the equations (12), (13):

$$\lg k = \lg A - \frac{E}{2.3RT} \quad (12)$$

$$E = -2.303 \cdot R \cdot [\Delta \lg k / \Delta (T^{-1})] \quad (13)$$

The method of inhibitors is used to reveal the route of the hydroperoxide decomposition when additives are introduced into polymer. According to the specified method the free radical acceptor, namely 2,6-di-*tert*-butyl-4-methylphenol (BHT), was introduced into the system, that acceptor, if particles of radical nature are in the system, reduces the rate of hydroperoxide decomposition or stops the reaction completely. The absence of the influence of a free radical acceptor of testifies the heterolytic character of hydroperoxide decomposition.

The important characteristic of stabilizing action of inhibitors is stoichiometric coefficient (v), corresponding to the quantity of hydroperoxide molecules destroyed by one molecule of an inhibitor (the catalysis factor). Stoichiometry of hydroperoxide consumption is defined in the range of 0.1-0.5 mole/l and the concentration of the investigated antioxidant 1×10^{-1}–1×10^{-4} mole/l according to the formula (14):

$$\nu = ([ROOH]_0 - [ROOH]_\infty) / [In]_0, \qquad (14)$$

where $[ROOH]_0$ and $[ROOH]_\infty$ are initial and end concentrations of hydroperoxide; $[In]_0$ is initial concentration of an inhibitor.

2.2.1.1. Full Esters of Phosphorous Acid (Trialkyl(Aryl)Phosphites)

Organophosphorous compounds (derivatives of three-and four-coordinated phosphorus atom) and, first of all, phosphorous acid esters (organic trialkyl(aryl)- and dialkyl(aryl)phosphites) are known as stabilizers of polyolefines, rubbers and vulcanizates, heterochain polymers, polyvinylchloride etc. [19-24].

It is established that the full aromatic and alkylaromatic esters of phosphorous acids possess the strongest antioxidizing action , they do not color polymeric materials during their processing and service performance, develop considerable synergistic effect in mixtures with phenols, they are well combined, have low volatility and oxidability by molecular oxygen.

Phosphites possess high nucleophilicity due to rather low ionization potential of a free pair of 3s-electrons and stability of forming sp^3-bonds. On the other hand, in reactions with strong nucleophiles a phosphorus atom can participate in the formation of new bonds by its free orbitals, being thus the electrophylic centre. Such bifility in reactions often reveals simultaneously and causes large reactivity and a large variety of reaction mechanisms and formed products. The important feature of the chemical nature of a phosphite molecule is that it has two nucleophilic centres (at oxygen and phosphorus atoms) also promoting the display of dual reactivity (ambidentedness) [25, 26].

It is these features of phosphites that make possible to consider phosphites as polyfunctional stabilizers.

Based on the high activity of full esters of phosphorous acid (phosphites) towards hydroperoxide bonds it is postulated [27, 28] that they can inhibit the polymer oxidation at the expense of hydroperoxide decomposition formed during the process of hydrocarbon substrata oxidation and thereby suppress degenerated branching of chains.

The practical confirmation of this is the phenomenon of nonadditive amplification of stabilizing efficiency of mixture of phosphites with phenols which are classical antioxidants of thermal-oxidative destruction of polymers [29, 30].

Analyzing this effect the authors of these papers came to the conclusion that the scheme of polymer oxidation in the presence of an inhibitor of free radical action and phosphite should include the termination of kinetic chains at an inhibitor and practically nonradical decomposition of polymer hydroperoxides by phosphites leading to the formation of stable molecular products:

$$InH + RO_2^{\cdot} \rightarrow In^{\cdot} + ROOH$$

$$(R'O)_3P + ROOH \rightarrow (R'O)_3P(O) + ROH.$$

Kinetically such scheme is confirmed by the shift of critical concentration of phenol antioxidants onto the area of low values in the presence of phosphites when inhibited oxidation of polypropylene and polyethylene takes place [29, 30].

It is suggested [27, 28] that reactions of organic phosphite oxidation by hydroperoxide proceed basically according to the ionic mechanism by the formation of intermediate ionic pairs of two types:

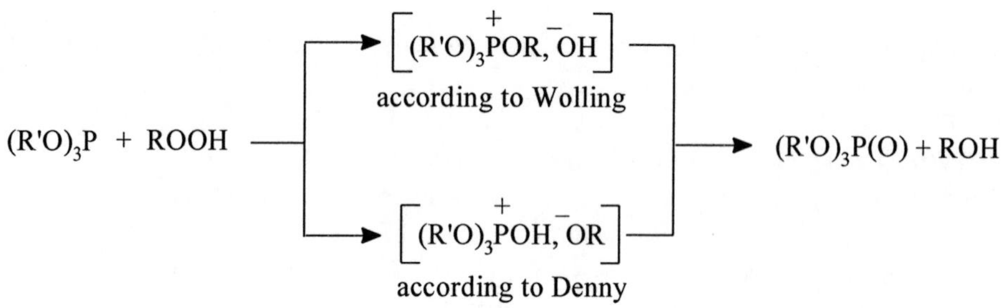

Scheme 2.1.

Trialkyl(aryl)phosphate and appropriate alcohol are the end products.

At present kinetic regularities of reactions of organic phosphites with hydroperoxides are investigated sufficiently well [31-35].

The approach based on using the combination of methods of kinetic polarography and radical traps appeared to be effective for the quantitative estimation of phosphites as hydroperoxide decomposers [33].

It is shown that the consumption of hydroperoxides in modeling reactions with phosphites comply with the first order equation concerning both components; bimolecular rate constants are calculated from these dependences, their values practically remain constant in the wide range of reaction rates (from 10 to 10^{-3} l/(mole·s)) at the stoichiometric ratio of components 1:1.

Activation energy is defined for phosphites of various structures, namely, 5-6 kcal/mole for aliphatic phosphites and 9-14 kcal/mole for aromatic phosphites. In these reactions aromatic phosphites possess smaller reactivity in comparison with aliphatic phosphites (table 2.1).

Table 2.1. Rate constants (293 K) and Arrhenius parameters of interaction of hydroperoxides with phosphites

Compound number	Phosphite	Solvent	Δt, K	ROOH	$k\times10^{-3}$, litre/(mole·s)	E, kcal/mole	A, litre/(mole·s)
(1)	$(C_2H_5O)_3P$	CH_2Cl_2	–	TBH[a]	56.0	–	–
(2)	$(n\text{-}C_3H_7O)_3P$	benzene	–	TBH	11.0	–	–
(3)	$(C_4H_9O)_3P$	benzene	283–323	CHP[b]	124.5	5.9	3.2×10^3
(4)	$(cyclo\text{-}C_6H_{11}O)_3P$	benzene	283–323	CHP	105.0	6.1	3.8×10^3
(5)	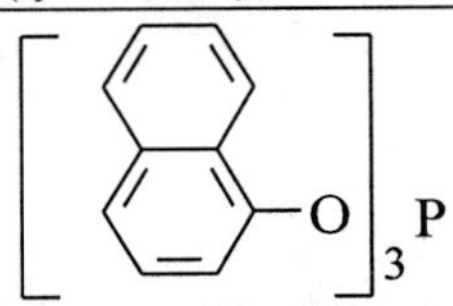	benzene	283–323	CHP	9.3	9.0	4.9×10^4
(6)	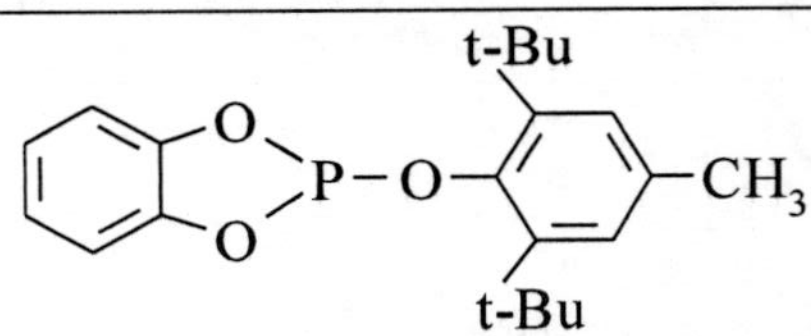	benzene	283–323	CHP	2.3	10.9	3.2×10^5
(6)	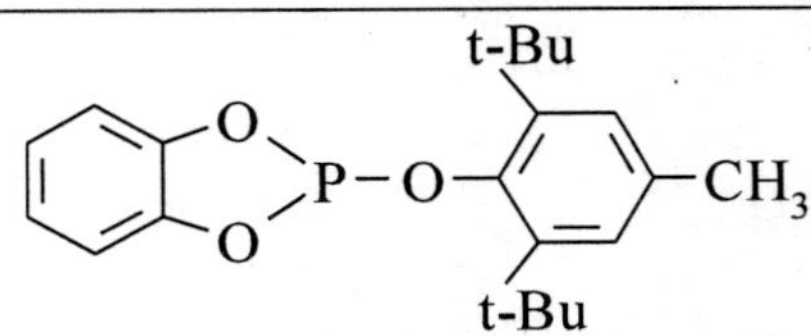	C_6H_5Cl	–	CHP	3.0 (calc.)	14.1[c]	6.0×10^7 [c]
(7)	$(C_6H_5O)_3P$	C_6H_5Cl	–	TBH	5.0	–	–
(7)	$(C_6H_5O)_3P$[d]	Heptane	308–328	TBH	10.7 (calc.)	8.1	1.2×10^4

Compound number	Phosphite	Solvent	Δt, K	ROOH	k×10-3, litre/(mole·s)	E, kcal/mole	A, litre/(mole·s)
(8)	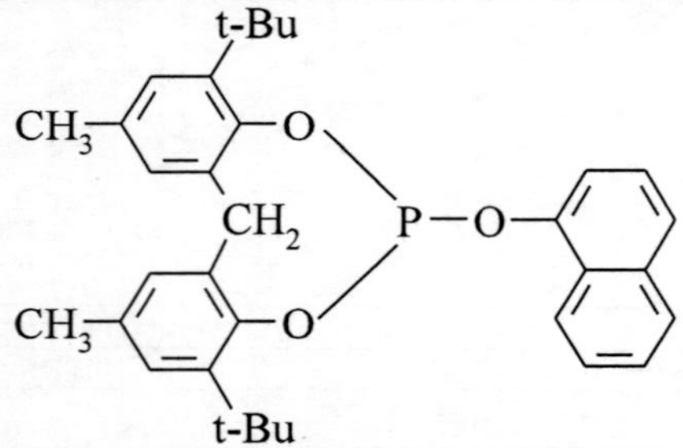	Benzene	283–323	CHP	2.1	14.0	5.8×10^{7}
(9)	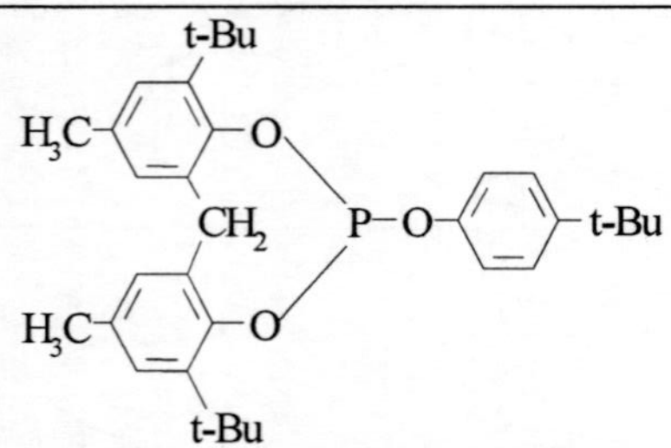	Benzene	283–323	CHP	1.9	13.8	3.8×10^{7}
(10)	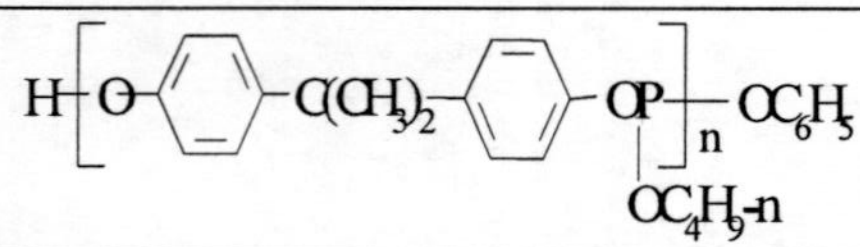	Benzene	283–323	CHP	27.2	13.1	1.6×10^{8}
(11)	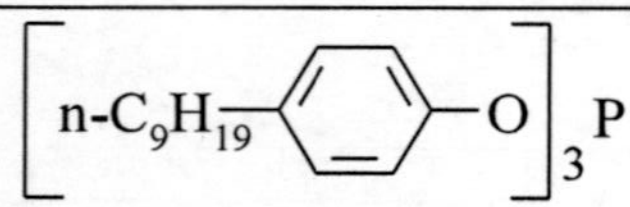	Benzene	283–323	CHP	36.6	9.5	4.6×10^{5}

Table 2.1. (Continued).

Compound number	Phosphite	Solvent	Δt, K	ROOH	k×10-3, litre/(mole·s)	E, kcal/mole	A, litre/(mole·s)
(12)	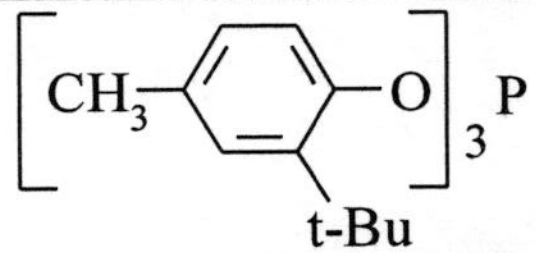	Hexane	283–323	TBH	3.5	–	–

[a] TBH – *tert*-butylhydroperoxide; [b] CHP – cumene hydroperoxide.
[c] E and A values were obtained in Reference [38]; [d] E and A values were obtained in Reference [39].

The comparison of reactivity of phosphites and typical hydroperoxide decomposers shows that the efficiency of phosphites is higher than the efficiency of dilauryl thiodipropionate, mercaptans and sulfides [36].

When the Kabachnik's σ correlation is used it is found that rate constants depend on the nature of substitutes R in phosphites $P(OR)_3$ [37]. It explains that electrodonor properties of aryloxy substituents are weaker than those of alkoxy derivatives.

The possibility of a homolytic route in the system *hydroperoxide – phosphite* is evaluated by the method ESR (electron spin resonance) according to the fixation of the consumption of stable nitroxyl radicals used as trapping of radicals formed in the course of reaction. The output of the radicals is low enough, from 0.02 % (for aliphatic phosphites) to 5 % (for aromatic phosphites) (table 2.2).

Table 2.2. Rate constants and probabilities (e) in reaction of TBH or CHP with phosphites (determined by nitroxyl trapping at 293 K, in the absence of O_2)

Compound umber	Phosphite	ROOH	Solvent	$k\times10^{-3}$, litre/(mole·s)	*e*
(1)	$(C_2H_5O)_3P$	TBH	Benzene	170	0.6×10^{-4}
(1)	$(C_2H_5O)_3P$	TBH	Styrene	170	1.0×10^{-3}
(2)	$(n\text{-}C_3H_7O)_3P$	TBH	Benzene	12	0.2×10^{-4}
(5)	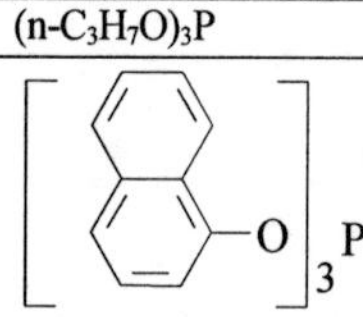	CHP	Benzene	6.6	0.6×10^{-2}
(6)	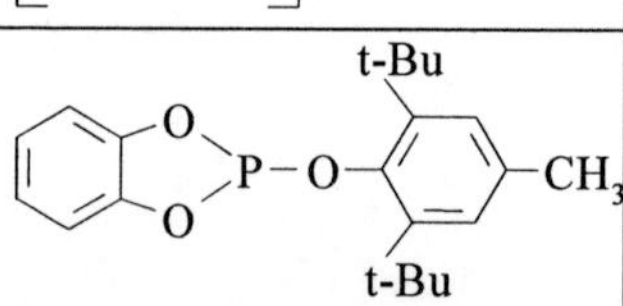	CHP	Benzene	1.3	4.5×10^{-2}
(6)	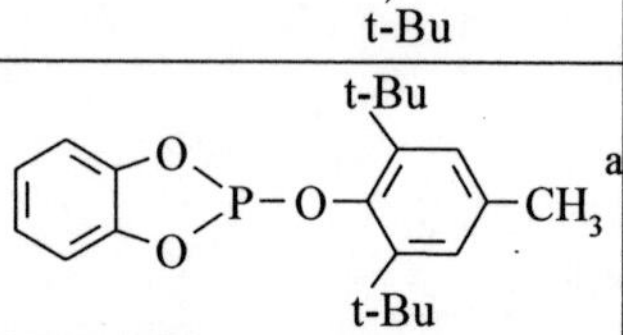 [a]	CHP	Styrene	–	$(5\text{-}10)\times10^{-2}$
(7)	$(C_6H_5O)_3P$	TBH	CH_2Cl_2	10.0	4×10^{-2}
(13)	$(C_6H_5O)_2P\text{-}OC_8H_{17}\text{-}i$	TBH	Benzene	30.4	2×10^{-4}
(14)	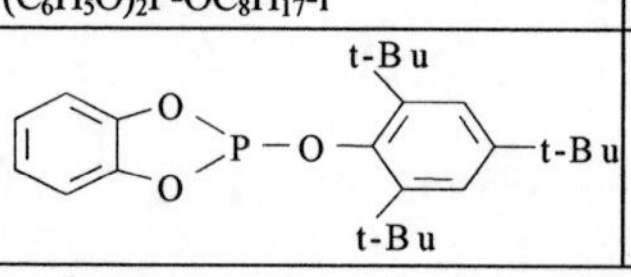	CHP	Benzene	5.0	3.3×10^{-2}

[a] Measured by rate of styrene polymerization at 348 K.

The scheme of forming radicals according to the publication [34] can be presented within the framework of latent-radical mechanism:

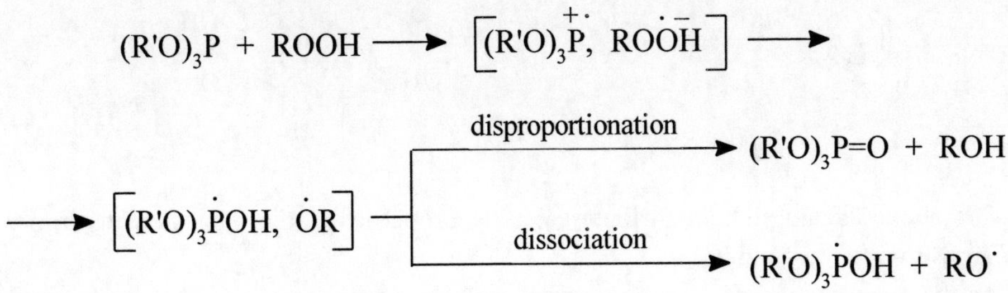

Scheme 2.2.

The research of the given reaction by the method of chemically induced dynamic nuclear polarisation (CIDNP) showed that alternative (nonradical and radical) reactions actually proceed by two parallel routes. However the radical route plays a smaller role though it becomes apparent in nuclear polarization [40].

As a whole, the data on interaction of hydroperoxides with phosphites provide the evidence to consider that this reaction will stipulate the nonradical decomposition of hydroperoxides (Scheme 2.1), formed in the process of thermal-oxidative destruction of polymers and, as a result, prevent the degenerate branching of oxidation chains.

At the same time the degree of efficiency of phosphite action as hydroperoxide decomposers can be various depending on the structure of phosphites. That is caused by the further conversions of phosphates formed according to Scheme 2.1.

Reactions of hydroperoxides with a number of cyclic phosphites during modeling polymer stabilization illustrate that [41-45].

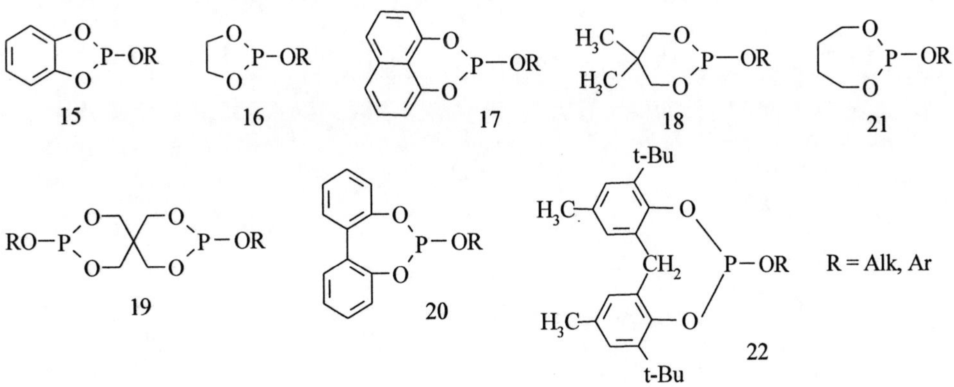

Phosphites with small cycles, in particular with pentamerous ones (15, 16) on the basis of pyrocatechin, glycol etc. are of special interest, they are characterized by high stabilizing efficiency in polymers and low-molecular hydrocarbon substrata.

A number of researchers attribute these particular properties of cyclic phosphites, in particular, *ortho*-phenylene-alkyl(aryl)phosphites, to their ability to be exposed to slight hydrolysis followed by the formation of *ortho*-phenylenephosphorous acid (23) [46] and *ortho*-phenylene- and hydroxyphenylenephosphorous acid (24, 25) [47].

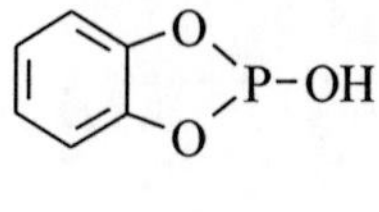

23

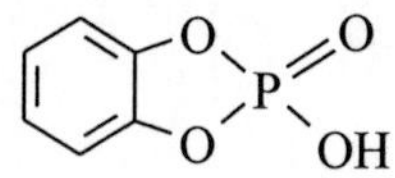

24

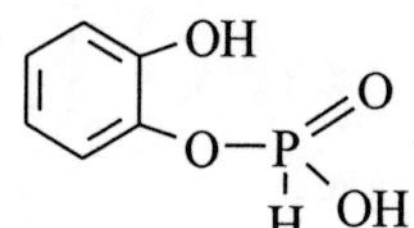

25

The above-named authors attribute catalytic properties in the process of hydroperoxide decomposition to these acids.

As there can be phenol (in case of aromatic esters), pyrocatechin, phosphorous acid etc. in oxidized system it is not excluded that the above-stated products in various combinations can promote the formation of synergism effect during the stabilization by *ortho*-phenylenephosphorous acid esters.

However a number of the experimental data do not confirm these hypotheses. K. Shvetlik et al. [41-45] conducted the detailed research of kinetics and mechanism of modeling reactions of cyclic phosphites with cumene hydroperoxides (CHP), it allowed to find out peculiarities of interaction of *ortho*-phenylenephosphite derivatives with hydroperoxides depending on conditions (the quantitative ratio of *phosphate – hydroperoxide*, temperature, the structure of initial phosphite).

It is ascertained by methods of NMR-spectroscopy, liquid chromatography and iodine metric titration that pentamerous cyclic phosphites react with CHP initially with forming the corresponding phosphate (26) and cumyl alcohol (Scheme 2.3). However, the disclosing of the cycle of formed unstable phosphate (27) takes place along with this process at the subsequent peroxidelysis (surplus of CHP) or hydrolysis (water presence) and a new product, openly chain phosphate of acid nature, hydroxyphenylphospate (28) is formed.

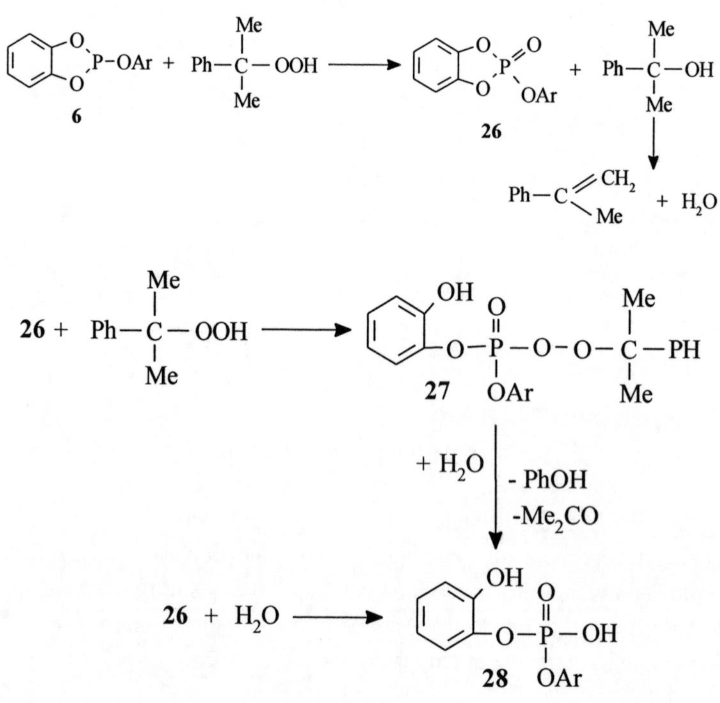

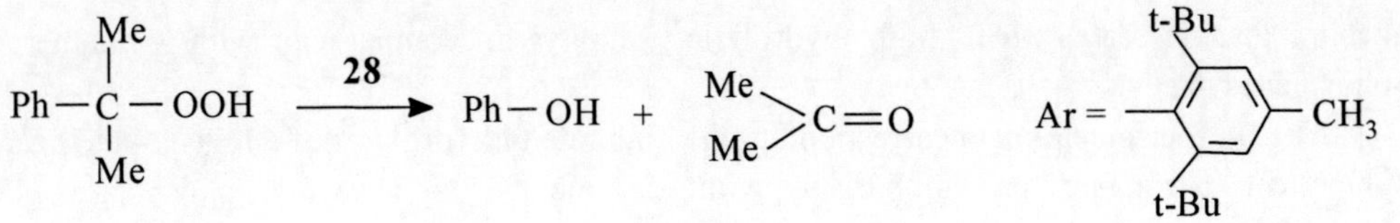

Scheme 2.3.

The suggested Scheme 2.3 is illustrated by figure 2.1.

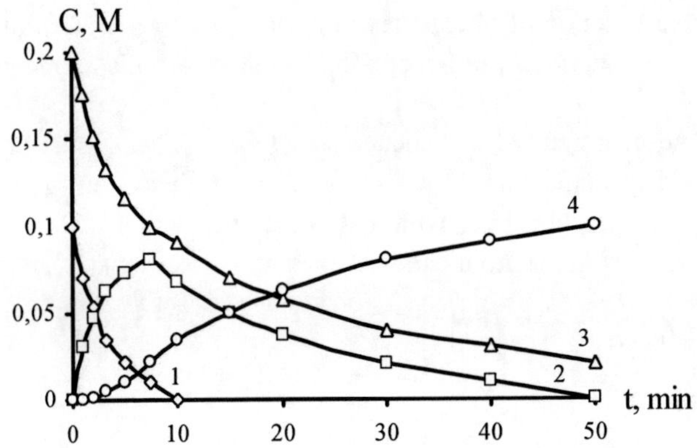

Figure 2.1. Products of reaction of cumene hydroperoxide (CHP) (0.2 mole) with 2,6-di-tert-butyl-4-methylphenyl-ortho-phenylenephosphite (6) (0.1 mole): 1 – compound (6), 2 – compound (27), 3 – CHP, 4 – compound (28) (323 K, chlorbenzene).

Scheme 2.3 illustrates that initially the interaction of aryl-*ortho*-phenylenephosphite (6) with cumene hydroperoxide takes place that corresponds to the curves of the consumption of these products. As the reaction proceeds there is the accumulation of the product (27) which turns into hydroxyphenylphosphate (28) after some time.

It is acid phosphate (28) that, according to the authors' opinion, is the catalyst of CHP decomposition into phenol and acetone, forming thus high stabilizing effect both in modeling hydrocarbon and polymer. Actually when hydroxyphenylphosphate (28) obtained by means of counter synthesis is introduced into oxidized polymer, the stabilizing effect, similar to the one when aryl-*ortho*-phenylenephosphite (6) is used as initial stabilizer, takes place (figure 2.1).

Cyclic phosphites with large 6-8 member cycles (17, 20-22) react with CHP stoichiometrically, forming only relevant phosphate and cumyl alcohol and negligible amount of acetophenone (less than 4 %) [42].

The nature of exocyclic residue in phosphites does not render any influence on the reaction mechanism; the formation of products of catalytic hydroperoxide decomposition at 348 K (phenol, acetone, methyl styrene) it is not observed either.

Besides, special experiments show that cyclic phosphates with 6-8 member cycles are stable as to the surplus of hydroperoxides at 348 K and 363 K and hydrolysis resistant. The catalytic decomposition of hydroperoxide in the presence of other cyclic phosphates does not occur, that indicates their high hydrolytic stability in comparison with pentamerous phosphates [42].

These experimental data correspond to the high rate of hydrolysis of ethylene- and *ortho*-phenylene phosphates in comparison with high-member and acyclic phosphates [48]. Phosphates with the large sizes of cycles are characterized by the rate constant of hydrolysis similar to acyclic esters of phosphoric acid [25, p.361]

Thus, during the process of polymer oxidation inhibition by arylene- and alkylenephosphites with minor cycles at their interaction with hydroperoxides the products of conversion are formed which are effective catalysts of nonradical hydroperoxide decomposition, and as a result they stipulate the high effects of stabilization.

It testifies to the function of phosphites of such type as generators of a new inhibitor in the process of thermooxidation of polymers, that is one of promising means of their effective stabilization.

To establish the quantitative dependence *structure – property* the correlation equations (15) – (17) [49-51], connecting stoichiometric coefficients (v) and rate constants of interaction of trialkyl(aryl)phosphites with hydroperoxides with Tafft's induction constants of substitutes (σ^*) at a phosphorus atom and with their steric constants R_s are deduced:

$$\lg k = a_0 + a_1 \Sigma\sigma^* + a_2 R_S \quad (15)$$

$$\lg v = a_0 + a_1 \Sigma\sigma^* + a_2 R_S \quad (16)$$

$$\lg k = a_0 + a_2 R_S \quad (17)$$

Steric constants R_S of a substitute are defined on the basis of additive equation (18):

$$R_S = 30 \times \lg(1 - \Sigma R_i^2 / 4r_i^2), \quad (18)$$

where R_i is a covalent radius of i-atom, r_i is its remoteness from reaction center.

Tafft's induction constants of substitutes at the nonhydrocarbon reaction center are calculated theoretically on the basis of equation (19):

$$\sigma^* = 7{,}840 \sum \Delta\chi R_i^2 / r_i^2, \quad (19)$$

where R_i is a covalent radius of i-atom, r_i is its remoteness from reaction center, $\Delta\chi_i$ is difference of electronegativities of i-atom and reaction center.

When calculating steric constants R_S for aryl substitutes the angle of rotation of the plane of an aryl ring round the bond $=Csp^2$–O was varied every 10^0. It turned out that the best conformity R_S(P) with the rate constant was observed at the value of rotation angle of the plane of aryl rings regarding the plane P-O-Csp^2 equal to 45 ± 10^0 that agrees well with the publication data and meets the least sterically loaded molecule conformation.

The correlation analysis in coordinates *lg k* – R_S(P) for the bulk of phosphites showed that it broke into 5 reactionary series (figure 2.2). In each reactionary series substances with the one-type electronic and spatial structure of molecules are united, and, in whole rather strict linear correlation ratios between logarithms of rate constants and total steric effect of substitutes at a phosphorus atom are observed in each reactionary series. Thus in all cases the rate of bimolecular reactions decreases with the increase of spatial requirements of substitutes. Extremely close within an experimental error sensitivity to steric effect (nearly identical inclination in coordinates *lg k – R_S*) for reactionary series A–D (figure 2.2) testifies that the mechanism of reactions for all specified series remains practically constant, and distinctions between them are caused by not taking into account the electronic and stereoelectronic effects of substitutes which vary considerably within the common studied.

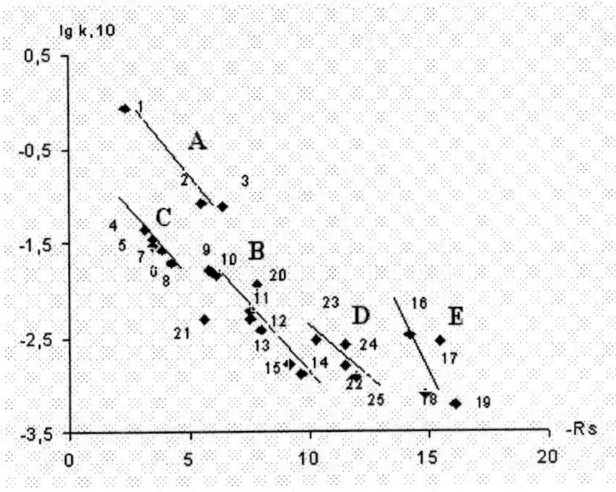

Figure 2.2. Dependence between logarithms of rate constants of the reaction of phosphites with cumene hydroperoxide and R_S-constants of substitutes at a phosphorus atom: 1 – compound (1); 2 – (3); 3 – (4); 4 – (29); 5 – (30); 6 – (31); 7 – (32); 8 – (33); 9 – (7); 10 – (11); 11 – (34); 12 – (35); 13 – (5); 14 – (6); 15 – (14); 16 – (36); 17 – (9); 18 – (8); 19 – (37); 20 – (38); 21 – (39); 22 – (12); 23 – (40); 24 – (41); 25 – (42).

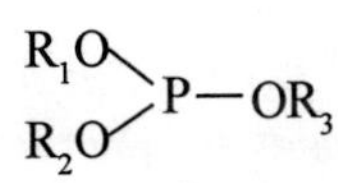

$R_1R_2 = CH_2CH_2CH_2$ **(29)**, $CH(CH_3)CH_2CH_2$ **(30)**, $CH_2C(CH_3)_2CH_2$ **(31)**, $CH(CH_3)CH(CH_3)$ **(32)**,

$CH(CH_3)CH_2$ **(33)**, 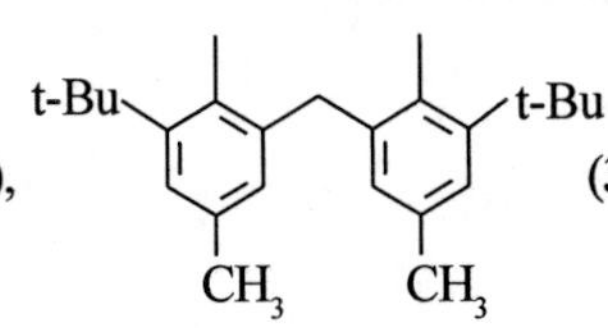 **(36)**, 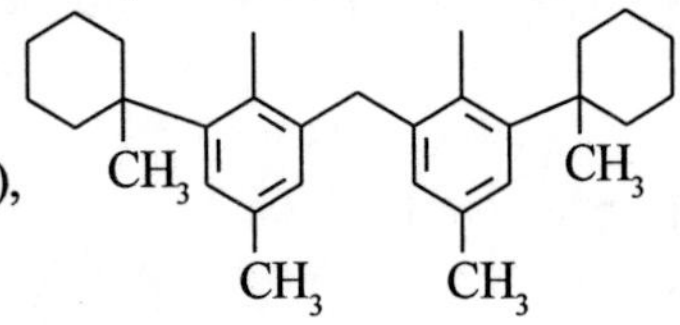 **(37)**;

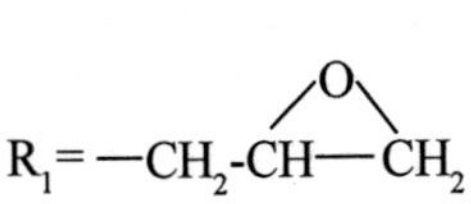 **(38, 39)**; $R_1=R_2=R_3=$ 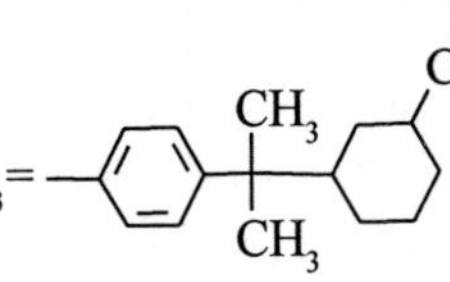 **(34)**, 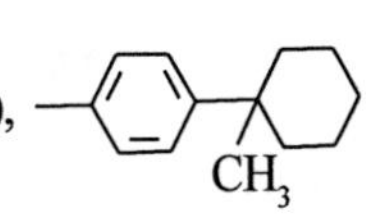 **(35)**,

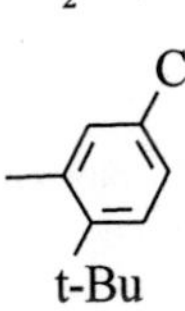 **(40)**, 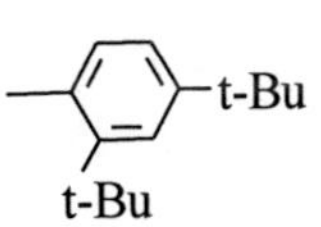 **(41)**, 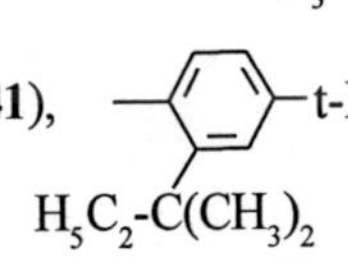 **(42)**;

$R_2 = C_6H_5$ **(38)**, $CH_2CH(C_2H_5)_2$ **(39)**; $R_3 = C_2H_5$ **(29-33)**, C_6H_5 **(36, 38, 39)**, 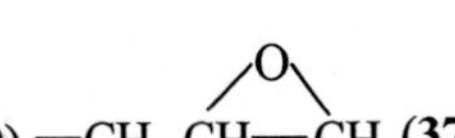 **(37)**

Series A includes trialkylphosphites (1, 3, 4) and is described by correlation equation (20)

$$\lg k = -(1.868 \pm 1.098) + (0.492 \pm 0.203)R_S; \qquad (20)$$

R=0.9243; S_0=0.319; N=3

Series B (equation 21) combines basically sterically non-overladen triarylphosphites (5-7, 11, 14, 34, 35) which contain no more than one *ortho*-substituted aryl substitute. They possess lower reactivity in comparison with trialkylphosphites, it is obvious that it is stipulated by electron-accepting nature of aryl substitutes, not allowing them to form single reaction series with trialkylphosphites in the said coordinates (straight line B is lower than straight line A, though practically it is parallel to it).

$$\lg k = (0.052 \pm 0.360) + (0.296 \pm 0.046)R_S; \qquad (21)$$

R=0.9979; S_0=0.030; N=7

Series C includes cyclic dioxyalkylene substitutes possessing apparent acceptor effect (compounds 29-33), that reduces the rate constant of interaction of phosphites with CHP (equation 22). Thus the sensitivity to steric effect remains practically invariable.

$$\lg k = (0.170 \pm 0.082) + (0.378 \pm 0.021)R_S, \qquad (22)$$

R=0.9941; S_0=0.041; N=6

As a whole, monotone decrease of reactivity alongside with the increase of electron-accepting nature of substitutes at a phosphorus atom at the transition from series A to series B and C is quite natural for derivatives of trivalent phosphorus and testifies to the nucleophilic (from standpoint of the phosphoric partner) mechanism of the reaction, the limiting stage of which is likely to be the oxidation of phosphate by cumene hydroperoxide

A very interesting situation was observed for compounds (12, 40-42) united into series D. These compounds are triarylphosphites, their difference from agents of series B is that each of three aryls has an *ortho*-substitute. Firstly, it leads to the large total load of the reactionary centre (a phosphorus atom), and secondly, to noticeable steric repulsion of three *ortho*-substitutes of aryl groups in a molecule of triarylphosphite. Such repulsion should lead inevitably to the increase of valence angles O-P-O in a phosphite molecule and, as a result, to the change of hybridization of atomic orbitals of phosphorus from pure p- (valence angle 90°) to sp^3-state (valence angle 109°). In turn, this change of nature of unshared electronic pair of a phosphorus atom leads to the increase of its nucleophilicity and reactivity of phosphites (equation 23). Thus, classical demonstration of stereoelectronic effect can be observed and three *ortho*-substituted radicals formally behave as donors in comparison with usual aryls.

$$\lg k = (0.146 \pm 0.168) + (0.231 \pm 0.015)R_S \qquad (23)$$

R=0.9959; S_0=0.019; N=4

The further growing of screening of the reactionary centre (significantly sterically overloaded molecules (8, 9, 36, 37)) increases even more the contribution of stereoelectronic effect according to the described above mechanism. Besides, it leads to an appreciable distortion of the structure of the activated complex, that is seen in the sharp increase (approximately three times) of the sensitivity of reaction rate to steric effect of substitutes (equation 24, series E).

$$\lg k = -(10.160 \pm 0.377) + (0.934 \pm 0.022) R_S \qquad (24)$$

$R = 0.9996$; $S_0 = 0.016$; $N = 4$

Thus, the results of the research demonstrate that reactivity of phosphorous acids esters in the reaction with hydroperoxide is checked by electronic and steric effects of substitutes at a phosphorus atom which is the reactionary centre. The increase of spatial load and electron-accepting nature of substitutes leads to the decrease of reaction rate. Strict linear dependences between logarithms of rate constants and steric R_S-constants of substitutes at phosphorus are carried out with excellent (in most cases) quality of correlation ($R > 0.99$) in series of substitutes which are of single-type according to their electronic influence.

2.2.1.2. Acid Esters of Phosphorous Acid (Dialkyl(Aryl)Phosphites)

Acid esters of phosphorous acid (acyclic dialkyl(aryl)- and cyclic alkylene(arylene) phosphites) are of considerable interest taking into consideration their stabilizing abilities. They can act as independent stabilizers, being highly reactive compounds in relation to active products of thermal-oxidative destruction of polymers, and also as components of mixtures with other types of stabilizers demonstrating synergistic effects [52-55].

Moreover, acid esters of phosphorous acid are capable to be generated in the course of conversion of full phosphites under oxidation inhibited by them and to take part in the common process of polymer stabilization [56].

Reactions of acid esters of phosphorous acid with cumene hydroperoxide (CHP) are investigated by means of kinetic polarography (Tab. 2.3) [57].

It is established that interaction of CHP with diaryl-, alkylene- and arylenephosphites proceed with appreciable rate at 373 K. The calculated rate constants correspond to the kinetic equation for trimolecular reactions at kinetic orders by hydroperoxide and acid, equal to 2 and 1 respectively [56, 58].

The interrelation of structure of acid esters of phosphorous acid with reactivity in respect of hydroperoxides is shown in a number of examples in table 2.3.

So, biphenylene- and diphenylphosphorous acids (44) and (47) possess low-activity, di-*i*-octyl phosphorous acid (50) doesn't react under conditions of testing. It is essential, that oligo(propylidenediphenyl)phosphorous acid (51) reacts with the greatest amount of hydroperoxide molecules, its stoichiometric coefficient ν equals to 800, that is caused, apparently by the presence of several reactionary centres in a molecule.

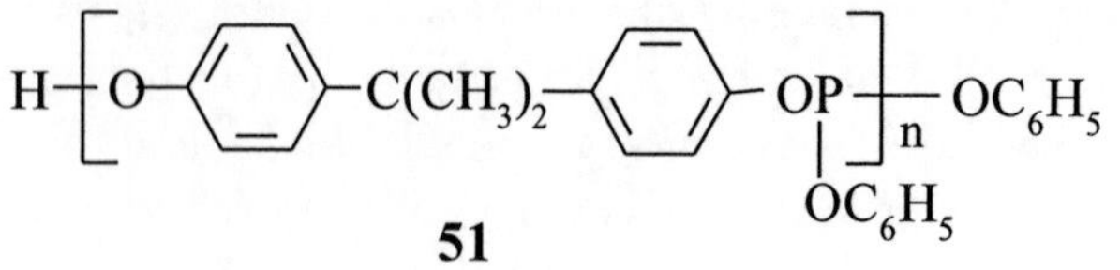

n=3-4

51

Table 2.3. Kinetic parametres of CHP decomposition reaction under the influence of acid esters of phosphorous acid (chlorobenzene, 373 K, $[CHP]_0$=0.1 mole/l), their constants of ionisation and values of chemical shifts of phosphorus atoms

Compound number	Compound	v_{CHP}	*k*, $l^2/(mole^2 \cdot s)$	pk_a, propanol	^{31}P, δ, ppm (J_{PH}, Hz)
(23)	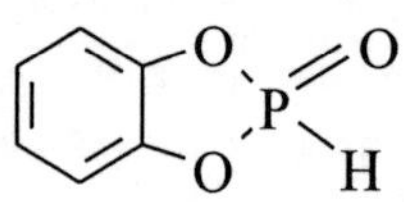	100.0	210.0	4.19 10.57	129 126
(43)	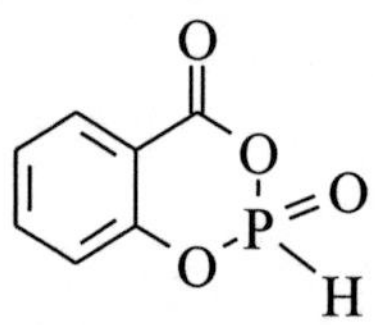	28.0	3.0	–	5 (680)
(44)	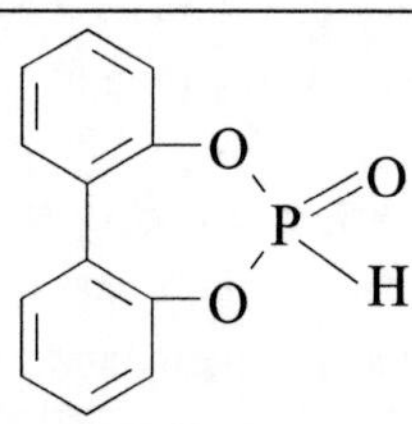	15.0	0.4	10.68	6 (720)
(45)	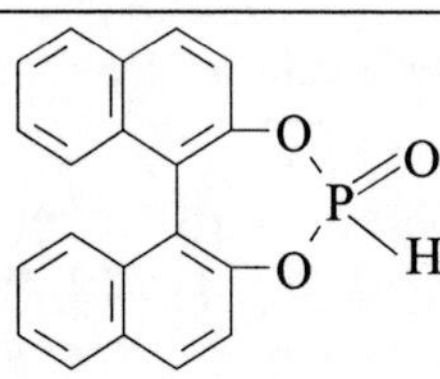	7.0	–	10.76	6 (730)
(46)	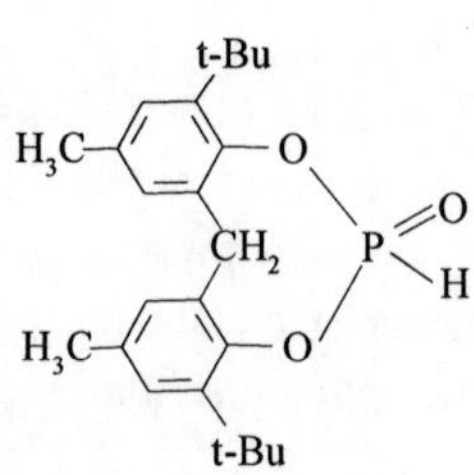	0.5	–	12.54	2 (735) 5 (700)

Compound number	Compound	νCHP	k, l2/(mole2·s)	pka, propanol	31P, δ, ppm (JPH, Hz)
(47)	$(C_6H_5O)_2P(O)H$	11.0	0.6	11.46	4 (740) 0 (746)
(48)	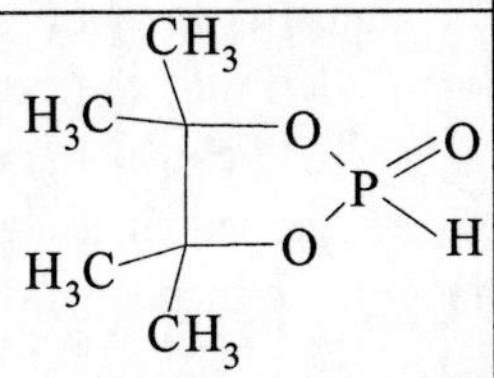	30.0	1.2	8.54	-20 (705)
(49)	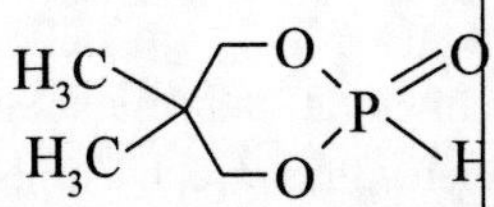	27.0	–	11.85	-9 (670) -5 (660)
(50)	$(i\text{-}C_8H_{17}O)_2P(O)H$	doesn't react			

Dyadic prototropic tautomerism is characteristic of acid esters of phosphorous acid [58]. According to it the reactivity of these compounds, which differs greatly depending on the structure of the substitute, can be connected with the possibility of compounds of the given group being in two forms (A) and (B):

$(RO)_2P(O)H \leftrightarrow (RO)_2POH$
(A) (B)

Tautomeric balance is caused by protondonor properties of a substitute at a phosphorus atom [59-61].

Values pk_a, obtained for alkylenephosphites refer to P^{III} form and testify that in these systems tautomeric balance is considerably shifted to more acid P^{III} form. Such rather easy transition in compounds with P^{III} atom is likely to be promoted considerably both by large acceptor effect of phospholane substitutes and by their spatial structure: angles x-R-x in pentamerous derivatives (~98°) are slightly less than in phosphacene systems (~104°), that naturally makes the specified transition easier. Thus, pentamerous alkylenephosphites are considerably stronger acids than hexamerous and acyclic analogues.

It is essential that the open structure of β-oxyalkylphosphite (**52**) formed as a result of opened pentamerous rings is responsible for the increased acidity of phosphorous acids of a number of 1,3,2-dioxaphospholanes [60]. This hydrolytic process is reversible and when the attempt is made to isolate the opened form the latter is exposed to recycling which results in releasing water and pinacophosphate.

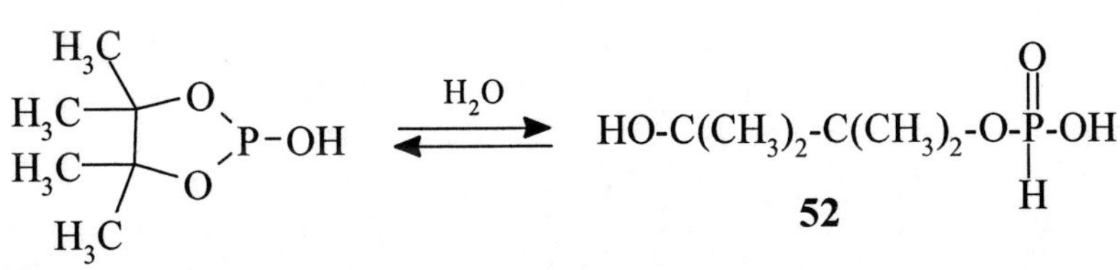

52

It is known that increasing the size of the ring from pentamerous to octatomic one leads to considerable reduction of proton-donor properties of cyclic phosphorous acids and makes them comparable according to acid properties with acyclic analogues [62]. The share of structure (A) increases in the same sequence in tautomeric balance A↔B, structure (A) becomes completely prevailing for an octatomic cyclic derivative (46) in comparison with acid (23), the data of NMR ^{31}P spectroscopy and values *pK* testify to that (Tab. 2.3) [63].

According to the adduced data the essential factor defining the reactivity of cyclic phosphorus acids in respect of hydroperoxides is the size of a cycle and its steric impediment as well as the quantitative content of the form with OH-bond which basically characterizes the degree of their acidity and is increased while changing from pentamerous to octatomic cycloanalog.

Comparing to the data on the ionization constants which are defined by the certain method of potentiometric titration, it is possible to conclude, that the acids characterized by large (above 8) values of *pK* do not react; hydroperoxides are decomposed only by the strongest acids of the above-stated series with *pK* less than 8 (tautomeric form (B), containing trivalent phosphorus).

Catalytic decomposition of hydroperoxide under the influence of strong acid is confirmed by its consumption during the stepped adding of acid (23) into the reaction system, as well as by high values of stoichiometric coefficients of a reaction (Tab. 2.3).

The scheme of the interaction with hydroperoxide, *ortho*-phenylenephosphorous acid (23) taken as an example, can be presented by the following sequence of reactions [47, 55, 64]:

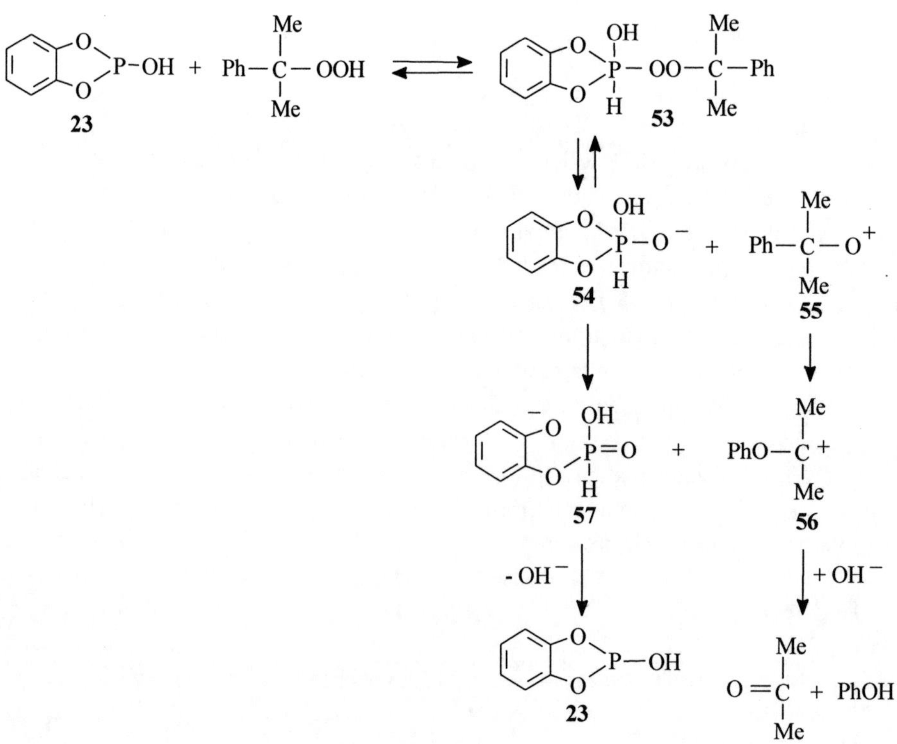

Scheme 2.4.

Thus, as a result of attack of *ortho*-phenylenephosphites (23) by hydroperoxide hydrophosphorane (53) is formed [65], which is likely to be able to dissociate into oxyphosphoranyle-anion (54) and dimethylbenzyloxy-cation (55) converting into carbcation (56). Delocalization of electron density in anion (54) leads to an anion of open structure (57), and further recycling with the formation of initial acid (23) ensures elimination of hydroxyl-anion. Carbcation (56) owing to the attack of eliminated hydroxyl turns into phenol and acetone.

Lower activity of diphenylphosphate (47) in comparison with *ortho*-diphenylphosphite (23) fits well into the given scheme, lower activity is explained by the fact that the probability of reverse process of substitution at a phosphorus atom is less possible at the termination of ester bond R–OP.

The probability of radical mechanism of reactions of acids with hydroperoxides is investigated. According to CNP ^{31}P (chemical nuclear polarization) data when hydroperoxide reacts with *optho*-phenylenephosphite (23) a series of very fast reactions with the participation of radicals takes place, and products having phosphate and phosphonate structures are formed (Scheme 2.5) [55].

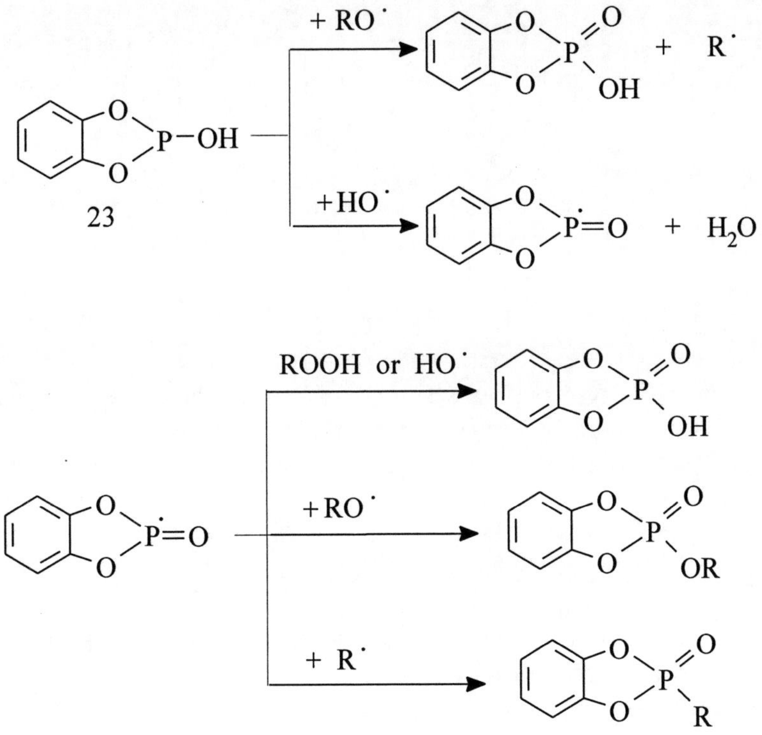

Scheme 2.5.

However, as the chemical polarization at link positions satisfying these products was absent, there are no signals found, and therefore, a share of radical mechanism is negligible.

The dominance of non-radical catalytic decomposition of hydroperoxide is testified as well by the absence of influence of typical acceptor BHT on the rate of hydroperoxide decomposition and by the detection of only 0.2 wt% products of radical decomposition of

CHP (cumyl alcohol, acetophenone), the detection is made by the method of gas-liquid chromatography.

The identification of intermediate and end products of reaction is conducted on the basis of analysis of NMR ^{31}P spectra [57].

Apart from the products shown in Scheme 2.4, there is the possibility of forming phosphorous acid and pyrocatechin due to hydrolysis of *ortho-* phenylenephosphite (23) by the water releasing in the course of reaction.

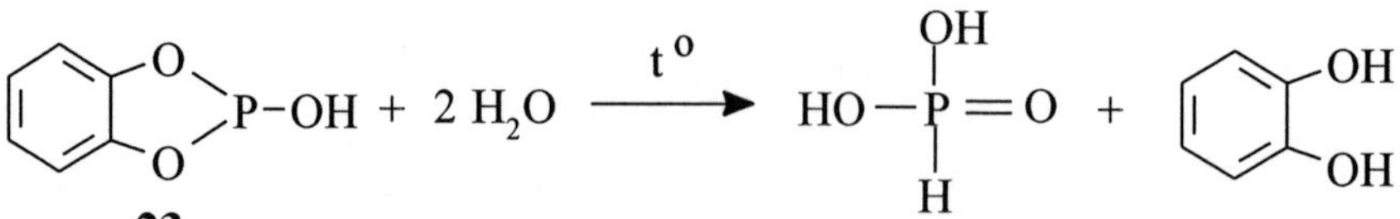

23

The formation of hydrospirophosphorane (58) during the experiment can take place according to the reaction which is modeled separately. The reaction of phosphorous acid with two molecules of pyrocatechin results in the same spirophosphorane (58).

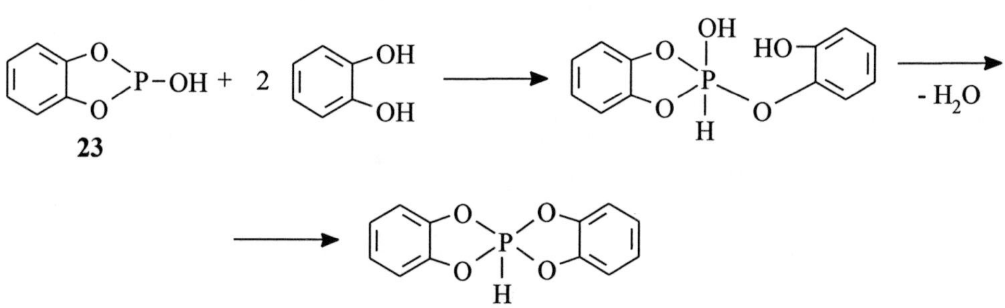

23 **58**

Parallel running of reactions leads to the formation of nonreactive organophosphorous compounds and to dying-away of catalytic decomposition of CHP.

When predicting efficient structures of acid esters of phosphorous acids in the processes of hydroperoxide decomposition it should be taken into consideration that dialkylphosphites will possess the most activity, pentacovalent intermediates promoting further regeneration of initial acid are formed very easily for these dialkylphosphites.

Acid esters of phosphorous acids are more efficient in nonradical hydroperoxide decomposition in comparison with full esters of phosphorous acids. When OH-group is substituted by OR (phosphites) there can be observed a sharp fall of the activity, that might be stipulated by stronger bond of RO–P in an anion (57) (Scheme 2.4), as a result the process of regeneration of phosphites is not ensured.

The specific character of the mechanism of reaction of acid esters of phosphorous acids is seen in the autocatalytic nature of the reaction, as a result there is considerable increasing of stoichiometric coefficient.

2.2.1.3. Phosphorus Dithioacids

Phosporus dithioacids having strongly pronounced acid properties [66] display high activity in reactions with hydroperoxides. This served as the basis to consider them from the position of the possibility to stabilize polymers due to suppression of degenerated branching of oxidation chain.

The reactions of cumene hydroperoxide (CHP) with a number of dithiophosphorous (59, 60), dithiophosphoric (61-67) and dithiophosphonic (68-74) acids were studied [67-69].

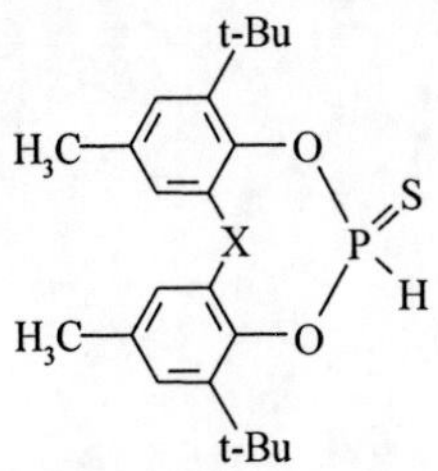

X = CH_2 (**59**), S (**60**)

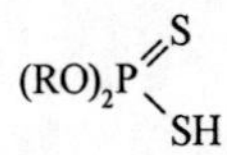

R = CH_3 (**61**), C_3H_7 (**62**), i-C_3H_7 (**63**), C_6H_5 (**64**)

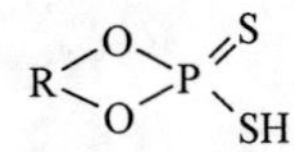

R = $-CH(CH_3)CH(CH_3)-$ (**65**), $-CH(CH_3)CH_2CH_2-$ (**66**), $-CH_2C(CH_3)_2CH_2-$ (**67**)

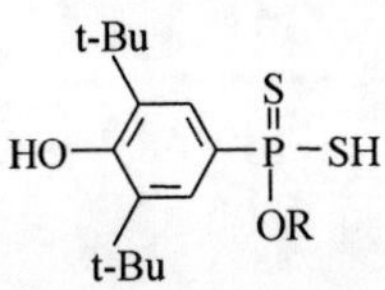

R = C_2H_5 (**68**), i-C_3H_7 (**69**), i-C_8H_{17} (**70**), cyclo-C_6H_{11} (**71**), $-CH_2CH{=}CH_2$ (**72**)

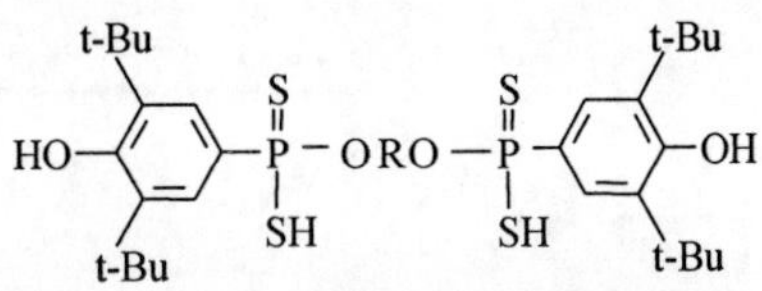

R = $(CH_2)_3$ (**73**), $(CH_2)_4$ (**74**)

Methods of polarography, ^{1}H, ^{13}C, ^{31}P NMR spectroscopy were used to consider the changing of reactivity of dithioacids of various structure including ones with sterically hindered phenol fragment in reactions with CHP [68, 69].

When the dependence of initial rates of CHP decomposition on the concentration of phosphorus dithioacids was investigated it was determined that reactions with dialkyldithiophosphoric (61-63) and diphenyldithiophosphoric (64) acids proceed without the induction period.

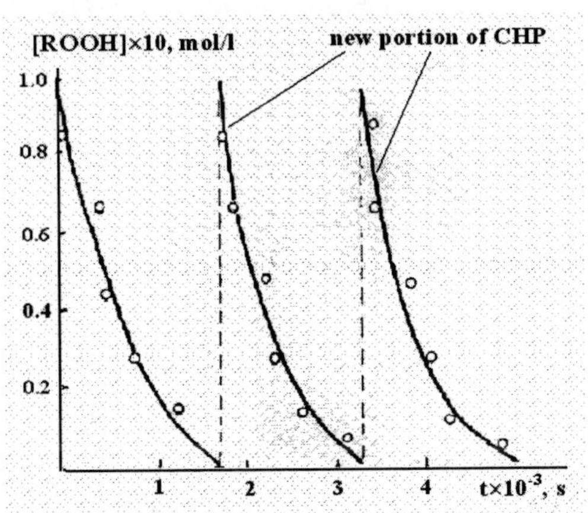

Figure 2.3. Kinetic curves of cumene hydroperoxide (CHP) decomposition under the influence of the compound (63) with the addition of a new portion of CHP after its decomposition (373 K, chlorobenzene, $[CHP]_0$=0.1 mole/l, [(63)]= 0.1 mole/l).

In case of dithiophosphonic acids (68-72) containing sterically hindered phenol fragments, a short "slow" zone appears on curves of CHP consumption depending on time (figure 2.4, curves 1-4; figure 2.5) where no more than 10 % of hydroperoxide is consumed (stepped character of a reaction).

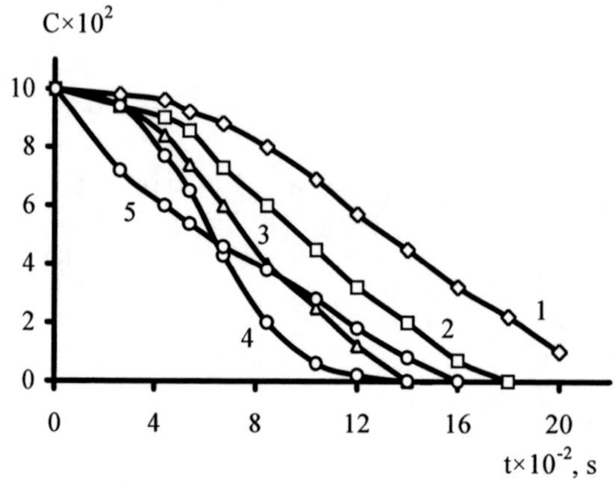

Figure 2.4. Kinetic curves of CHP decomposition (C, mole/l) under the influence of O-alkyl-(3,5-di-tert-butyl-4-hydroxyphenyl)-dithiophosphonic acids (DTPC): 1 – compound (68), 2 – compound (70), 3 – compound (71), 4 – compound (72), 5 – compound (73) (373 K, chlorbenzene, $[CHP]_0$=0.1 mole/l; [DTPC]=1×10^{-3} mole/l).

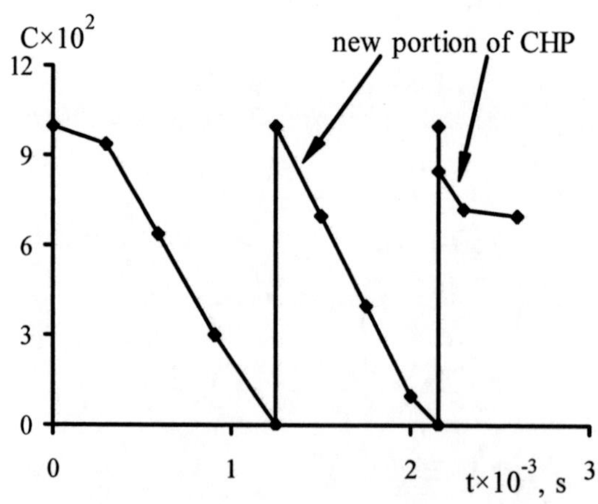

Figure 2.5. Kinetic curves of CHP decomposition (C, mol/l) under the influence of dithiophosphonic acid (71) with the addition of a new portion of CHP after its decomposition (373 K, chlorbenzene, $[CHP]_0$=0.1 mole/l; [(71)]=1×10^{-3} mole/l).

If during the reaction a new portion of hydroperoxide is added after its full exhaustion, a "slow" zone on kinetic curves it is not observed (figure 2.5). That is why it is possible to consider that in the course of the interaction products are formed which are active with respect to hydroperoxides capable of continuing a reaction.

Investigating the dependence of initial rates of CHP decomposition on the concentration of acids allowed to determine that reactions with thiophosphorous (59, 60) and dithiophosphoric acids (61-67) have the first order regarding the acid and the second one regarding hydroperoxide; for dithiophosphonic (68-72) acids both phases of reaction are described by the equation of the general third order and zero one regarding CHP respectively.

Values of rate constants, activation energy and preexponents of these reactions are given in table 2.4 [57].

Studying gross-consumption of cumene hydroperoxide in reactions with dithiophosphoric (61-65) and dithiophosphonic (68-71) acids showed that they are characterized of abnormal high values of stoichiometric coefficients *ν* (table 2.5) that testifies to the catalytic decomposition of hydroperoxide in the presence of both types of acids [68, 69].

Table 2.4. Kinetic characteristics (rate constants k, activation energy E, preexponents A) of the reaction of cumene hydroperoxide (CHP) decomposition under the influence of phosphorus dithioacids (293-373 K, chlorbenzene, $[CHP]_0 = 0{,}1$ mole/l; $[In]_0 = 0{,}1\times10^{-2}$ mole/l)

Compound number	"Slow' zone	"Fast" zone		
	k_s, s^{-1} at 373 K	k_f, $l^2/(mole^2 \cdot sec)$ at 373 K	$E_{акт}$, kJ /mole	*lg A*
(62)	N/A	12.30	53.5	8.7
(63)	N/A	16.70	54.9	9.6
(65)	N/A	13.30	89.5	13.6
(68)	Not calculated	14.40	62.2	9.9
(70)	2.60×10^{-3}	3.20	-	-
(71)	3.97×10^{-3}	5.05	-	-

Table 2.5. Stoichiometric coefficients (ν) of reactions of phosphorous dithioacids with cumene hydroperoxide

Compound number	*ν* (373 K)	Compound number	*ν* (373 K)
(61)	15000	68	1800
(62)	18000	69	2000
(63)	20000	70	2600
(64)	25000	71	2500
(65)	22000		

Stoichiometric coefficients are interrelated with *pK* values of acids. The maximum coefficients are observed for diphenyl- (64), 2,3-butylene- (65), diisopropyl- (63) dithiophosphoric acids which are the strongest among the investigated phosphorus dithioacids. For example, *pK* for the specified dithiophosphoric acids amounts to 3.90-3.62 while for dithioacids with sterically hindered phenol fragment (68-71) this value is in the range of 4.37–4.12. The obtained data assume the acid nature of catalysis in the reactions of phosphorus dithioacids with hydroperoxides.

The suggested scheme of acid decomposition of CHP includes a cationic chain reaction where cation (75) and carbocation (76) are chain carriers, similar to the scheme given in the publication [70]:

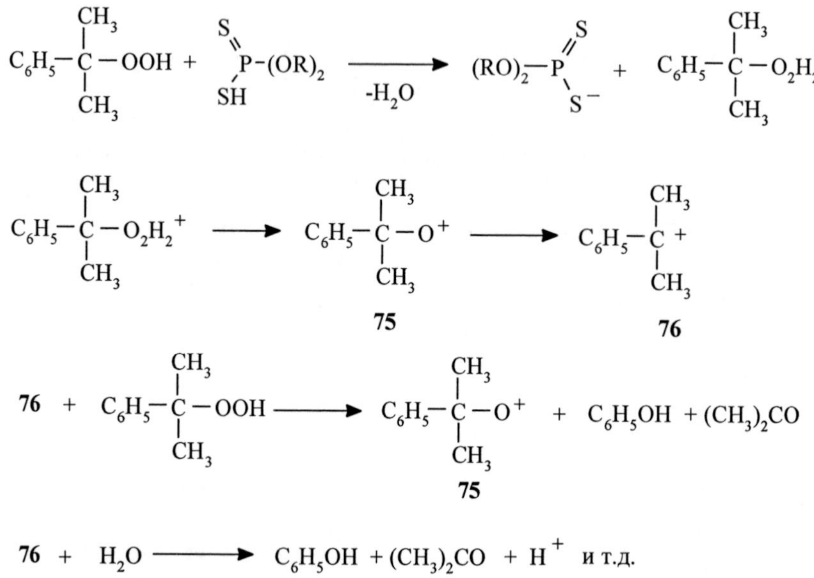

Scheme 2.6.

The data about the reaction of diisopropyldithiophosphoric acid (**63**) with CHP obtained by the method of NMR ^{31}P spectroscopy testify that under the conditions of rather large concentrations both for dithioacid (0.5 mole/l), and for CHP (1 mole/l) the oxidation process proceeds in several minutes resulting in the formation of sulfides [56, 69]. The formation of bis(diisopropylthiophosphoryl)disulfide (**77**) was confirmed by the results of IR-spectroscopy and presented by the Scheme 2.7:

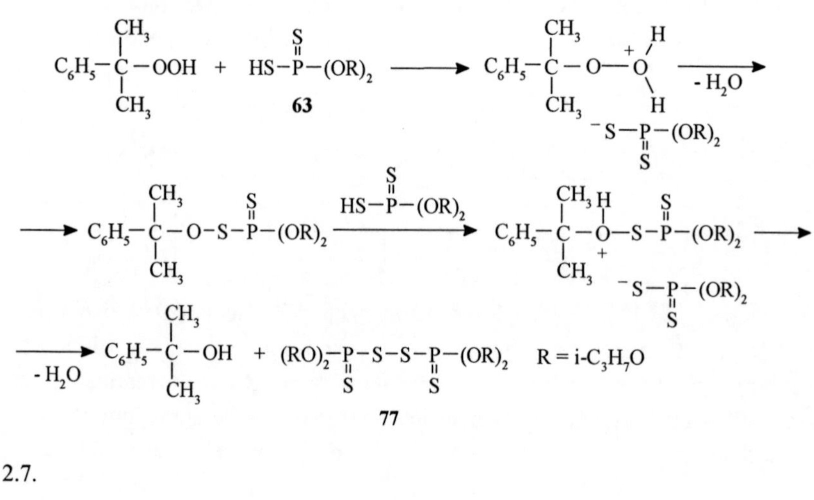

Scheme 2.7.

Apart from the main reaction of oxidation the side reaction of forming corresponding trisulfides (78) proceeds, that is proved by NMR ^{31}P.

$$(RO)_2P(S)SH + H_2O \longrightarrow \underset{79}{(RO)(OH)P(S)SH} + ROH$$

$$79 + (RO)_2P(S)SH + C_6H_5{-}C(CH_3)_2{-}OOH \longrightarrow \underset{80}{(RO)_2P(S)SSP(S)(OR)(OH)} + H_2O + C_6H_5{-}C(CH_3)_2{-}OH$$

$$80 + H_2O \longrightarrow \underset{81}{(RO)_2P(S)SSH} + (RO)P(S)(OH)_2$$

$$(RO)_2P(S)SH + 81 + C_6H_5{-}C(CH_3)_2{-}OOH \longrightarrow \underset{78}{(RO)_2P(S)SSSP(S)(OR)_2} + H_2O + C_6H_5{-}C(CH_3)_2{-}OH$$

As phosphorus dithioacids are characterized by high reactivity both in electrophilic and in nucleophilic processes [66], they are likely to be able to interact rather easily with products of heterolytic and homolytic decomposition of hydroperoxide, these products were found out by means of gas-liquid chromatography, spectra of NMR ^{1}H and ^{13}C.

As a result of oxidizing conversions of dithio- and thiocompounds sulfuric acid can be formed in the reaction mixture too, sulfuric acid is capable to be an effective catalyst of hydroperoxide decomposition [69].

While discussing the mechanism of the process of hydroperoxide decomposition by dithioacids it is necessary to take into consideration the possibility of reaction proceeding according to radical way.

Thus the addition of BHT into the initial composition results in decreasing the reaction rate in case when dialkyldithiophosphoric acids (63) and (64) are used, along with this the relative share of the radical route makes approximately 40% and 20% respectively. The essential contribution of proceeding reactions of CHP with phosphorus dithioacids according to the radical route is detected by method of CNP ^{31}P [57, 69].

In this connection for phosphorus dithioacids the occurrence of the free-radical mechanism will reduce their efficiency in the processes of polymers inhibition despite high stoichiometric coefficients in reactions with CHP.

However, in case of phosphorus dithioacids containing a sterically hindered phenol fragment (68-72) there was no influence of BHT on the rate of CHP decomposition detected. The nonradical nature of the process of hydroperoxide decomposition under the action of phosphorus dithioacids (68-72) is testified by the products of its decomposition: GLC (gas-liquid chromatography) method detected less than 1% of α-methylstyrene which is a typical product of radical decomposition of CHP.

Thus the reactivity of various phosphorus acids in reactions with hydroperoxides is connected with their acidic nature and the contribution into the catalytic decomposition of hydroperoxides is likely to be brought by all acids formed in the process of reaction. The largest stoichiometric coefficients for strong acids is caused by the reason that they have time to react with a large number of hydroperoxide molecules before they are consumed in secondary reactions of forming by-products.

For defining quantitative interrelation *structure – reactivity* the comparison of values of stoichiometric coefficients (v) and reaction rate constants (k) of phosphorus dithioacids with hydroperoxides with stoichiometric constants of substitutes R_S both at a phosphorus atom of and at thione and thiol sulfur atoms according to the equations (15) – (19) was conducted [71].

Satisfactory correlations of values lg$v$ and lg$k$ and the structure of phosphorus dithioacids allow to draw the conclusion that the reaction centre in reactions of phosphorus dithioacids with hydroperoxides is a thiol sulfur atoms.

The obtained results confirm the suggested scheme of interaction of cumene hydroperoxide with phosphorus dithioacids (Scheme 2.6).

In coordinates lg$v$ – R_S (-S-) dithiophosphoric (61-64) and dithiophosphonic (68-71) acids produce two separate dependences (Equations 25, 26 respectively) with close angular coefficient.

$$\lg v=(3.578\pm0.074)-(0.196\pm0.022)\,R_S\,(\text{-S-}) \quad (25)$$
$R=0.9845$; $S_0=0.017$; $N=5$

$$\lg v=(1.865\pm0.043)-(0.297\pm0.088)\,R_S\,(\text{-S-}) \quad (26)$$
$R=0.9586$; $S_0=0.029$; $N=4$

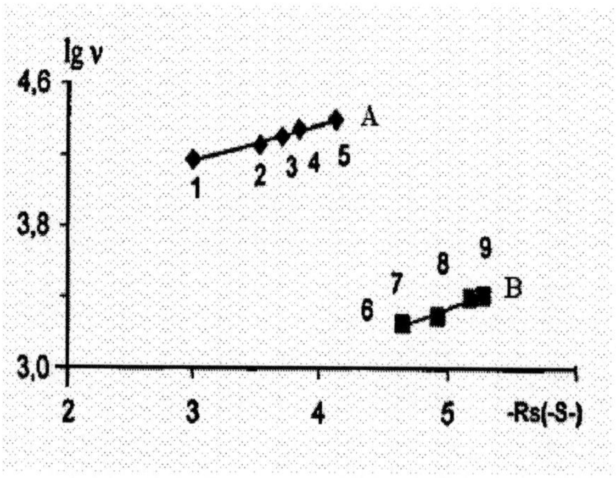

Figure 2.6. Dependence of stoichiometric coefficient ν of interaction of CHP on steric effect of the substitute (R_S) at thiol sulfur atoms of phosphorus dithioacids. A: 1 – compound (61), 2 – (62), 3 – (107), 4 – (65), 5 – (64); B: 6 – compound (68), 7 – (69), 8 – (70), 9 – (71).

Good parallelism (figure 2.6) of straight lines (25), (26) allowed to describe the interrelation of experimental parameters and the nature of substitutes using one common dependence (27) taking into consideration steric and electronic effects of variable substitutes.

$$\lg v=3.826+(0.391\pm0.106)\Sigma\sigma^{*}-(0.276\pm0.124)\,R_S\,(S) \quad (27)$$
$R=0.9513$; $S_0=0.085$; $N=9$

The analysis of the obtained correlation equations (25-27) showed that stoichiometric coefficient v increases with the growth of steric effect of substitutes at thiol sulfur atoms and with the increase of their electron-accepting nature.

The rate constant of interaction of phosphorus dithioacids with cumene hydroperoxide, on the contrary, decreases with the increase of steric effect of substitutes. In this case both dithiophosphates (61-64) and dithiophosophonates (68-71) are described by common one-parameter dependence (28) with high correlation coefficient.

$$\lg k=(2.821\pm0.162)+(0.431\pm0.33)\, R_S\,(S) \quad (28)$$

$$R=0.9912;\ S_0=0.064;\quad N=6$$

The detected high activity of phosphorus dithioacids in the processes of hydroperoxide decomposition, connected with catalytic nature of the reaction, is the potential prerequisite of their activity in the inhibition of polymer oxidation. It is necessary to underline, that phosphorus dithioacids with a sterically hindered phenol fragment unlike dialkyldithiophosphoric acids decompose hydroperoxides without forming radicals.

2.2.1.4. Aryldithiophosophonates and Dithiophosphates of 3d-Metals

The specific feature of phosphorus dithioacids is their ability to interact with ions of some metals, basically transition metals, leading to the formation of stable complexes (ML_2) [72-76]. A number of reactions of metal dithiophosphates with hydroperoxides of *tert*-butyl and cumene are described [77-79].

Aryldithiophosphates and aryldithiophosophonates of 3d-metals, including those that contain sterically hindered phenol fragment are of interest from the point of view of stabilization.

At the same time, the replacement of an alkoxyl substitute at a phosphorus atom (i.e. bonds O-P) by the substituted aromatic radical (bond C-P) influences the efficiency of metalcomplex action on hydroperoxide. According to the experimental data the presence of a sterically hindered phenol fragment in the structure of dithiophosophonate leads to increasing rate constant of the process (table 2.6).

Table 2.6. Kinetic and stoichiometric coefficients of the decomposition reaction of cumene hydroperoxide (CHP) under the action of dithiophosphates and dithiophosophonates of 3d-metals ($[CHP]_0$=0.1mole/l, 373 K)

№	Formula	v	1 stage			2 stage		
			k, $l^2/(mole \cdot s)$	E, kJ/mole	lg A_0	$k \times 10^{-3}$, s^{-1}	E, kJ/mole	lg A_0
(82)	$[(C_6H_5O)_2P(S)S]_2Ni$	11000	2.52	–	–	1.060	–	–
(83)	$[(ArO)_2P(S)S]_2Ni$	10800	Not defined			1.000	74.5	6.8
(84)	$[(i\text{-}PrO)_2P(S)S]_2Ni$	9000	1.66	–	–	0.876	–	–
(85)	$[(i\text{-}PrO)(Ar)P(S)S]_2Zn$	10800	2.60	84.09	12.227	0.390	57.30	4.836
(86)	$[(i\text{-}PrO)(Ar)P(S)S]_2Ni$	12000	9.00	59.70	9.398	2.083	17.10	2.490

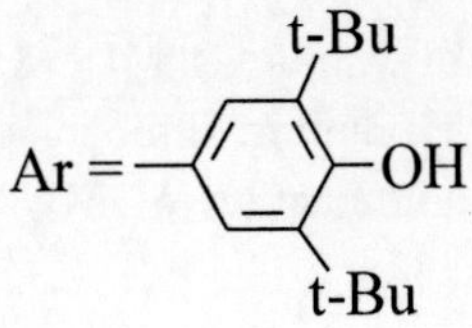

Kinetics of CHP decomposition in the presence of zinc and nickel dithiophosphonates with a sterically hindered phenol fragment (85, 86) in the range of temperatures 293-373 K is characterized by "stepped" change of concentration of cumene hydroperoxide during the time period (figure 2.7).

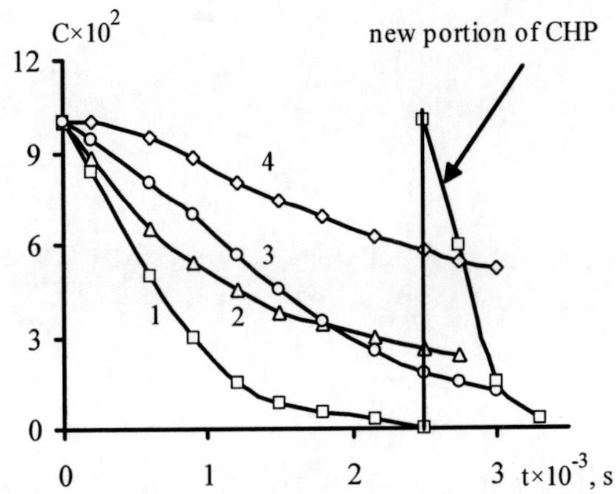

Figure 2.7. Kinetic curves of CHP decomposition under the action of ML_2: 1 – NiL_2 (86); 2 – ZnL_2 (85), 3 – NiL_2 (86) + BHT, 4 – ZnL_2 (85) + BHT (373 K, chlorbenzene, $[CHP]_0$=0.1 mole/l; $[ML_2]=1\times10^{-3}$ mole/l; $[BHT]=5\times10^{-2}$ mole/l).

Up to 15% of CHP is consumed at the rate $W=k_s\cdot[CHP]^2\cdot[ML_2]$ for zinc dithiophosophonate (85) at the short-term "slow" zone of the reaction; then there is a zone of "fast" hydroperoxide decomposition ($W=k_f\times[ML_2]$); at the final stage the rate of CHP consumption reduces again.

The "slow" zone of a reaction is stated slightly for nickel dithiophosophonate (86), containing an analogue sterically hindered fragment at a phosphorus atom.

The difference of the mechanisms of interaction of dithioacid complexes of nickel and zinc with hydroperoxide carried out under similar conditions was defined in the publications [80-83].

The introduction of pyridine into the reaction system "ML_2 – CHP" both at the first and the second stage of the reaction practically results in their complete stop. On the basis of that it can be suggested that active compounds here as in case of CHP decomposition by phosphorus dithioacids are the products of acid nature.

Superstoichiometric interaction of reagents testifies to catalytic nature of the process of hydroperoxide decomposition. The obtained values of coefficients ν(373 K) are in the range of 10000 – 12000 (table 2.6).

It should be noted that values ν of complexes ML_2 (85, 86) exceed for an order the values of stoichiometric coefficients of corresponding dithioacids acting as ligands.

In principle the mechanism of interaction of ML_2 with CHP can be represented by Scheme 2.8 [81]:

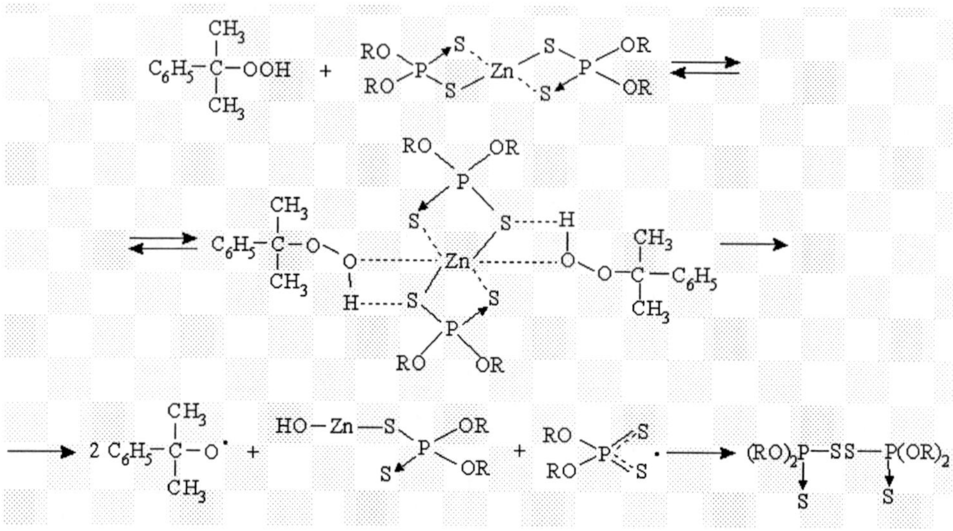

Scheme 2.8.

The introduction of radical acceptor BHT (in the ratio 1:2 regarding CHP and 50:1 regarding ML_2) into systems "ML_2 – CHP" before the beginning of the reaction results in decreasing the rate of hydroperoxide consumption. In case of CHP decomposition by zinc dithiophosophonates **(85)** the slowing down of the process is noticeably greater. Both radical and nonradical routes are possible in the course of interaction of ML_2 with hydroperoxide [82].

The evidence of the mixed route of the process of CHP decomposition, where ML_2 is a catalyst, is the detection(as products of CHP decomposition) of phenol and acetone [81] which are typical of its heterolytic decomposition, as well as of α-methylstyrene, which is a product of homolytic decomposition of CHP, by method of vapor-liquid chromatography (table 2.7).

Table 2.7. The composition (%) of products of cumene hydroperoxide decomposition in the presence of ML_2 (chlorbenzene, 373 K)

Compound number	(86)	(85)	(83)
Phenol	30.4	18.0	33.1
Acetone	39.8	25.1	40.2
α-butylstyrene	9.0	29.5	8.0
Acetophenone	0.9	2.0	0.8
Dimethylphenylcarbinol	0.65	2.53	0.60
Cumylphenol	0.33	0.90	0.30
Dimer of α-methylstyrene	0.33	2.39	0.30
ΣX[a]	0.95	0.95	0.94

[a] The sum of unidentified components.

The interaction of NiL_2 (83) proceeds according to mixed route under analogous conditions. At kinetic curves of hydroperoxide (CHP) decomposition distinct initial «slow» zone of the reaction appears under action of di(3,5-di-*tert*-butyl-4-hydroxyphenyl)nickel phosphate (83) (figure 2.8)

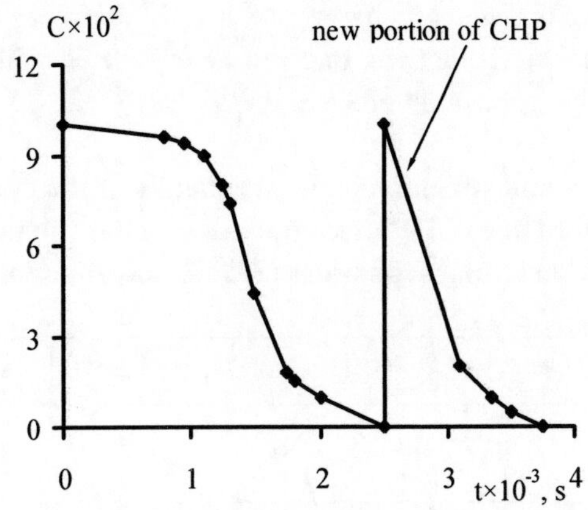

Figure 2.8. Kinetic curves of cumene hydroperoxide (CHP) decomposition under the action of NiL_2 (83) (373 K, chlorbenzene, $[CHP]_0$=0.1 mole/l; $[ML_2]=1\times10^{-3}$ mole/l).

The addition of a new portion of hydroperoxide into the reaction mixture at the moment of its complete consumption results in continuation of the reaction without the second «slow» zone (figure 2.8). Thus, the products of initial compound conversion are responsible for superstoichiometric hydroperoxide decomposition in the given reaction, as in case of phosphorus dithioacids.

Metal dithiophosophonates reveal higher catalytic ability in reactions with CHP if compared with respective dithiophosphates (table 2.6). However, the replacement of labile hydrogen atom in SH-group of acid for a metal ion can result in undesirable changing of radical mechanism into radical one [83].

2.2.1.5. Aryl- and Benzoylthiocarbamides and Their Complexes with C 3d-Metals

Thiocarbamides of the general formula

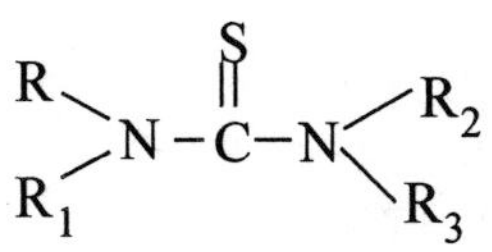

$$R = R_1 = R_2 = R_3 = H, \text{Alk}, \text{Ar}$$

generally participate in the reactions, stipulated by their nucleophil activity, sulfur atom or nitrogen one being a reaction center [84].

Thiocarbamide derivatives are known to be used as noncoloring heat stabilizers [85]. However, according to the analysis of publications systematic research of the regularities of stabilizing actions for them, including qualitative features has not been conducted.

Antioxidant properties of thiocarbamide derivatives are studied in a number of publications of the last years [86-89].

So, kinetic investigations indicate that interaction of cumene hydroperoxide with arylthiocarbamides (87-94) starts to proceed with noticeable rate at the temperature above 323 K [86].

Gross-consumption of cumene hydroperoxide (CHP) is characterized by super stoichiometric coefficients ν (table 2.8), that can be connected with the catalytic nature of cumene hydroperoxide decomposition process.

Table 2.8. Kinetic and stoichiometric parameters of the reaction of cumene hydroperoxide (CHP) decomposition under the influence of arylthiocarbamides (373 K, iso-propanol)

Compound number	Compound	ν_{CHP}	k×10, l/(mole·s)
(87)		40	0.45
(88)	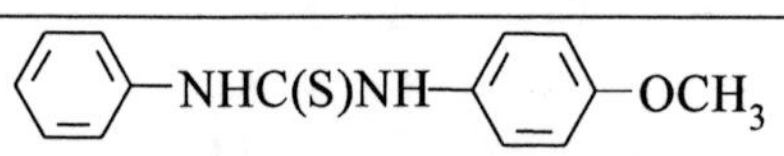	40	0.42
(89)	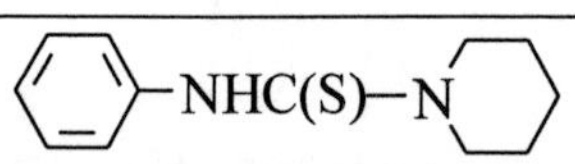	50	0.47
(90)	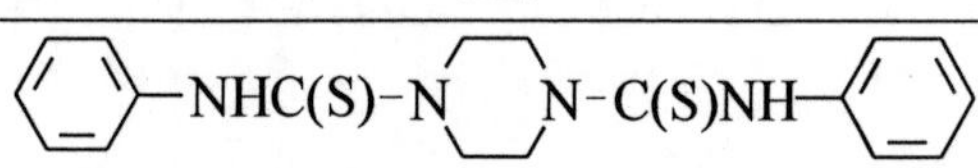	140	0.55
(91)	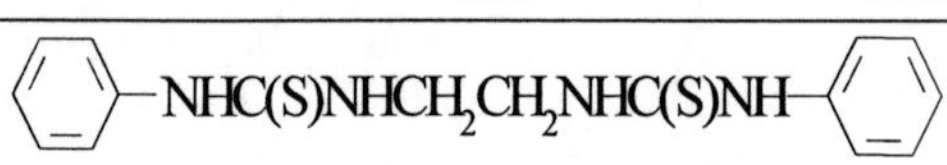	180	0.58
(92)	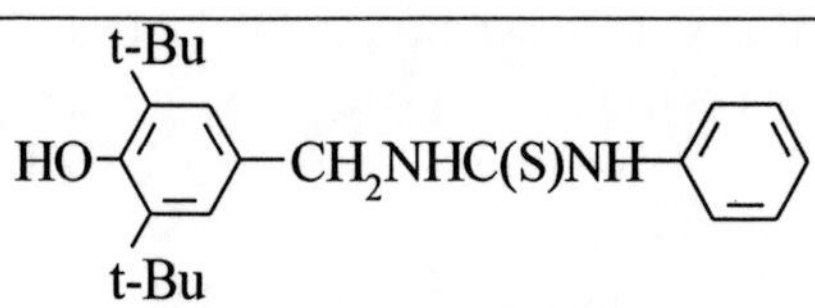	250	0.75
(93)	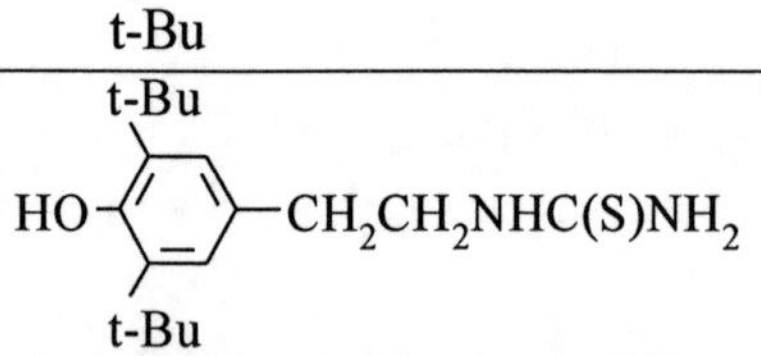	250	0.73
(94)	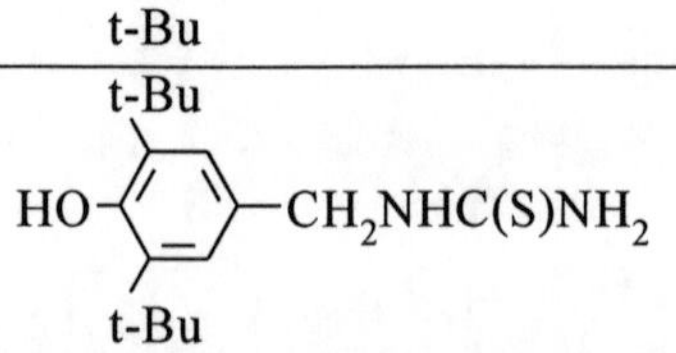	500	0.87
(95)	$H_2NC(S)NH_2$	10	0.41

Induction periods are absent at the curves of CHP consumption depending on time. It testifies to that the initial compounds are primary reagents, active in respect of CHP (figure 2.9).

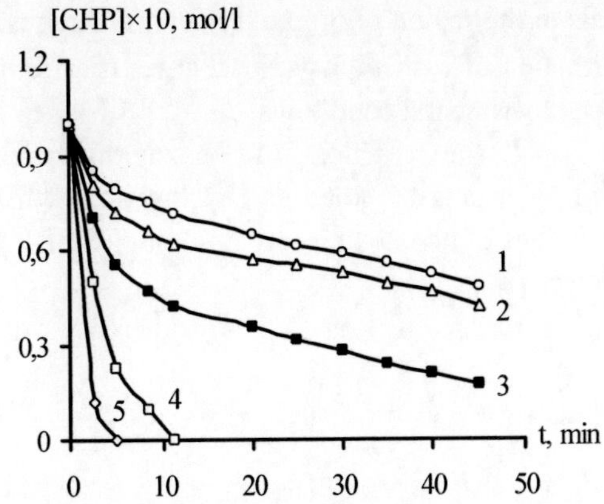

Figure 2.9. Kinetic curves of CHP decomposition under the influence of arylthiocarbamides (ATC): 1 – (87) and (88), 2 – (89); 3 – (90); 4 – (92) and (93); 5 – (94); (373 K, iso-propanol, $[CHP]_0$=0.1 mole/l; [ATC]= 0.1 mole/l).

It is determined that the reactions have the first order concerning both hydroperoxide and arylthiocarbamides.

To determine the possible homolytic route of interaction of CHP with arylthiocarbamides radical acceptor, namely 2,6-di-*tert*-butyl-4-methylphenol (BHT), was introduced into the reaction mixture. During that process CHP decomposition rate didn't change distinctly (figure 2.10) that testified to the predominance of heterolytic nature of the process.

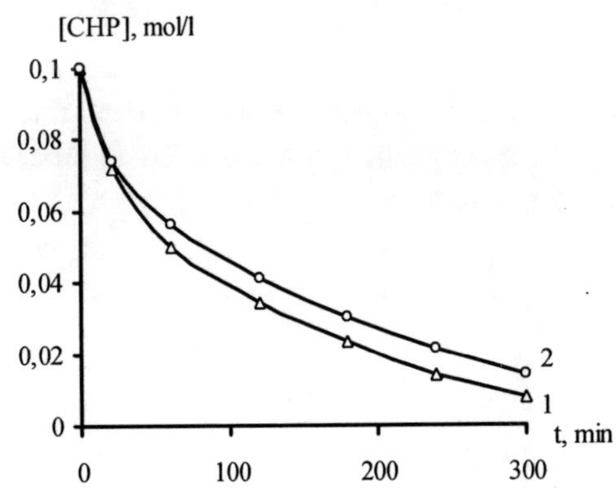

Figure 2.10. Kinetic curves of CHP decomposition under the influence of arylthiocarbamide (94): 1 – compound (94); 2 – compound (94) + BHT; (373 K, iso-propanol, $[CHP]_0$=0.1 mole/l; [(94)] = 5×10^{-3} mole/l; [BHT] = 5×10^{-2} mole/l).

This conclusion is confirmed by NMR ^{1}H spectroscopic investigation of the reaction of CHP decomposition under the influence of N-phenyl-N′-piperidylthiocarbamide (89). According to the typical chemical shifts in NMR spectra it was determined that phenol and acetone are formed and accumulated in the course of the reaction (heterocyclic route of a reaction).

The absence of signals in the region of 5 ppm, typical for the resonance of vinyl protons allows to exclude the formation of α-methylstyrene that testifies to the heterolytic nature of CHP decomposition under experimental conditions.

Kinetic investigations of the reaction of benzoylthiocarbamides (96-105) with hydroperoxides conducted by means of polarography determine that unlike reactions with arylthiocarbamides the reactions of benzoylthiocarbamides with CHP proceed with noticeable rate even at room temperature [87].

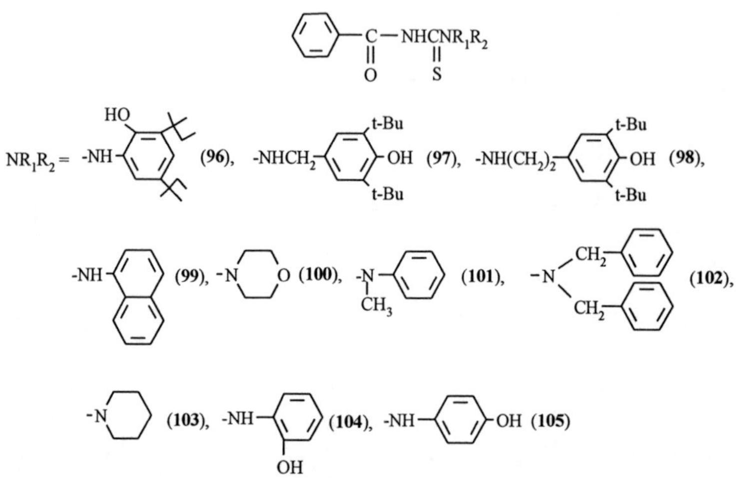

As in case of arylthiocarbamides induction periods were not observed during the experimental process (figure 2.11).

The reaction orders are defined according to the dependence of starting rate of the process on the initial concentration of reagents. The determined fraction order of the reaction points at the complexity of the process, the process is likely to proceed through intermediates which are active towards hydroperoxide.

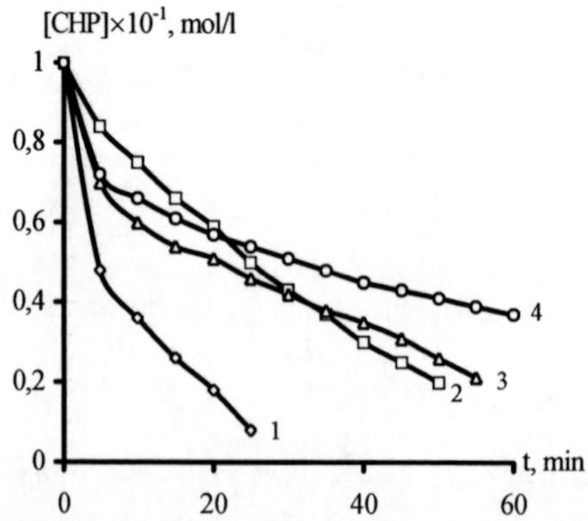

Figure 2.11. Kinetic curves of CHP decomposition under the influence of benzoylthiocarbamide (BTC): 1 – compound (104), 2 – (102), 3 – (101), 4 – (103); (373 K, chlorbenzene, $[CHP]_0=1\times10^{-1}$ mole/l; $[BTC]=1\times10^{-2}$ mole/l).

According to the data from table 2.9, the interaction of CHP with benzoylthiocarbamide derivatives is characterized by low values of activation energy.

Table 2.9. Kinetic and stoichiometric parameters of the reaction of cumene hydroperoxide (CHP) decomposition by benzoylthiocarbamide derivatives (96-105), ([CHP]$_0$ = 1×10^{-3} mole/l, chlorbenzene, 303-373 K)

Compound number	k, $l^{1.5}\cdot mole^{-1.5}\cdot s^{-1}$				E_{act}, kJ /mole	lgA_0	ν (323 K)	ν (373 K)
	303 K	323 K	343K	373 K				
(96)	0.27	0.69	0.93	1.85	22.84	3.53	200	12000
(97)	0.48	0.80	1.04	1.67	16.10	2.51	400	8000
(98)	0.32	1.19	1.8	2.56	26.29	3.26	200	8000
(99)	0.33	0.67	1.11	1.67	22.94	3.63	1000	13000
(100)	0.44	0.67	0.93	1.36	18.05	2.75	200	9000
(101)	–	–	–	0.57	–	–	200	1000
(102)	–	–	–	0.83	–	–	400	10000
(103)	0.24	0.42	0.67	1.11	19.94	3.84	100	5000
(104)	–	–	–	2.78	–	–	300	10000
(105)	0.31	0.69	1.01	1.58	21.34	3.613	360	10000

In whole, benzoylthiocarbamide derivatives are characterized by higher rate constants than arylthiocarbamides, as well as superstoichiometric decomposition coefficients in reactions with CHP (table 2.9). It gives grounds to consider that benzoylthiocarbamide derivatives will possess relatively higher antioxidant activity in the processes of stabilization of polymers than arylthiocarbamides.

Kinetic parameters of model reactions of hydroperoxides with benzoylthiocarbamides were the basis of correlation analysis of dependence of reactivity upon the structure of benzoylthiocarbamides. The high degree of correlation of kinetic parameters and calculated values of substitute constants for a nitrogen atom (N^3) indicates that the most probable reaction center in the investigated compounds of general formula $C_6H_5C(O)$-N^1H-$C^2(S)$-$N^3R'R''$ is the given nitrogen atom (equations 29-32).

$$\lg k_{30}=(-0.048 \pm 0.012)\Sigma\sigma^* + (0.244 \pm 0.040)R_S + (1.087 \pm 0.257); \quad (29)$$
R=0.9749, S_0=0.041, N=5

$$\lg k_{50}=(0.010 \pm 0.006)\Sigma\sigma^* + (0.160 \pm 0.019) R_S + (-1.213 \pm 0.121); \quad (30)$$
R=0.9984, S_0=0.018, N=5

$$\lg k_{70}=(0.017 \pm 0.007)\Sigma \sigma^* + (0.124 \pm 0.023) R_S + (0.79 \pm 0.023); \quad (31)$$
R=0.9808, S_0=0.023, N=4

$$\lg k_{100}=(0.031 \pm 0.008)\Sigma \sigma^* + (0.096 \pm 0.028) R_S + (0.593 \pm 0.179); \quad (32)$$
R=0.9732, S_0=0.028, N=5

The obtained data indicate that the reaction rate of CHP decomposition under the influence of benzoylthiocarbamide derivatives depends both on electron and on steric (to a greater extent) effect of substitutes. The rate constant of interaction of benzoylthiocarbamide with CHP decreases alongside with the increase of steric effect of substitutes at a nitrogen atom (N^3). Temperature increasing the influence of steric factor decreases, that is confirmed by the decrease of the coefficient at R_S-constant in equations (29-32).

Thiocarbamide derivatives are known to bind metal ions into complexes (ML_2). At the same time metal ions in complexes change the nature of distribution of electron density of ligands, hence their reactivity [90, 91].

In fact arylthiocarbamide metal complexes, in contrast to ligands forming these metal complexes, start to interact with hydroperoxides at room temperature. When the temperature increases the reaction rate of interaction grows considerably. Induction periods are absent at the curves of cumene hydroperoxide (CHP) consumption depending on time (figure 2.12). This, as in case of arylthiocarbamide, testifies that initial compounds are primary reagents, active ones in respect to CHP.

Conducting the reaction in the presence of radical acceptor BHT results in decreasing the reaction rate by ≈10-15% both for CuL_2 and NiL_2 (figure 2.12) that is an evidence of a certain share of homolytic route in the general process.

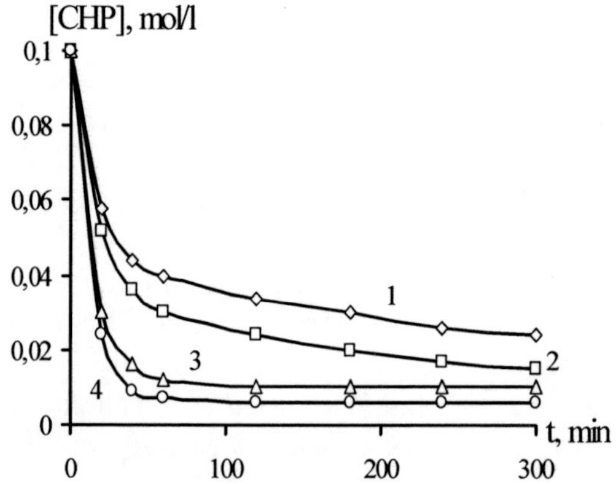

Figure 2.12. Kinetic curves of cumene hydroperoxide (CHP) decomposition under the influence ML_2 (106, 107): 1 – NiL_2(106); 2 – NiL_2 (106)+ BHT; 3 – CuL_2(107); 4 – CuL_2 (107)+ BHT; L=Ph-NHC(S)NH-Ph-p-OCH_3 (88); (373 K, isopropanol, the method of iodometry; $[CHP]_0$=0.1 mole/l; $[ML_2]=1\times10^{-4}$ mole/l; $[BHT]=5\times10^{-2}$ mole/l).

Heterolytic route of the reaction being the main one is confirmed by the investigation of the interaction of metal complex CuL_2 (108) (where L=Ph-NH-C(S)-NH-Ph) with CHP in dimethyl sulfoxide at 323 and 373K, the initial ratio CHP – metal complex being 500:1, the investigation was conducted by using the method of NMR ^1H.

As in case with thiocarbamide derivatives, in NMR ^1H spectra there were no peaks detected in the area of 5.0 ppm, which are typical for vinyl protons of α-methylstyrene, and the occurrence of peaks which are typical for protons of phenol ring (~7.0 ppm), that indicates the heterolytic nature of the process.

The rate of interaction of metalcomplex (108) with cumene hydroperoxide was increasing as compared with the rate of a similar reaction with ligand (figure 2.13).

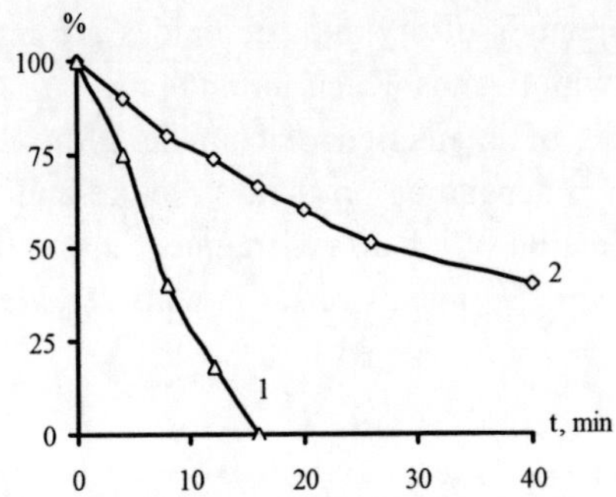

Figure 2.13. Changing of integral intensity of OH-group of cumene hydroperoxide (1.0 mole/l) during the period time: 1 – metalcomplex (108) (0.005 mole/l); 2 – ligand (87) (0.2 mole/l) (DMSO-d_6, 373 K, area ≈11.0 ppm, method of NMR ^1H spectroscopy).

The complete decomposition of hydroperoxide in the presence of metalcomplex (108) took place during 15 minutes, whereas the decomposition of CHP by ligand took more time, more for an order. Moreover, if in case of thiocarbamide derivative at 323 K in DMSO-d_6, the reaction doesn't take place at all, for metalcomplexes it proceeds with noticeable rate.

Defining stoichiometric parameters ν of the reaction of cumene hydroperoxide decomposition under the influence of metal complex compounds, given in table 2.10 indicates that stoichiometric coefficients ν_{CHP} increase more than for an order in comparison with initial arylthiocarbamides after introducing a metal ion and depend on metal nature. Moreover stoichiometric coefficient ν of CHP decomposition under the influence of complexes formed by ions of Cu(II) is 2-3 times larger if compared with complexes formed by ions of Ni(II), that can be explained by their different electronegativity.

Table 2.10. Stoichiometric coefficients of the reaction of cumene hydroperoxide decomposition by metalcomplexes ML_2 (373 K, iso-propanol)

Compound number	M(II)	Ligand (L)	ν_{CHP} (L)	$\nu_{CHP}(ML_2)$
(106)	Ni	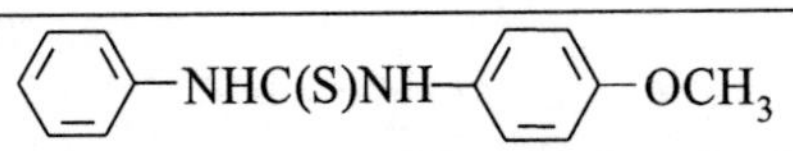	40	500
(107)	Cu	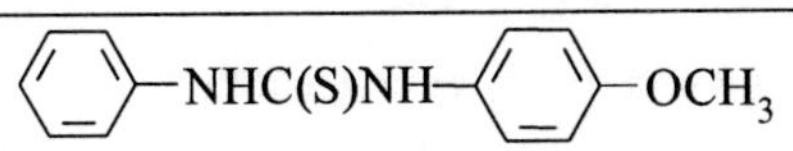	40	1200
(108)	Cu	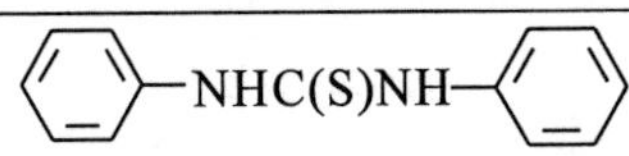	40	1600
(109)	Ni	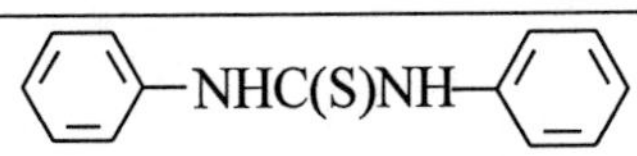	40	550
(110)	Ni		50	300
(111)	Cu	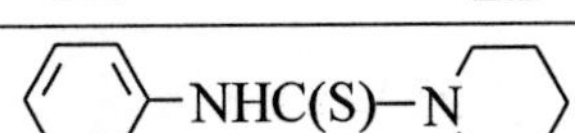	50	1000

Thus, coordination compounds of arylthiocarbamides are more efficient hydroperoxide decomposers in comparison with ligands which form them.

In whole, the regularities of interaction of cumene hydroperoxide with derivatives of thiocarbamides are determined depending on their structure and the presence of this or that functional group. The introduction of a benzoyl fragment into a thiocarbamide molecule and the formation of complexes with 3d-metals result in considerable strengthening of reactivity regarding hydroperoxide.

2.2.2. Reactions with Peroxide Radicals

Reactions of inhibitors with $RO_2^{\cdot}$-radicals were investigated under conditions of model initiated oxidation of low-molecular hydrocarbon (styrene, ethylbenzene), capable to interact with oxygen per chain mechanism without forming hydroperoxides [5].

The efficiency of oxidation inhibition was characterized by rate constant k_7 of the reaction of an inhibitor with radicals $RO_2^{\cdot}$ (or the ratio k_7/k_2, where k_2 is a constant of chain transferring on substrate) and by the stoichiometric coefficient of inhibition f, defining the efficiency of stabilizing effect of the substance [5].

2.2.2.1. Full Esters of Phosphorous Acid (Trialkyl(Aryl)Phosphites)

Organophosphorous compounds are involved in a lot of free-radical conversions of hydrocarbon substrates [1-5]. Autooxidation of such compounds as phosphites is of radical-chain nature and make numerous and important group of organic reactions [23, 24, 92].

The idea of participation of organic (aromatic) phosphites in terminating kinetic chains of polymer oxidation is based on the data about reactions with low-molecular $RO^{\cdot}$ and $RO_2^{\cdot}$-paradicals [27, 28, 93-96]. The experimental proof of this concept are regularities detected during the investigation both of model interactions of phosphites with free radicals of α,α-diphenyl-β-picrylhydrazyl [97, 98] and of inhibited oxidation of low-molecular hydrocarbons and polymers [30, 99].

At the same time the presence of critical concentrations, linear dependence of the induction period of polymer oxidation on aromatic phosphites concentration etc were detected [29, 30]. The detected linear dependence of the induction period on the concentration of aromatic phosphites can be explained explicitly, if the mechanism of terminating kinetic chains of oxidation by means of substitution of active peroxide macroradicals $RO_2^{\cdot}$ for low-activity, resonant-stabilized phenoxyl radicals are taken.

In this case the general scheme of inhibited oxidation can be represented in the following way:

Taking the interaction of phosphites with $RO_2^{\cdot}$-radicals as the main reaction (7) at $k_2 < k_7$ (key condition of stabilization) we get the ratio $\tau = [P]_0 2k_0$ (where k_0 is the constant of initiation rate), that is τ depends linearly on the concentration of phosphites. This scheme is confirmed by kinetically detected critical concentrations of investigated phosphites [30], that is, such concentrations, the exceeding of which results in changing the oxidation process from nonstationary to stationary one.

$$RH + O_2 \xrightarrow{k_0} R^{\bullet} + HO_2^{\bullet} \qquad (0)$$

$$R^{\bullet} + O_2 \xrightarrow{k_1} RO_2^{\bullet} \qquad (1)$$

$$RO_2^{\bullet} + RH \xrightarrow{k_2} ROOH + R^{\bullet} \qquad (2)$$

$$RO_2^{\bullet} + (R'O)_2P(OAr) \xrightarrow{k_7} O{=}P(OR')_2(OR) + ArO^{\bullet} \qquad (7)$$

Scheme 2.9.

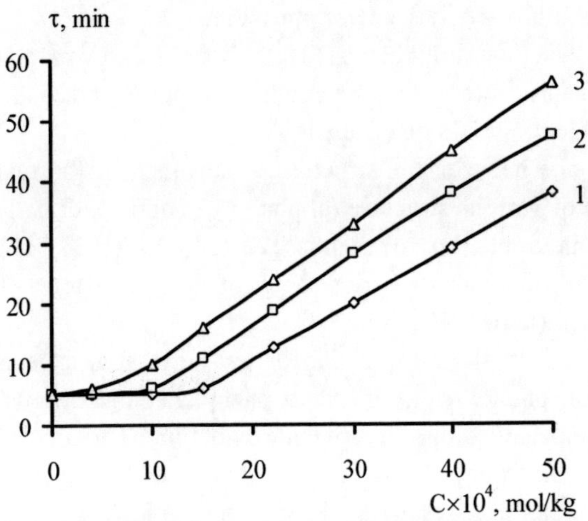

Figure 2.14. Dependence of the induction period (τ) of polypropylene oxidation on the concentration of phosphites: 1 – compound (112), 2 –(6), 3 –(14); (473 K, oxygen pressure 33.3kPa).

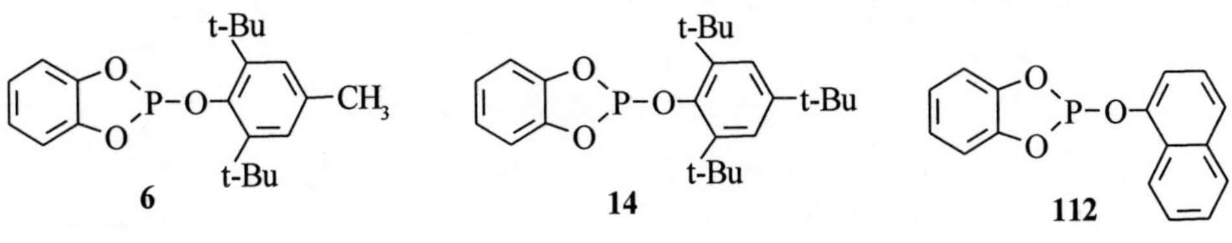

6 14 112

According to Scheme 2.9 the efficiency of phosphorous acid esters as oxidation inhibitors depends on the rate and mechanism of their reactions with peroxide radicals (chain-developing agents) during the oxidation of polymers.

Phosphites, which accept peroxide radicals sufficiently fast in comparison with competitive hydrocarbon substrate, can efficiently terminate chains of oxidation. However as phosphites are oxidized by $RO_2^{\bullet}$-radicals forming phosphates and $RO^{\bullet}$-radicals foremost, then it is the interaction of alkoxyradicals with phosphites that is the decisive factor [92-96, 100-103]:

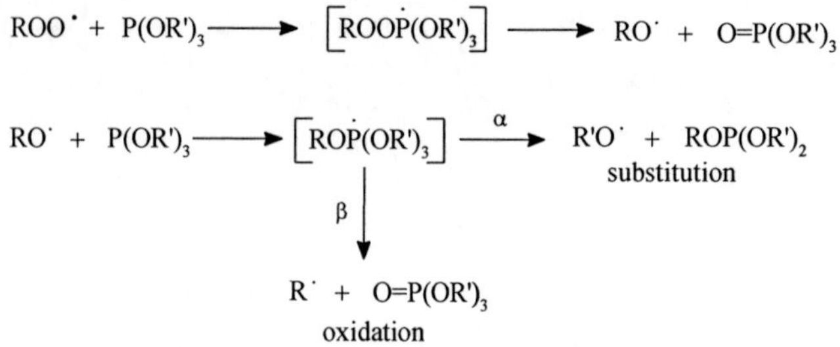

Scheme 2.10.

According to Scheme 2.10 these reactions proceed through intermediate phosphoranyl radical. Its decomposition can be carried out through one or two routes – α and β, the nature of which defines the inhibiting efficiency of phosphites.

According to the reactions in Scheme 2.10, only phosphites reacting with alkoxyradicals along the route α, that is, along the substitution reactions with the formation of chains terminating aroxyl radicals should be efficient.

The interdependence between the structure, mechanism of the reaction and anti-oxidant activity of phosphites of various structure (aliphatic, aromatic, sterically-hindered acyclic and cyclic) was studied in a number of works [22, 24, 100-103]. The investigations were conducted under conditions of oxidation both of polypropylene (table 2.11) and low-molecular hydrocarbons (table 2.12).

According to table 2.11 the introduction of *tert*-butyl and, especially, isobornyl groups into *ortho*-position of phenol ring of *ortho*–phenylenephosphorous acid ester increases sharply the induction period of polypropylene oxidation. The investigation of kinetics of oxygen absorption during propylene oxidation in the presence of phosphites indicated that ineffective phosphites didn't possess critical concentration [30].

Table 2.11. Induction periods of polypropylene oxidation in the presence of phosphites (473 K, oxygen pressure 33.3kPa) [30]

Compound number	Phosphite (0.05 mole/kg)	Induction period, min
(6)	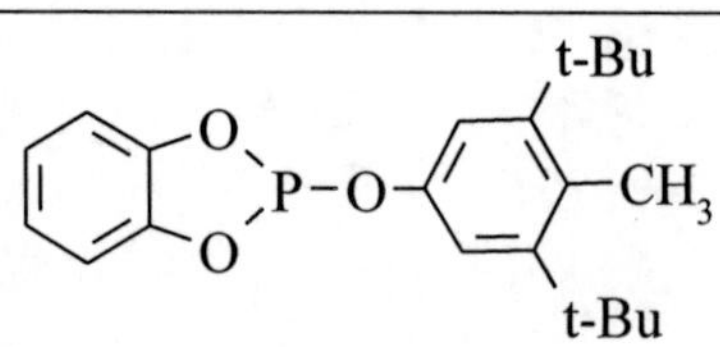	260
(11)	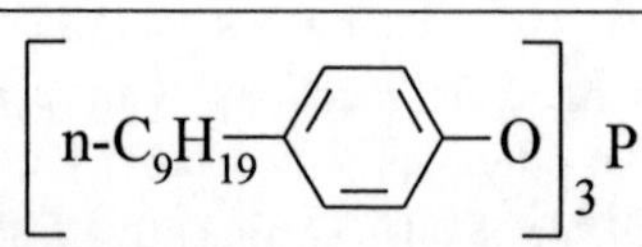	30
(13)	$(C_6H_5O)_2P\text{-}OC_8H_{17}\text{-}i$	35

Compound number	Phosphite (0.05 mole/kg)	Induction period, min
(14)	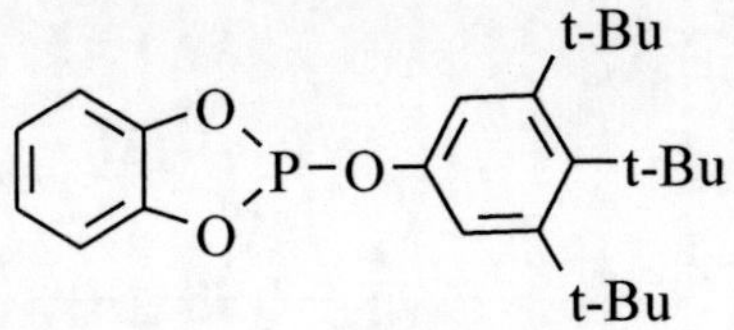	400
(112)	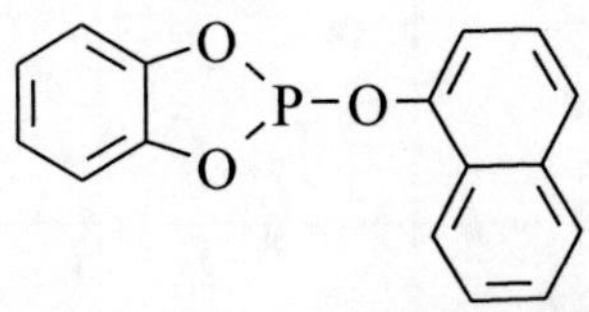	190
(113)	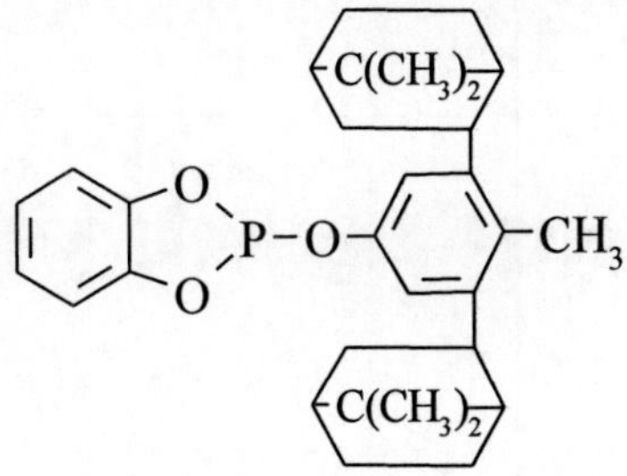	1000
(114)	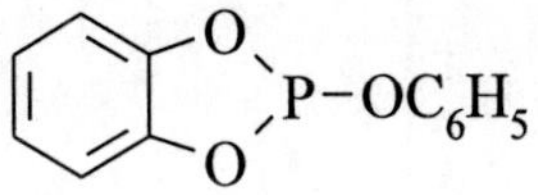	60
(115)	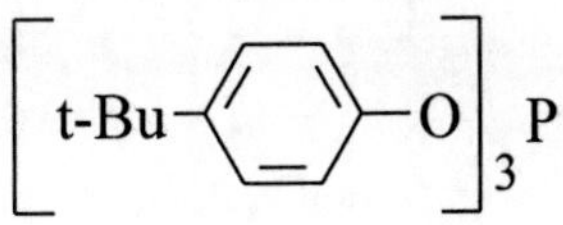	90

Table 2.12. Kinetic parameters of the initiated oxidation of phosphites (338 K, oxygen pressure 760 torr, o-dichlorobenzene, initiator – AIBN) [24]

№	Phosphite	$R_i \times 10^8$, mole·litre^{-1}·s^{-1}	([-dP]/dt)×10^4, mole·litre^{-1}·s^{-1}	n	$k_p/k_i^{1/2}$, litre$^{1/2}$·mole$^{-1/2}$·s$^{-1/2}$
(1)	$(C_2H_5O)_3P$	480.0 4.8 2.4 0.5	– 7.8 6.4 2.1	0.2×10^4 1.5×10^4 2.6×10^4 4.2×10^4	3.8
(3)	$(C_4H_9O)_3P$	12.0 2.4 1.2	14.4 5.8 3.7	1.2×10^4 2.4×10^4 3.1×10^4	4.0
(4)	$(C_6H_{11}O)_3P$	24.0 12.0 2.4	9.2 5.8 2.2	0.4×10^4 0.5×10^4 0.9×10^4	1.6
(7)	$(C_6H_5O)_3P$	480.0 240.0 48.0	3.1 1.5 0.3	0.6×10^2 0.6×10^2 0.6×10^2	–
(12)	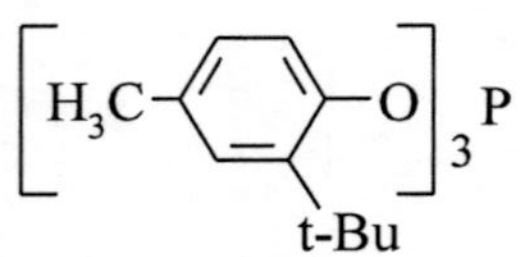	240.0	0.5	0.2×10^2	–
(13)	$(C_6H_5O)_2POC_8H_{17}$-i	276.0 120.0	3.6 1.2	1.3×10^2 1.0×10^2	–
(14)	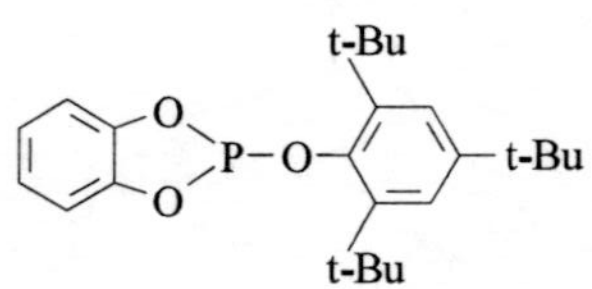	240.0	0.2	0.1×10^2	–

№	Phosphite	Ri×108, mole·litre-1·s-1	([-dP]/dt)×104, mole·litre-1·s-1	n	kp/ki1/2, litre1/2·mole-1/2·s-1/2
(36)	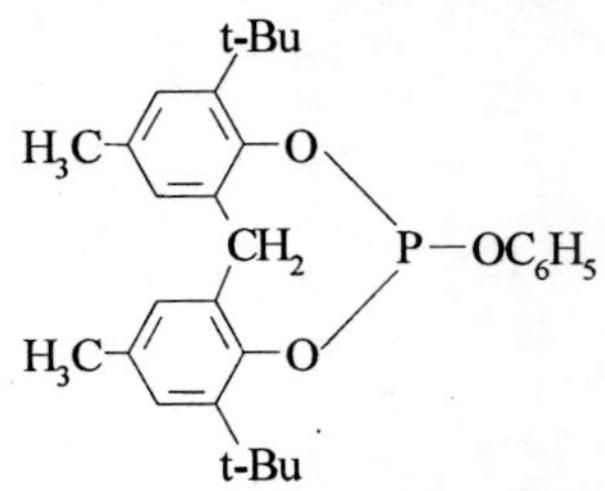	240.0	0.4	0.2×10^2	–
(116)	$C_6H_5OP(OC_8H_{17}\text{-}i)_2$	240.0 102.0 24.0	2.4 1.8 0.9	1.0×10^2 1.5×10^2 3.8×10^2	0.2

The chain length value was obtained by extrapolation of the observed linear dependence $v \sim R_i^{-1/2}$; R_i – rate of initiation; [P] – phosphite concentration; k_p – rate constant of chain oxidation; k_i – rate constant of initiation reaction; n – chain length.

The observed regularities can be conditioned by the following factors. The interaction of aliphatic phosphites, in particular, of trialkylphosphites, with peroxide radicals proceeds through intermediate alkoxy radicals $RO^{\cdot}$, the latter participate actively in the development of oxidation chain. Strong secondary chain terminators (for example, aroxyl radicals) do not form, that's why the polymer oxidation proceeds with very high rate even if constant k_7 for trialkylphosphites (triisopropylphosphites, in particular) is higher than constant k_2 of the transmission of the chain to substrate. Hence, phosphites of the given structure are not capable to perform functions of primary antioxidants, as their oxidation is of chain nature and proceeds up to complete exhaustion of phosphites ("autoxidation" of phosphites).

In the publications [104-107] the radical mechanism of these reactions was proved by the method of CNP and the causes of the low polarization efficiency (chain reaction with long chains) were explained. For phosphites the analysis of the rate of their chain oxidation as a function of initiation rate and their concentrations was conducted.

As it is seen from table 2.12, chain length (*n*) of oxidation shortens considerably from 10000 (for trialkylphosphites) to 5-130 units (for aromatic phosphites), particularly when sterically hindered phenoxyl fragments are introduced into a phosphite molecule. When aliphatic phosphites and aromatic ones are compared the change of mechanism of oxidation chain termination is observed for the latter, bimolecular termination is changed into linear termination of chains at secondary phenoxyl radicals.

The latter, as it was shown above, can be formed under substitution reactions (route α, scheme 2.10) from kinetically unstable phosphoranyl radicals. It is through this phenomen that chemical polarization of nuclei ^{31}P is not detected while aromatic phosphites go through chain oxidation.

Sterically unhindered arylphosphites and arylenephosphites without *ortho*-sabstitutes also reveal slight antioxidant effect similar to one which aliphatic phosphites possess.

Sterically hindered arylphosphites, which react with alkoxyl radicals by substitution forming chain-terminating aroxyl radicals (route α), can react as efficient antioxidants, similar to classical inhibitors (amines, phenols). However as the rate constants of interaction of these phosphites with peroxide radicals are 100 times less than the rate constants of phenol, to get the equivalent effect the concentrations of organic phosphites should be higher respectively.

Much attention is paid to the investigation of the mechanism of stabilizing action of cyclic phosphites in the papers of K. Shvetlik et al. [100-103]. It is shown that cyclic phosphites with sterically hindered aryl exogroups under conditions of thermally initiated oxidation of polypropylene show efficiency owing to the forming of phosphites, alongside with oxidation products (route β, scheme 2.10), and substitution products by means of β-splitting (route α). Liberated sterically hindered aroxyl radicals bind chain-developing peroxide radicals $RO_2^{\cdot}$.

Alkyl(aryl)-*ortho*-phenylenephosphites (6, 114, 117, 118) containing pentamerous cycle are characterized by high antioxidant effects under conditions of initiated oxidation of tetralin at 333 K. It is observed that they act as effective antioxidants at later stages of oxidation process [101].

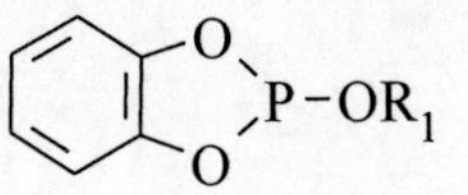

R_1= i-Pr (**117**), t-Bu (**118**), Ph (**114**),

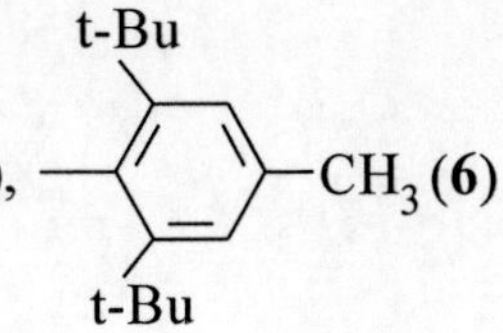

The authors connect the observed effects with the formation of a new product due to oxidizing conversions of initial *ortho*-phenylenephosphorous acid esters, which proceed through the forming of primary unstable alkyl(aryl)-*ortho*-phenylenephosphate (119) by analogy with the reactions given in Scheme 2.5 (Part 2.2.1.2). Subsequent opening of its cycle results in forming acid hydroxyphenyl-alkyl(aryl)phosphate (120), which is responsible for high stabilizing effect due to the interaction with hydroperoxides included, the interaction proceeds evidently through the mechanism of proton catalysis [96, 100-103].

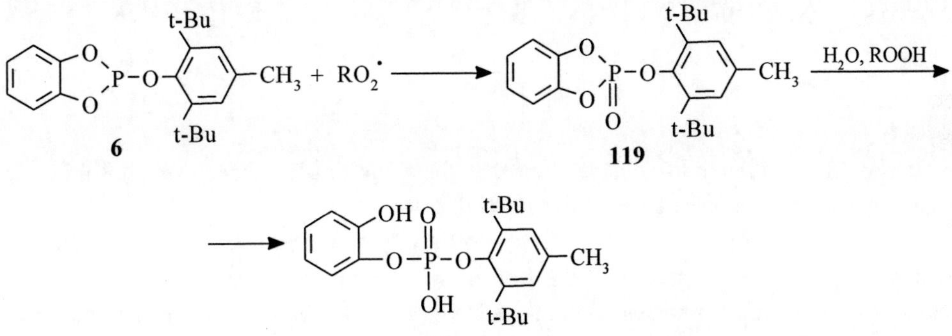

Studying the product (120) in the process of initiated oxidation of polypropylene and tetralin determines its high stabilizing efficiency, exceeding antioxidant activity of initial *ortho*-phenylene-alkyl(aryl)phosphites (6) and BHT (table 2.13) [101, 102].

Table 2.13. Induction periods of thermally initiated oxidation of polypropylene (453 K) and initiated oxidation of tetralin by AIBN ($[AIBN]_0=5\times10^{-3}$ M, 333 K, chlorbenzene) [101]

Compound number	Induction period of oxidation (τ), min	
	Polypropylene [a]	Tetralin [b]
(6)	230	0
(7)	72	0
(120)	280	>500
(121)	146	15
(122)	210	5
(123)		0
(124)		5
(125)		10
(126)		0
Pyrocatechin		230
BHT	210	245

[a] [In] = 0.05 mole/kg; [b] [In] = 0.0005 mole/kg.

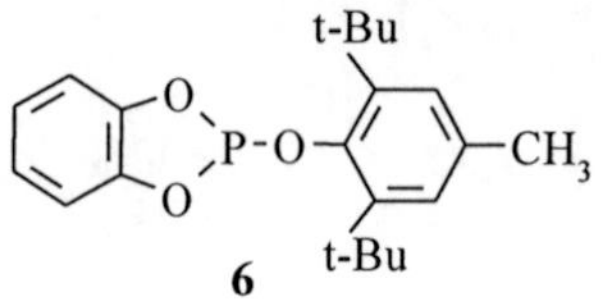

6

$(PhO)_3P$

7

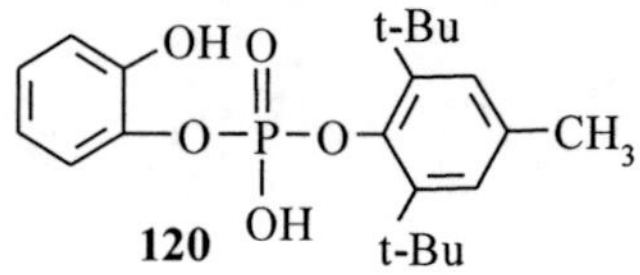

120

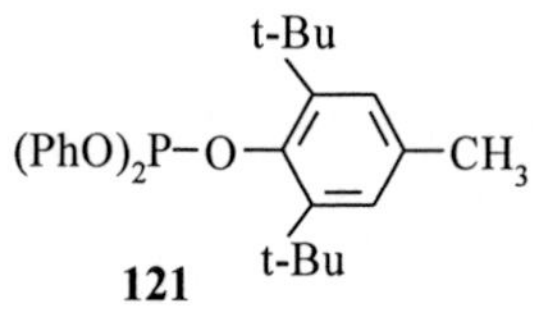

121

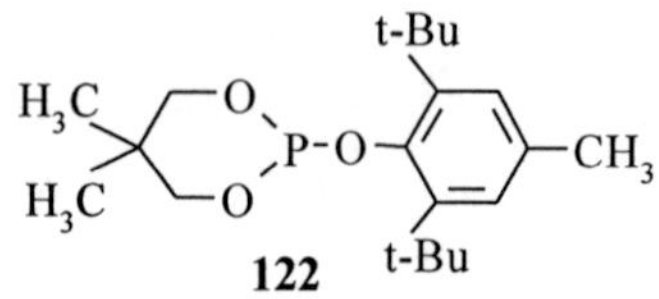

122

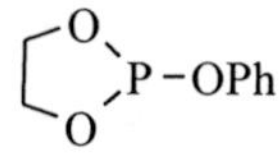

123

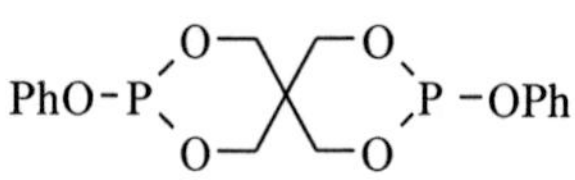

124

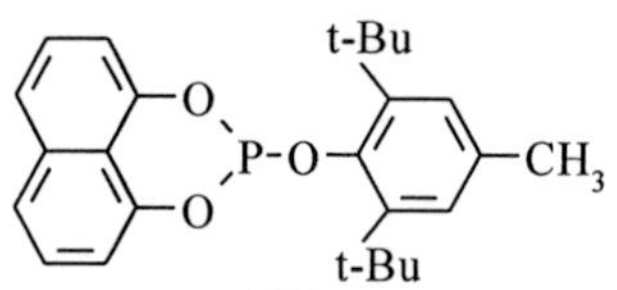

125

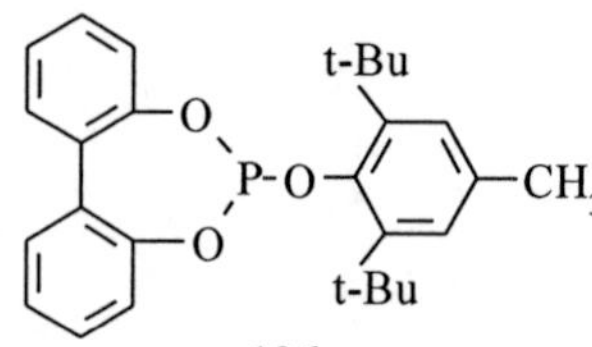

126

It should be taken into consideration that the inhibiting efficiency of phosphites may depend on chemical nature of the stabilizable substrate.

Thus, triarylphosphites react with cyano-*i*-propyloxyl- and *tert*-butoxyl radicals by various means. When reacted with the former triarylphosphates are formed due to the β–cleavage of intermediate phosphoranyl radicals, while reacted with the latter (that is with *tert*-butoxyl radical) *tert*-butyl phosphates are formed according to the route of α–decomposition (Scheme 2.10). It is evident that it is the result of the fact that the bond $C(Me)_2C$–O is weaker than the bond $(Me)_3C$–O, and its termination competes well with the P-OR-bond cleavage in a phosphoranyl radical [101].

On the whole, organic phosphites as inhibitors of free radical oxidation of polymers can be divided into the following main groups.

The first group includes aliphatic and sterically unhindered aromatic acyclic and cyclic phosphites which react according to the oxidation reaction (route β, Scheme 2.10), and that's why they can not inhibit efficiently free radical processes of thermal-oxidative destruction.

The second group includes acyclic and cyclic phosphites with sterically hindered aryl substitutes, which interact with an alkoxyl radical according to the substitution reaction (route α, Scheme 2.10). At the same time sterically hindered aroxyl radical is liberated which terminates the chains of oxidation.

The specific group of pentamerous cyclic phosphites is formed, they are capable of forming a new product, namely acyclic hydroxyphenylphosphate (**127**) having acidic nature under the influence of various oxidizing agents (ROOH, $RO^{\cdot}$, $RO_2^{\cdot}$). As it is defined experimentally, it is this product that determines the high results of inhibition of hydrocarbon substrate oxidation. The possible mechanism of its inhibiting action due to catalytic decomposition is suggested:

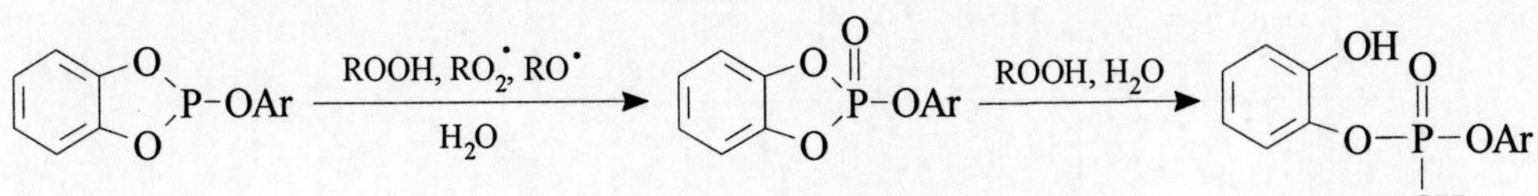

127

Kinetics of reactions of macroradicals RO and RO_2 with phosphites is of exceptional interest for the theory of polymer stabilization. The kinetic regularities of inhibition of polymers oxidation by phosphites are discussed in the publication [24].

In figure 2.15 the kinetic curve of oxygen uptake at the inhibited oxidation of polypropylene by phosphites is given, it is typical of model hydrocarbons (styrene, tetralin, benzaldehyde) and polyolefines As it is given in figure 2.15, aromatic phosphites inhibit the initiated oxidation, confirming by it that the reaction of peroxy radicals with an inhibitor (phosphites in the given case) proceeds very vigorously under oxidation.

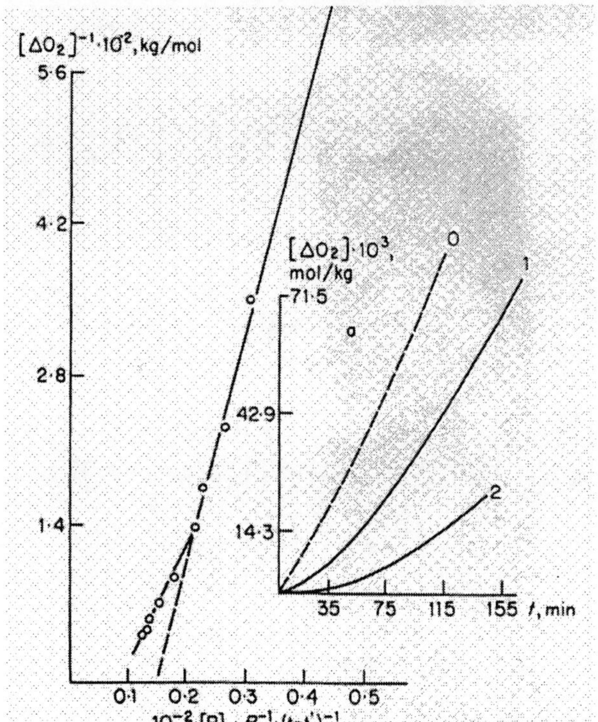

Figure 2.15. (a) Kinetic curve of the initiated oxidation of polypropylene ($R_i = 3.8\times10^{-7}$ mole·kg^{-1}·s^{-1}, 358 K, oxygen pressure 650 torr) in the presence of 1×10^{-2} (1) or 3×10^{-2} (2) mole·kg^{-1} tri(phenyl)-phosphite (7) (dotted line in the absence of (7); (b) its linear anamorphosis in the equation (B) co-ordinates.

The parameters of inhibition defined from the kinetics of oxygen uptake in various systems are given in table 2.14 [24].

Table 2.14. The quantitative characteristics of reactions of peroxy radicals with organic phosphites (method of initiated and inhibited oxidation: by the absorbed O_2)

Composition	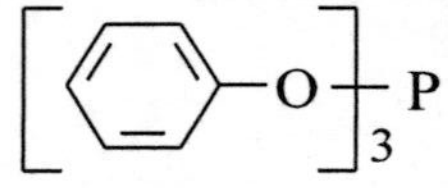 7		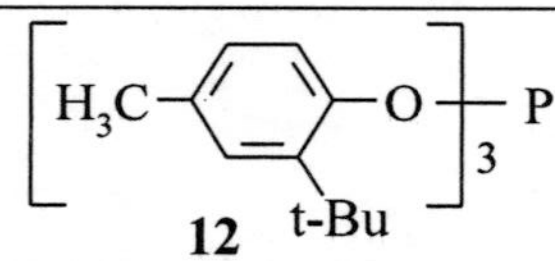 12		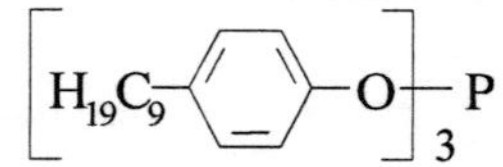 11		Oxygen pressure, torr	R_i
	f	$lg\ k_7$	f	$lg\ k_7$	f	$lg\ k_7$		
Solid polyethylene at 390 K	0.63	3.63 [a]	0.97	3.83 [a]	Not measured		700	1.6×10^{-6} mole/kg·s
Molten polyethylene at 390 K	Does not inhibit		0.63	3.43 [a]	Not measured		700	2.8×10^{-6} mole/kg·s
Solid polypropylene at 358 K	6.6×10^{-2}	3.85 [a]	0.12	4.49 [a]	Not measured		650	3.8×10^{-7} mole/kg·s
Styrene + chlorobenzene (50%) at 323 K	1×10^{-2}	4.86	6.3×10^{-2}	4.76	0.27	4.76	760	5×10^{-9} mole/kg·s
Tetralin + chlorobenzene (50%) at 323 K	1.5×10^{-2}	5.00	4.9×10^{-1}	5.49	Not measured		760	3.6×10^{-8} mole/litre·s
Benzaldehyde (0.2 mol/litre) + chlorobenzene (50%) at 323 K	0.5	6.20	2.0	6.48	3.4	6.74	760	6×10^{-8} mole/litre·s

[a] rate constant is k_{eff}; the measured ratio k_7/k_2 is given.

As it can be seen from table 2.14 phosphites terminate kinetic chains of oxidation of styrene and tetralin more efficiently than the typical chain terminator BHT. The rate constants for the reaction of RO_2–radicals with phosphites and BHT in solid polypropylene are nearly the same, but in molten polymer they are considerably lower than rate constants (k_7) for the known antioxidants as amines and phenols. The rate constants of chain termination (k_7 or k_7/k_2) increase when there are sterically hindered phenoxyl groups in phosphite molecules; simultaneously the stoichiometric coefficient of inhibition increrases. Direct quantity measurements of phosphites efficiency are confirmed by the assumption [22], that new more efficient stabilizers of polyolefine should be found in the series of aromatic phosphites with sterically hindered (or electron-donating) groups (substitutes).

To identify the products of conversion of aromatic phosphites during polyolefine oxidation (e.g. polyethylene) the experiments were interrupted at the end of inhibited oxidation, formed low-molecular products extracted CCl_4. According to IR-spectra in the oxidated polymer there were C-O-P and aromatic bonds, while the products of inhibitor conversion extracted from polyethylene contained P=O and C-O-P fragments. This result can be interpreted by the given above Scheme 2.9: phosphites are oxidized changing into phosphates by macroradicals $RO_2^{\cdot}$ (Reaction (7)), while the second phosphate molecule joins quickly to the formed macroalkoxyl radical $RO^{\cdot}$, that is followed by the elimination of phenoxyl radical (Reaction (8)).

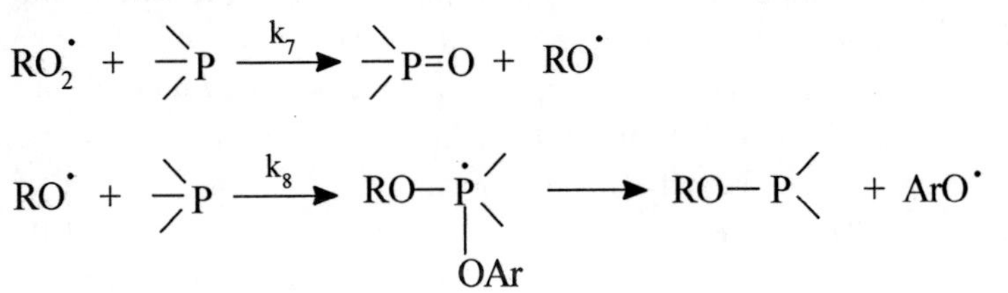

(7)

(8)

Later phenoxyl radicals can participate in the reaction of termination with macromolecules $RO_2^{\cdot}$ (if possible with the least value of f=1) or in the reaction of selfrecombination (f=0.5) [14].

Sharp decrease of inhibiting properties of phosphite (12) (table 2.14) in molten polyethylene is likely to be connected with reduced diffusion because of structural impediments which are conditioned by the specific properties of solid phase [24]. Apart from that resulting phenoxyls are not inert ones and can take part in the initiation of new chains in polymer. Coefficient f decreases more sharply in liquid phase (table 2.14). In latter case "autoxidation" of phosphite is of importance leading to the additional consumption of an inhibitor. In the course of oxidation inhibition both of styrene and tetralin the coefficient f increases (increases from $(1\text{-}1.5)\times10^{-2}$ for triphenylphosphite (7) to $(5\text{-}6)\times10^{-2}$ for sterically hindered phosphite (12). On the other hand such changing of f proportionally leads to the inverse value (n^{-1}) of the chain length of "autooxidation" of the phosphites themselves [$(0.6\times10^2)^{-1}$ and $(0.2\times10^2)^{-1}$, respectively], in dichlorbenzene (see table 2.12). Finally the consumption rate (R_P) of an inhibitor in styrene is found to be higher than the initiation rate R_i. In other words, phosphite is consumed according to chain mechanism, chain length being $n=0.1\times10^2$. This value is very close to the value found independently during initiated oxidation of the same phosphite.

Thus, if the efficiency of phosphites at inhibited oxidation is characterized by relatively high values of rate constant k_7, the duration of their stabilizing action depends on the value of their competitive chain "autooxidation" which defines the value of stoichiometric coefficient f. When combined these tightly bound processes define the specific behavior of organophosphorous inhibitors. The knowledge of these specific features is likely to predict in the future the optimality conditions of stabilization of polymers during their storing and usage. In particular, if "autooxidation" of phosphites is inhibited, the duration of their stabilizing action increases substantially.

The identification of the key function of phosphites as antioxidants determines the means of solving the problem that is how to raise this function and increase the efficiency of phosphites as inhibitors of chain developing. The problem was solved by binding phosphites with ions of transition metal (M) into kinetically stable complexes [108-112].

The considerable increase of efficiency of aromatic phosphites as inhibitors of initiated oxidation of low-molecular hydrocarbon (e.g. styrene, tetralin) and polyolefins (e.g. polypropylene) in the presence of acetylacetonates of transition metals (M_{acac}, where M=Co^{+2}, Cr^{+3}, Ni^{+2}, Vo^{+2}, Fe^{+3} etc) is defined. The main characteristics of this phenomenon can be summarized by the following.

In the presence of acetylacetonates of transition metals the dependence of induction period on concentration of phosphites is of "critical" nature (figure 2.16b). In other words, the period of inhibited oxidation increases practically unrestrictedly when the "critical" concentration of phosphites is exceeded. It is likely to be connected with the formation of a new more efficient inhibitor, namely a complex of metal acetylacetonate with phosphite which accepts efficiently peroxyradicals. Using the value of "critical" concentration of phosphite the rate constant of inhibited oxidation of styrene in the presence of phosphite complex was obtained. This value exceeds considerably the similar constant both for free phosphite and metal acetylacetonate and for the well-known phenol inhibitor BHT. The inhibited action of phosphite bound by means of coordination with a metal ion is confirmed as well by computer modeling of kinetics of the common process of styrene oxidation [109-110].

Fast distinct catalytic process of substrate oxidation in the presence of metal ions changes into slow linear termination of chains under the influence of the complex *phosphite – metal ion* (figure 2.16a, curves 3 and 4). The investigation of kinetics of this process allowed to calculate the quantity parameters of the observed result. The analysis of these parameters defines that catalysis of polypropylene oxidation by metal ions is prohibited most efficiently by phosphites at ratio metal/phosphite $[M]/[P] \leq 6\times10^{-2}$, that is when the concentration of metal in polymer doesn't exceed 0.02 wt %, and concentration of phosphite is close to the concentration equal to 0.5-1.0%, which is usually used. At high concentrations of metal ions ($[M]/[P] \geq 6\times10^{-2}$) the efficiency of synergistic composition depends on catalytic activity of metal component during the process of polymer autooxidation. Phosphite, however, doesn't influence noticeably on the rate of catalyzed oxidation (figure 2.16, curves 1 and 2), e.g. by chrome acetylacetonate (Cr_{acac}). The compositions where metalorganic component possess weak catalytic activity (e.g nickel acetylacetonate, Ni_{acac}) are the most efficient.

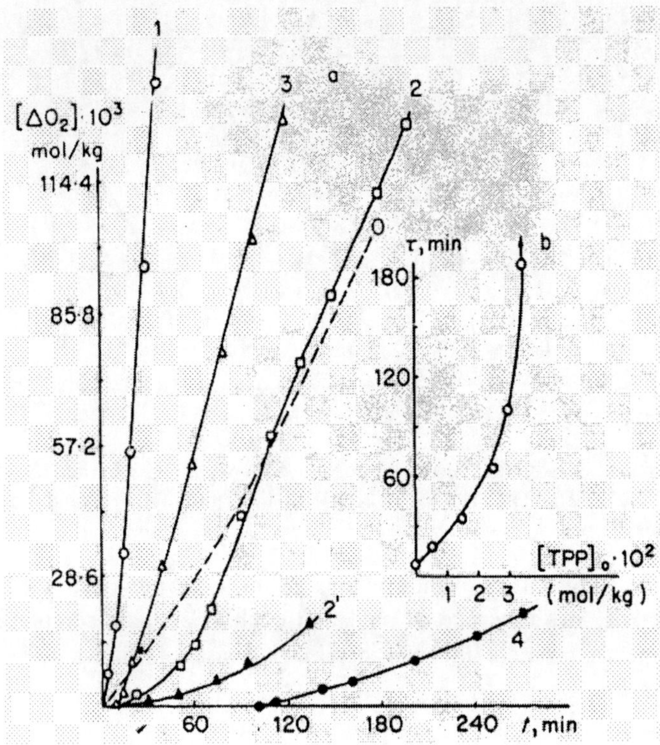

Figure 2.16. (a) Kinetic O_2 absorption curves by solid polypropylene (358 K; oxygen pressure 650 torr; $R_i = 3.8\times10^{-7}$ mole·kg^{-1}) in presence of 3×10^{-3} mole·kg^{-1} (1) or 6×10^{-4} mole·kg^{-1} (3) of Co_{acac2}; of the mixture 3×10^{-3} mole·kg^{-1} Co_{acac2} + 3×10^{-2} mole·kg^{-1} tris(phenyl)phosphite (2); of the mixture 6×10^{-4} mole·kg^{-1} Co_{acac2} + 3×10^{-2} mole·kg^{-1} tris(phenyl)phosphite (4); 3×10^{-2} mole·kg^{-1} of tris(phenyl)phosphite (2'); 0 – the curve of the initiated oxidation in the presence of metal complex and phosphite; (b) the dependence of the observed induction period upon the initial concentration of tris(phenyl)phosphite in polypropylene oxidation with initiator and or 6×10^{-4} mole·kg^{-1} Co_{acac2}.

The values of relative "synergistic" activity of metal ions in compositions with phosphites in styrene, tetralin and solid polypropylene were defined [64]. On the whole, these compositions reveal similar tendency concerning strengthening the efficiency, regardless of chosen substrate that is in all cases the nature of synergism does not depend on the type of oxidized hydrocarbon.

Stoichiometric coefficient *f* of chain termination by the complex of metal acetylacetonate with phosphite does not yield to the similar coefficient *f* for free phosphite in its importance. Based on that fact and other data the conclusion can be made that only complexes of metals with phosphites having antioxidant activity are kinetically active during the inhibition of substrate oxidation.

Defined regularities are confirmed by the results of the increasing of inhibiting action of phosphites when polymers are autooxidized. The particular example is the presence of residua of polymerization catalyst in commercial polyolefines which can catalyze polymer oxidation.

It is also found [64] that derivatives of titanium and vanadium can synergetically increase the efficiency of organophosphorous compounds in low density polyethylene. Apart from that induction periods for polyethylene both of high and low density stabilized by phosphites increase 2-7 times (figure 2.17).

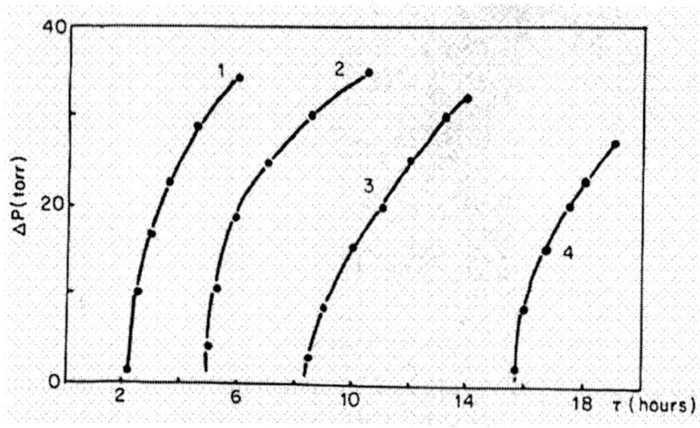

Figure 2.17. The change of induction period for polyethylene oxidation, inhibited by phosphite (9) (1) in the presence of: 0.05% $TiCl(OC_4H_9)_3$ (2), $[(C_4H_9O)(C_4H_9)(O)PO]_2TiCl_2$ (3) and the mixture of $TiCl_2(OC_2H_5)_2{\cdot}C_2H_5OH$ with $TiCl_2(OC_2H_5)$ (4).

The general behavior model of the complex *metal–phosphite* given above gives grounds to predict the activity of this group of oxidation inhibitors: stable complexes with organic phosphites as ligands promote the considerable increase of the value of stoichiometric inhibition coefficient *f*.

2.2.2.2. Acid Esters of Phosphorous Acid (Dialkyl(Aryl)Phosphites)

The reactions of homolytic substitution at hydrophosphoryl hydrogen when phosphonyle radicals are formed are nearly always the basis of homolytic conversions of hydrophosphoryl compounds.

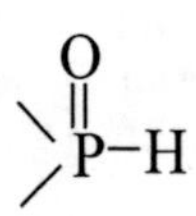

Rate constants of a number of reactions of such type are described and defined in publications [114, p.210-220].

In particular, the rate constant of hydrogen atom detachment from dialkylphosphorous acids by polystyrene radical at 333 K amounts to less than 1×10^{-2} l/(mole·s). The relative rate constant of hydrogen atom detachment from dialkylphosphites by phenyl radical *k* make (2.0-5.4)$\times10^{-3}$ l/(mole·s). From the point of value *k* dialkylphosphites are comparable with one of the most active donors of hydrogen, such as Ph_2NH, Ph_3CH, Ph_3SiH. The reactions of *tert*-butyloxyradical with a number of hydrophosphoryl compounds are known.

The rate constant of interaction with free radicals for dialkylphosphites exceed considerably the similar constants for organophosphorous compounds without bond P-H.

In the publication [115] the kinetics of hydrogen atom detachment for a number of phosphorus acids was investigated, and it was shown that spatial accessibility of bond P-H is of determinative importance in changing the ability to react with sterically hindered radical.

It was also established by the method of ESR (electron spin resonance) that cyclic diarylphosphites can easily give away hydrogen to peroxide radicals [116]:

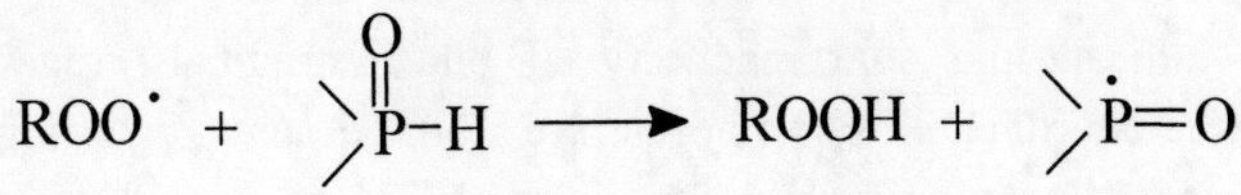

The possibility of forming phosphonyl radicals was confirmed in the paper [117], where phosphonradicals forming in model conditions of phosphonradical oxidation were identified by ESR using traps of radicals and phosphonradical stability depending on the structure was characterized.

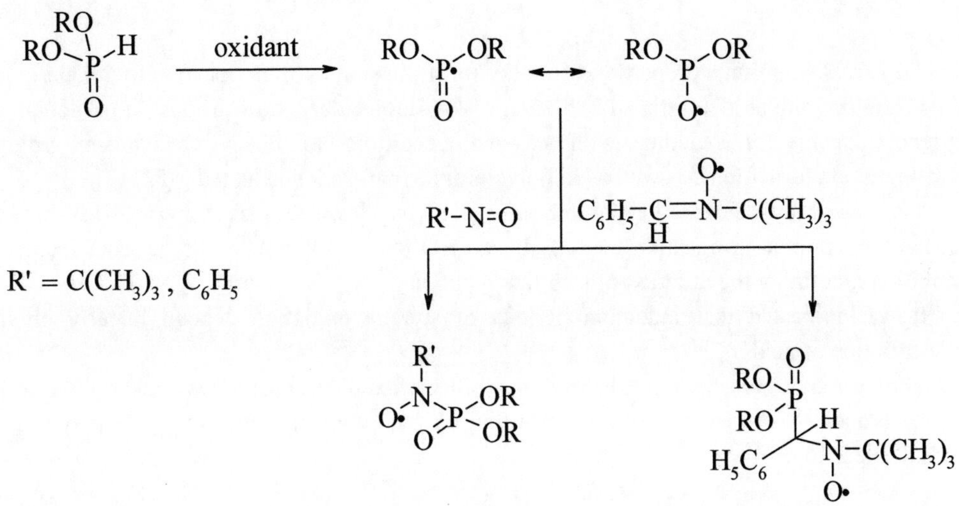

Diarylphosphites (46, 128) with bulky *tert*-butyl groups in *ortho*-position towards ester oxygen are the most efficient, dialkyl- and diphenylphosphites (47, 129) are noticeably less active.

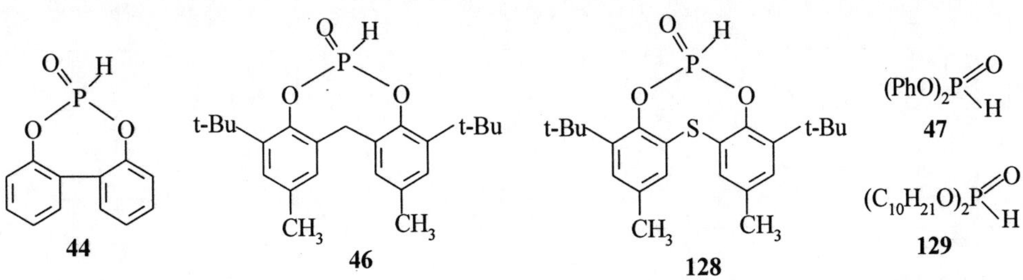

44 46 128 47 129

Parallelism of stabilizing activity and the ability of phosphites to react with nitroxyl and diphenylpicrylhydrazyl radicals was determined [117]. For the latter rate constants k (l/(mole·s)) at 293 K in benzene are equal to: 0.104 for the compound (128), 0.032 for the compound (44), 0.0154 for compound (129). The rate constant is equal to 4.2×10^{-4} l/(mole·s) for reactions of combining (47) at 328 K in toluene.

The method of CNP ^{31} is successfully used for the investigation of radical reactions with the participation of phosphonyl radicals [118, 119].

Spectra data indicate the dual reactivity of phosponradicals: regarding oxygen the derivatives of three-coordinated phosphorus are formed and regarding phosphorus the derivatives of four-coordinated phosphorus are formed.

Thus, phosphonyl radical (C) is ambident, as well as dialkylphosphite anion (D).

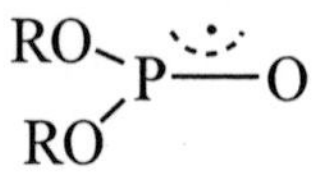

C

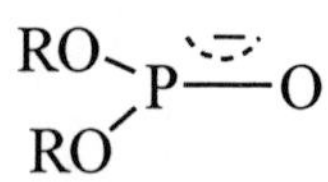

D

To reveal inhibiting properties of dialkyl(aryl)phosphites their reactions with peroxide radicals during the model initiated oxidation of low-molecular hydrocarbons, in particular, of styrene (capable of reacting with oxygen according to chain mechanism without hydroperoxide formation), as well as in polymer matrix were investigated [55, 117].

The efficiency of oxidation inhibition in the given case was characterized by the rate constant of inhibition reaction with radicals $RO_2^{\cdot}$(k_7) and stoichiometric coefficient of inhibition f, defining the duration of stabilizing action of the substance.

It was indicated that induction periods of styrene oxidation depend linearly on the concentration of acids.

The results of investigations defined the ability of studied phosphorus acids to inhibit the radical process of styrene oxidation depending on the nature of substitutes at a phosphorus atom (table 2.15).

Table 2.15. Kinetic and stoichiometric parameters of inhibiting action of phosphites during initiated oxidation of styrene and polypropylene (initiator – AIBN)

Compound number	Styrene, 323 K			Polypropylene, 358 K		
	f	$k_7 \cdot 10^{-4}$, l/(mole·s)	$k_7/k_2 \cdot 10^{-2}$	f	$k_7 \cdot 10^{-3}$, l/(mole·s)	$k_7/k_2 \cdot 10^{-3}$
(23)	1.35	81.00	70.43	–	5.85	39.00
(44)	1.35	1.40	1.22	1.2	3.38	22.52
(45)	1.35	0.18	4.00	1.2	3.81	25.31
(46)	0.80	6.16	0.16	0.5	0.04	0.25
(47)	0.60	4.54	0.62	0.6	2.70	18.00
(130)	1.25	0.71	5.39	0.8	2.93	19.55
(131)	0.11	1.25	3.95	–	–	–

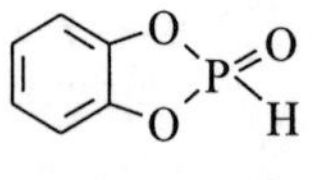

23

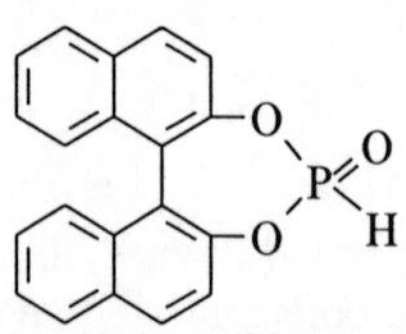

31

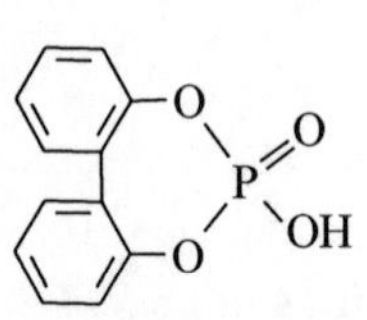

130

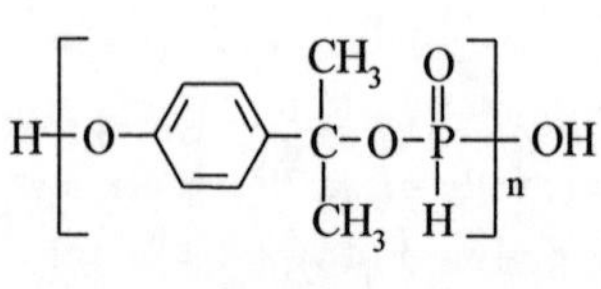

131

Kinetic parameters of the reaction of inhibition (k_7, k_7/k_2) indicate that inhibiting ability of cyclophosphorous acids (23, 44-46) decreases according to the increase of the cycle size in an acid molecule. In particular, the most efficiency, as it is seen from table 2.15, is characteristic for *ortho*-phenylenephosphorous acid (23). There is also the decrease of reactivity in reactions with peroxyradicals during the transition from cyclic to acyclic acids of phosphorus. The acid (47) is a clear demonstration of that. Stoichiometric coefficients of inhibiting action for diarylphosphorous acids are close, on the average, to 1, that is 1-2 orders more than the inhibition coefficient of respective full esters of phosphorous acids.

On the whole, reactivity of phosphorous acids in homolytic reactions of inhibition of the chain oxidation process of hydrocarbons and polymers is closely connected with the dissociation energy of the bond P–H and with the stability of formed phosphonyl radicals and their further conversions.

2.2.2.3. Dialkyl(Diaryl)-3,5-Di- Tert-Butyl-4-Hydroxybenzylphosophonates

Dialkyl(diaryl)-3,5-di-*tert*-butyl-4-hydroxybenzylphosophonates (132-137) are known as stabilizers of polymers [120-122].

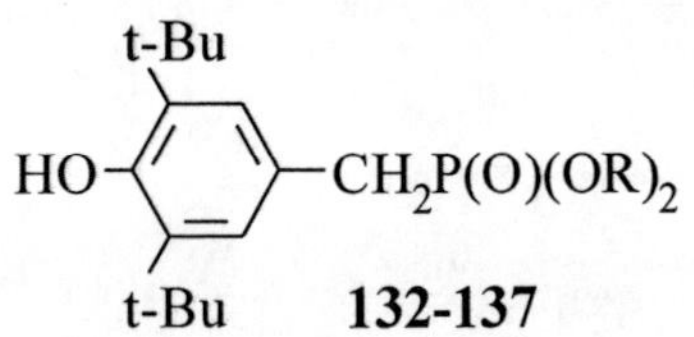

132-137

R = CH_3 (**132**), C_2H_5 (**113**), i-C_3H_7 (**134**), C_4H_9 (**135**), C_6H_5 (**136**), C_8H_{17} (**137**)

As consistent with the structure of phosophonates (132-137) their protective action can be connected, first of all, with the participation of a sterically hindered phenol fragment (contained in the composition of their molecules) in the accepting peroxide radicals. At the same time radicals $\dot{O}ArCH_2P(O)(OR)_2$ are formed. The efficiency of inhibiting action of phosophonates is defined by the stability of the forming radical in many respects. The modeling oxidation of these compounds by lead oxide was conducted to determine the ability of phosophonates (132-137) to generate similar radicals and to estimate their stability [120]. The forming of stable radicals $\dot{O}ArCH_2P(O)(OR)_2$ was stated by ESR (electron spin resonance). The constants of ultrafine interaction and the activation energy of the process were defined for some of them (table 2.16).

Table 2.16. The constants of ultrafine interaction and the activation energy (E_a) of the phosophonate oxidation reactions (solvent is propanol)

Compound number	E_a, kilocalorie/mole	$\alpha\beta^H$ (Э)*	α^P (Э)**	$\alpha\beta^H$ (Э)	α^P (Э)
		616 K		506 K	
(132)	8.0	2.70	14.25	2.53	15.22
(133)	3.8	2.63	14.70	2.49	15.43
(135)	6.3	2.64	14.55	2.48	15.40
(136)	8.9	2.63	15.25	2.55	16.05

* $\alpha\beta^H$ – the splitting constant from a hydrogen atom;
** α^P – the splitting constant from a phosphorous atom.

The presented data define that the constants of ultrafine interaction of an unpaired electron with the phosphorus nucleus practically do not depend on the length of alkyl substitute at a phosphorus atom, but increase during the transition to phenyl one. This fact complies with the conception about the difference of the nature of electron effects of alkyl and aryl substitutes.

The quantitative assessment of the ability of phosophonate (133) to inhibit the processes of free radical oxidation of hydrocarbons by accepting peroxide radicals was conducted by manometric method and by chemiluminescence (figure 2.18) [120].

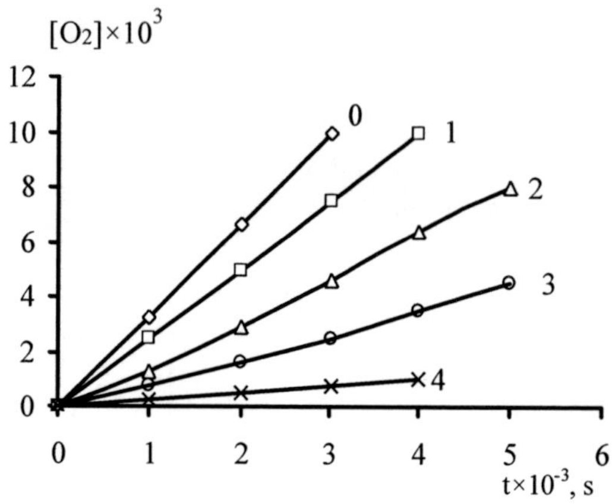

Figure 2.18. Kinetic curves of oxygen uptake ([ΔO_2], mole/l) by styrene in the presence of diethyl-3,5-di-tert-butyl-4-hydroxybenzylphos-phonate (133) in the concentration: 0 – without (133); 1 – 0.5×10^{-5} mole/l; 2 – 0.3×10^{-4} mole/l; 3 – 0.12×10^{-3} mole/l; 4 – 0.5×10^{-3} mole/l; (initiation rate $W_i = 0.5\times10^{-9}$ mole/l·s, 323 K).

It was found that values f and k_7 depend little on the nature of alkoxyl substitute at a phosphorus atom of phosophonates. At the same time aromatic substituted phosphonates are characterized by higher f and k_7 values (table 2.17).

Table 2.17. Kinetic and stoichiometric parameters of inhibiting action of 3,5-di-tert-butyl-4-hydroxybenzylphosophonates during initiated oxidation of ethylbenzene (333 K, AIBN, initiation rate $W_i = 1\times10^{-9}$m·s^{-1}, [In]=1×10^{-5}mole/l) and styrene (323 K, AIBN, W_i= 5×10^{-9}mole/l·s)

Compound number	Ethylbenzene		Styrene	
	$k_7\times10^5$, l/(mole·s)	f	$k_7\times10^5$, l/(mole·s)	f
(132)	1.5	1.0		
(133)	1.5	1.0	1.1	1.0
(135)	1.7	1.0		
(136)	2.0	1.5		
BHT	1.0	1.5	1.3	1.4

The parameters of inhibiting action of phosphonates concerning peroxide radicals are commensurable with the ones for BHT. However, indubitable priority of these compounds consists in impossibility of forming quinoid structures during oxidation, it is an essential drawback of most phenol stabilizers. It is known that even insignificant amounts of compounds of quinoid structures impart undesirable coloring to polymer [123]. Thus, the given phosophonates can be not only efficient antioxidants, but color stabilizers for polymers as well.

2.2.2.4. Phosphorus Dithioacids and Their Complexes with 3d-Metals

Unlike dialkylarylphosphites dialkyldithiophosphorus acids (61-64) display more considerable antioxidant action during initiated oxidation of styrene, that can be explained by forming low-activity sulfur-centered $(RO)_2PS_2^{\cdot}$-radical in oxidable substrate [114, p.257, 124].

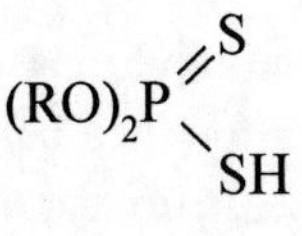

R = CH_3 **(61)**, C_3H_7 **(62)**, i-C_3H_7 **(63)**, C_6H_5**(64)**

Inhibiting properties of phosphorus dithioacids increase considerably during the introduction of sterically hindered phenol fragment into a molecule (compounds 68-74), the phenol fragment having characteristics of a strong inhibitor of free-radical oxidation processes (table 2.18) [123]. The stoichiometric coefficient of inhibition *f* of dithiophosphonic acids (68-72) increases to 1.5-2 units, and for the bis(dithiophosphonic) acids (73, 74) coefficient *f* practically doubles [55].

Table 2.18. Kinetic and stoichiometric parameters of inhibiting action of phosphorus dithioacids during initiated oxidation of styrene and polypropylene (initiator – AIBN)

Compound number	Styrene (323 K)			Polypropylene (358 K)		
	f	$k_7 \times 10^{-4}$, l/(mole·s)	$k_7/k_2 \times 10^2$	f	$k_7 \times 10^{-3}$, kg/(mole·s)	$k_7/k_2 \times 10^3$
(63)	0.35	120.00	104.35	0.30	6.89	37.93
(64)	0.22	70.00	61.65	–	–	–
(68)	2.00	3.76	3.26	–	–	–
(69)	1.50	3.00	2.61	–	–	–
(70)	2.00	3.78	3.29	1.35	2.34	15.60
(71)	2.00	1.30	1.13	1.30	0.84	5.62
(72)	1.65	2.21	1.92	–	–	–
(73)	3.60	1.25	1.09	–	–	–
(74)	3.50	1.20	1.05	1.80	1.03	6.84

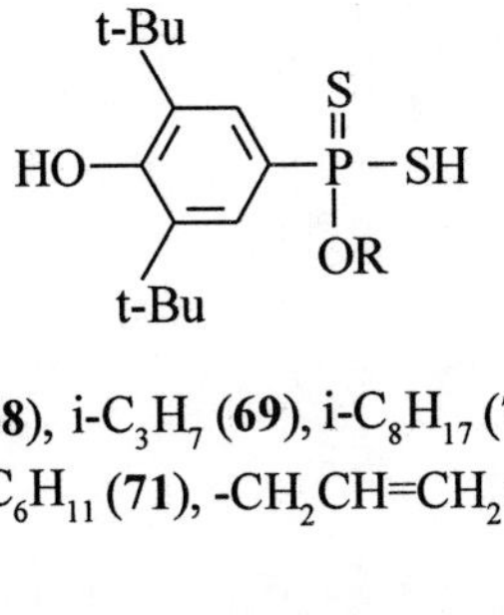

R = C_2H_5**(68)**, i-C_3H_7 **(69)**, i-C_8H_{17} **(70)**,
cyclo-C_6H_{11} **(71)**, -CH_2CH=CH_2**(72)**

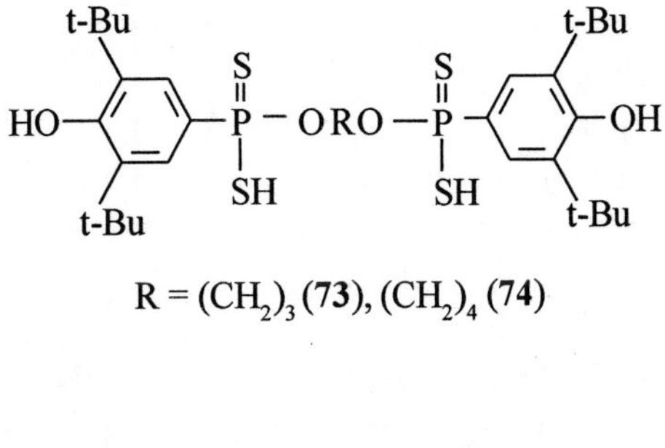

R = $(CH_2)_3$ **(73)**, $(CH_2)_4$ **(74)**

The rate constants (k_7) of initiated solid-phase oxidation of isotactic polypropylene, inhibited by acids of phosphorus, are 1-2 times lower in comparison with the ones during the styrene oxidation (table 2.18), that is likely to be connected with diffusion and structural impediment, the decreasing of molecular mobility of medium, resulting from the specific character of solid phase [125]. At the same time the general tendency of changing efficiency of action of phosphorus dithioacids is preserved.

The data about the high inhibiting action of phosphorus dithioacids containing a sterically hindered phenol fragment in model reactions indicate that their efficient stabilizing action is expected during polymer oxidation.

The characteristic of phosphorus dithioacids is their ability to interact with metal ions leading to the formation of stable complexes (ML_2) [126-131].

As compared with phosphorus dithioacids acting as ligands, the inhibiting efficiency of complexes ML_2 is characterized by higher parameters. The reasons for such conclusions are the values of inhibiting coefficients *f* (table 2.19), obtained during the oxidation of organic substrates. The results of "activation" of typical oxidation inhibitors by metal ions are given in the publication [132].

Table 2.19. Kinetic and stoichiometric parameters of inhibiting action of phosphorus dithioacids and their metal complexes during styrene oxidation initiated by AIBN (323 K, [styrene]=4.35 mole/l, [In]=5×10^{-4} mole/l, [AIBN]= 2.5×10^{-3} mole/l)

Compound number	Formula[a]	f	$k_7\times10^{-4}$, l/(mole·s)
(84)	$[(i\text{-}PrO)_2P(S)S]_2Ni$	0.8	0.50
(138)	$[(i\text{-}PrO)_2P(S)S]_2Zn$	0.9	1.60
(68)	Ar(EtO)P(S)SH	2.0	3.76
(139)	$[Ar(EtO)P(S)S]_2Zn$	3.6	3.10
(140)	$[Ar(EtO)P(S)S]_2Ni$	3.3	2.47
(69)	Ar(i-PrO)P(S)SH	1.5	3.00
(85)	$[Ar(i\text{-}PrO)P(S)S]_2Zn$	3.6	3.13
(86)	$[Ar(i\text{-}PrO)P(S)S]_2Ni$	3.3	2.45
	BHT	2.0	1.30

[a] Ar =

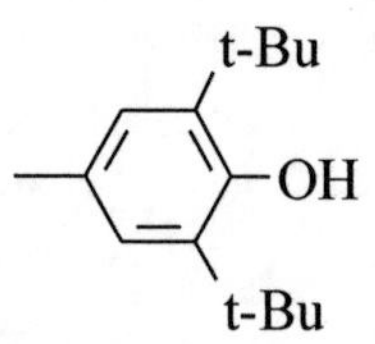

While comparing kinetic and stoichiometric parameters of styrene oxidation, initiated by dithiophosphates and dithiophosphonates of 3d-metals it was determined that the introduction of sterically hindered phenol fragments into their molecules resulted in considerable strengthening of stabilizing ability of complexes MeL_2 (table 2.20). The increasing of activity of compounds (85), (86) and (139) in the given process can be explained by the occurrence of an additional reactive center (phenol fragment), participating in accepting peroxide radicals.

2.2.2.5. 3,5-Di-Tert-Butyl-4-Hydroxybenzylamines

Stabilizing mixtures including amine and phenyl antioxidants and displaying synergetic effects are widely used [133-137].

The mechanism of stabilizing action of such compositions is determined by the regeneration in the substrate oxidation process of the most efficient inhibitor [138-140]:

$$InH + RO_2^{\bullet} \longrightarrow In^{\bullet} + ROOH$$

$$In^{\bullet} + ArOH \longrightarrow InH + ArO^{\bullet}$$

As the rate constant of the interaction of arylamine with a peroxide radical, as a rule, is considerably higher than for sterically hindered phenols due to the exchange reaction

$$Am^{\bullet} + ArOH \longrightarrow AmOH + ArO^{\bullet}$$

it is possible, firstly, to restore an efficient inhibitor and, as a result, to prevent the reaction of the oxidation chain transfer by an active amine radical [136]. Secondly, a stable aroxyl radical is formed capable of terminating oxidation chains.

Such approach turned to be the basis for obtaining polyfunctional stabilizers, namely arylamines, having the common formula

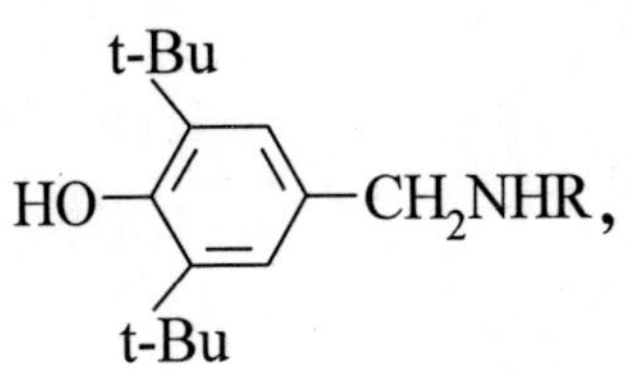

containing two types of reactive centers in a molecule, i.e. spatially hindered phenol and arylamine (generally, secondary one) groups, each of them can participate in the termination of kinetic chains of oxidation by means of accepting peroxide radicals. The combination of such interactions may promote the occurrence of intramolecular synergism [142].

The results of investigations of solid-phase initiated oxidation of polypropylene in the presence of arylamines (141-145) indicate that in the initial period the concentration of the introduced antioxidant falls sharply, and after some time it becomes stable (goes into plateau) (figure 2.19). The observed experimental result is probably determined by the fact that under conditions of inhibited oxidation initial arylamine generates another efficient inhibitor through an intermediate active radical.

The quantitative estimation of the ability of compounds (141-145) to inhibit the oxidation of organic substrates by accepting peroxide radicals was conducted by means of chemiluminescence. It was determined that these compounds are more efficient in these processes than phenol antioxidant BHT.

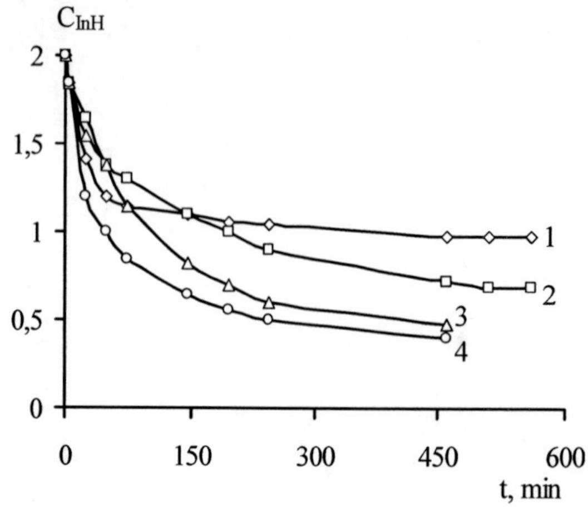

Figure 2.19. Dependence of inhibitor concentration on the time of polypropylene oxidation: 1 – compound (144); 2 – (145); 3 – (142); 4 – (143), (initiator – azoisobutyric acid dinitrile, 403 K, air, $[In]=2.0\times10^{-2}$ mole/kg).

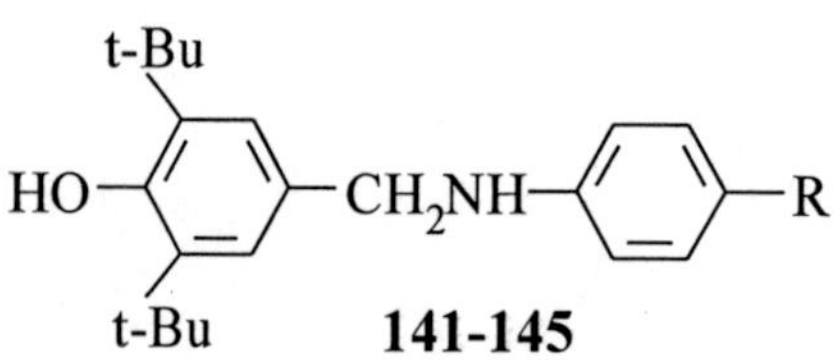

141-145

R = OH(**141**), CH_3 (**142**), OCH_3 (**143**), $NHC(O)CH_3$ (**144**), NH_2(**145**)

Table 2.20. Kinetic and stoichiometric parameters of inhibiting action of the compounds (141-145) during the initiated oxidation of cumene (333 K, initiator – AIBN, initiation rate 7×10^{-8} m/s)

Compound number	(141)	(142)	(143)	(144)	(145)	BHT
f	3.51	2.54	2.38	2.63	2.97	1.34
$k_7\times10^{-4}$, l/(mole·s)	2.09	0.84	0.92	0.89	1.23	1.43

According to table 2.20, the efficient rate constants of interaction of arylamines, containing a sterically hindered phenol fragment, with peroxide radicals reach the level of the action of phenol antioxidants.

From the point of practical stabilization of polymers N-(3,5-di-*tert*-butyl-4-hydroxybenzyl)benzthiazolthione-2 (146) is of interest, it is tertiary amine, containing sterically hindered phenol fragment in the molecule, the fragment is admittedly active regarding the accepting peroxide radicals [142].

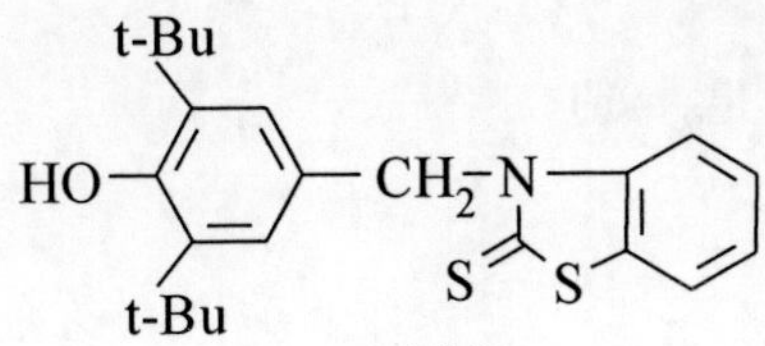

146

The obtained reaction parameters, that is, the stoichiometric coefficient of inhibition f=1.5, the rate constant of the reaction with a peroxide radical k_7=1.44×10^4 l/(mole·s), (k_7/k_2=1.25×10^2) indicate that amine (146) is comparable on efficiency with BHT.

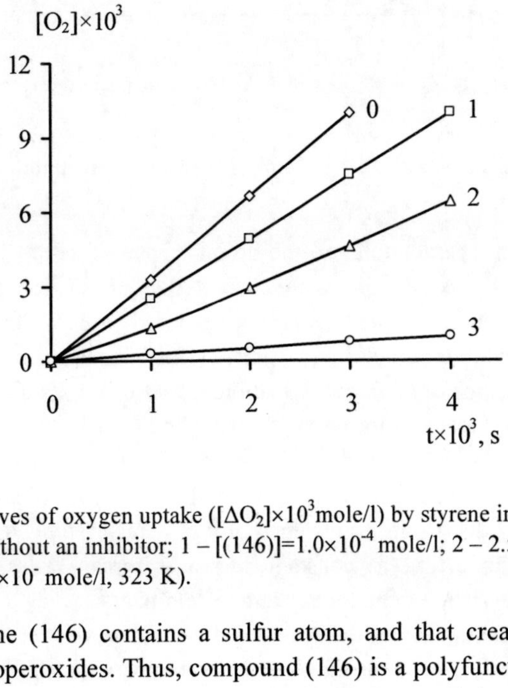

Figure 2.20. Kinetic curves of oxygen uptake ($[\Delta O_2]\times10^3$mole/l) by styrene in the presence of compound (146): 0 – without an inhibitor; 1 – [(146)]=1.0×10^{-4} mole/l; 2 – 2.5×10^{-4} mole/l; 3 – 5.0×10^{-4} mole/l; (initiation rate 5×10^{-} mole/l, 323 K).

In addition, amine (146) contains a sulfur atom, and that creates prerequisites for its interaction with hydroperoxides. Thus, compound (146) is a polyfunctional stabilizer, capable of participating both in inhibiting free-radical oxidation and in suppressing degenerate branching of oxidation chains.

2.2.2.6. Thiocarbamide Derivatives

Thiocarbamide derivatives (arylthiocarbamides and their metal complexes, benzoylthiocarbamides) were investigated under conditions of initiated oxidation using model low-molecular hydrocarbons and in polymers.

The assessment of reactions of arylthiocarbamides with peroxide radicals was conducted under conditions of initiated oxidation of styrene. It was shown that the intensity of styrene oxidation depends on the concentration of additives, the inhibition efficiency grows with the increase of their content. The illustrations for some of them are given in figure 2.21.

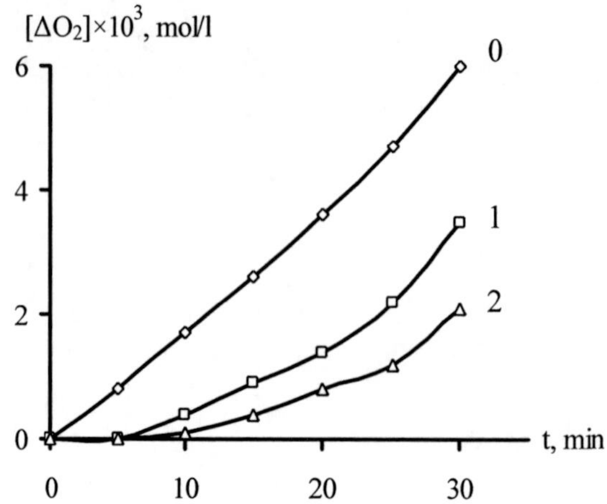

Figure 2.21. Kinetic curves of oxygen uptake by styrene in the presence of compound (92) in concentration: 1 – 0.25×10^{-2} mole/l; 2 – 0.5×10^{-2} mole/l; 0 – without an inhibitor (initiation rate 5×10^{-9} mole/l, 323 K).

According to kinetic parameters of inhibition given in table 2.21, the reactions of arylthiocarbamides with peroxide radicals are characterized by sufficiently high rate constants of the reaction exceeding similar constants for amine and phenol stabilizers.

Low values of the chain termination coefficient (less than 1) are probably connected with the fact that under conditions of inhibited oxidation of hydrocarbons there is the conversion of initial stabilizers into products promoting the destruction of substrates.

The increasing of the stoichiometric inhibition coefficient f is observed when a sterically hindered phenol fragment is introduced into a thiocarbamide molecule, i.e there is an increase of antioxidant efficiency due to the occurrence of a new reactive center, e.g. for compounds (92) and (94). At the same time k_7 constant decreases for an order approaching constant values which are characteristic of phenol and amine stabilizers.

The conducted mathematical analysis of the dependence of the efficiency of inhibiting action of arylthiocarbamides on their structure in reactions with peroxide radicals indicated that σ^*- and R_S are constants describing steric effect of substitutes at a nitrogen atom (N^1) of substituted thiocarbamides $R\text{-}N^1H\text{-}C(O)\text{-}N^3R'R''$, they correlate well with the value lg k_7 (equation 33):

$$\lg k_7=(0.160 \pm 0.044)\Sigma\sigma^* + (1.028 \pm 0.090)R_S+(9.909\pm0.387); \quad (33)$$
$$R=0.9926,\ S_0=0.131,\ N=5$$

The obtained dependence indicates that the rate constant of interaction of arylthiocarbamides with peroxide radicals decreases along with the decrease of steric effect of substitutes at a nitrogen atom (N^1).

Table 2.21. Kinetic parameters of reactions of arylthiocarbamides with peroxide radicals (styrene, 323 K, [AIBN] = 2.5×10^{-3} mole/l)

Compound number	Formula	f	k_7, l/(mole·s)	k_7/k_2
(88)		0.2	6.0×10^6	5.3×10^4
(89)		0.1	5.0×10^6	4.4×10^4
(92)	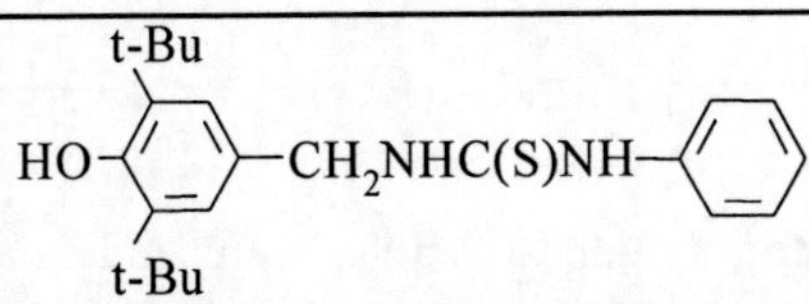	0.6	3.2×10^5	2.8×10^3
(94)	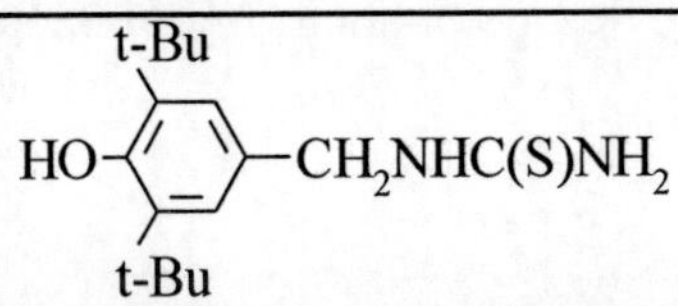	0.5	3.1×10^5	2.6×10^3
(93)	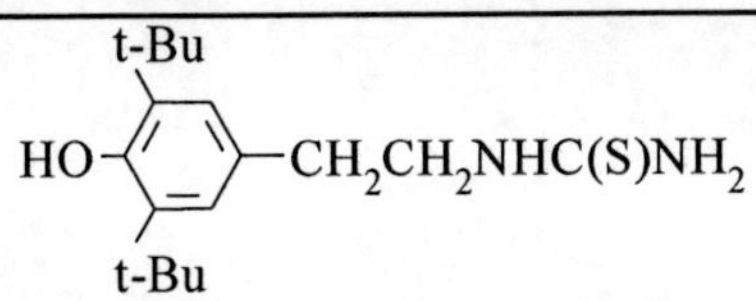	0.4	3.0×10^5	2.6×10^3
	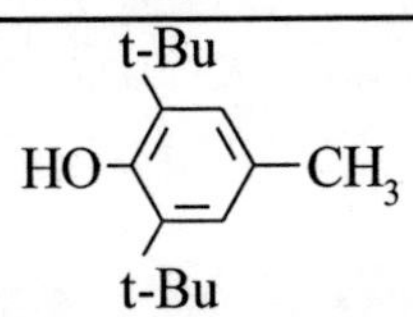	1.4	2.5×10^4	2.2×10^2
	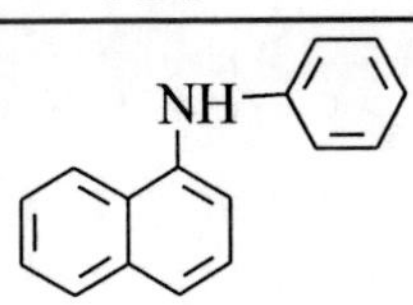	1.0	1.3×10^5	1.14×10^3

Thus, the obtained results indicate that there is statistically significant correlation of reactivity of arylthiocarbamide derivatives with their structure in reactions with peroxide radicals, this can be the well-grounded basis for predicting structures of efficient inhibitors of oxidation processes.

The strengthening of stabilizing action of thiocarbamides by means of complex formation with metal ions is determined.

Reactions of complex formation of aryl- and benzoylthiocarbamides with a number of compounds of 3d-metals result in the forming stable compounds by means of coordination through the nitrogen and sulfur atoms (through the sulfur and oxygen atoms in case with benzoylthiocarbamides), the structure of which is proved by the methods of infrared and electron spin resonance spectroscopy, and by the data of an elemental analysis.

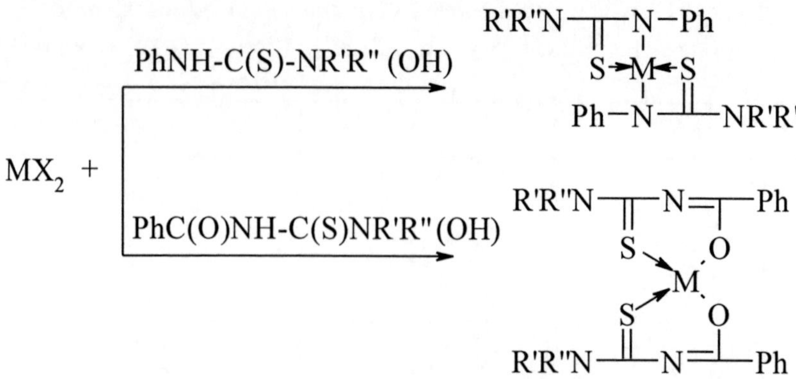

Metal complexes (ML_2) of investigated arylthiocarbamides were studied as inhibitors of styrene oxidation initiated by AIBN in the range of concentrations from 1.0×10^{-2} to 0.25×10^{-2} mole/l (figure 2.22, 2.23).

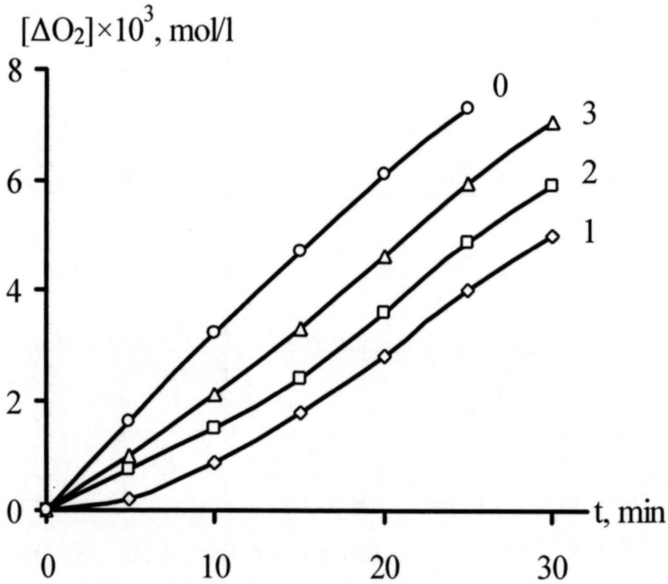

Figure 2.22. Kinetic curves of oxygen uptake by styrene in the presence of metal complex (147) in the concentration: 1 – 1.0×10^{-2} mole/l, 2 – 0.5×10^{-2} mole/l, 3 – 0.25×10^{-2} mole/l, (initiation rate 5×10^{-9} mole/l, 323 K).

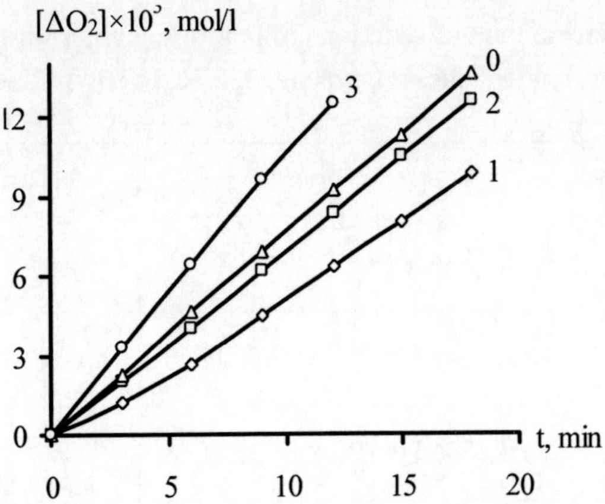

Figure 2.23. Kinetic curves of oxygen uptake by styrene in the presence of metal complex (107) in the concentration: 1 – 1.0×10^{-2} mole/l; 2 – 0.5×10^{-2}; 3 – 0.25×10^{-2} mole/l, (initiation rate 5×10^{-9} mole/l, 323 K).

Kinetic investigations showed that complexes of arylthiocarbamides with ions Ni(II) and Co(II) in the certain range of concentrations reveal properties of inhibitors of initiated styrene oxidation, at the same time they are more active inhibitors than the initial arylthiocarbamides (table 2.22). At the same time the efficiency of their inhibiting action depends on the concentration and nature of metal. For example, complexes of arylthiocarbamides with ions Cu(II) display characteristics both of an inhibitor and an initiator, at concentrations of 1.0 – 0.5×10^{-2} mole/l they behave as inhibitors (figure 2.22), when concentration decreases to 0.25×10^{-2} mole/l they behave as initiators (figure 2.23).

A large number of benzoylthiocarbamides in reactions with $RO_2^{\cdot}$-radicals during solid-phase oxidation of polypropylene and polyethylene initiated by benzoyl peroxide is studied (figure 2.24, 2.25). The increasing of induction periods alongside with the increasing of inhibitor concentration is illustrated by compound (91) (figure 2.25).

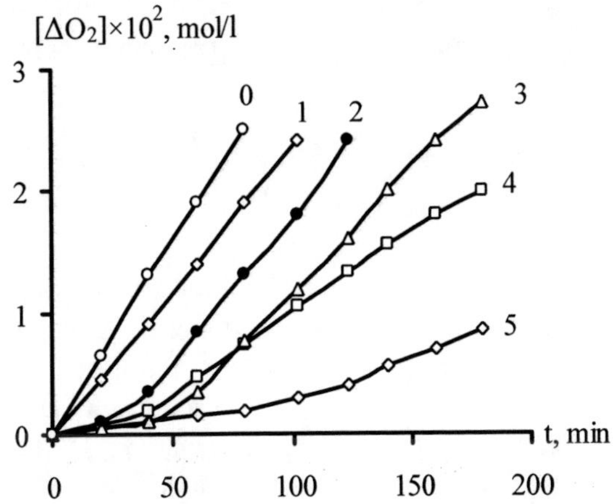

Figure 2.24. Kinetic curves of oxygen uptake by polypropylene in the presence of 1×10^{-4} mole/kg inhibitor: 0 – without inhibitor; 1 – compound (96); 2 – (97); 3 – (99); 4 – (98); 5 – (100); (358 K, initiator – benzoyl peroxide).

Table 2.22. Kinetic parameters of reactions of arylthiocarbamides and metal complexes ML_2 with peroxide radicals (styrene, 323 K, [AIBN]=2.5×10^{-3} mole/l)

M(II)	Ligand (L)	k_7, l/(mole·s)
–		6.0×10^6
Ni	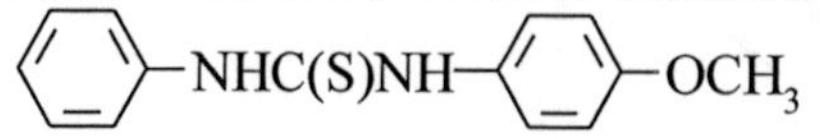	5.1×10^7
Co	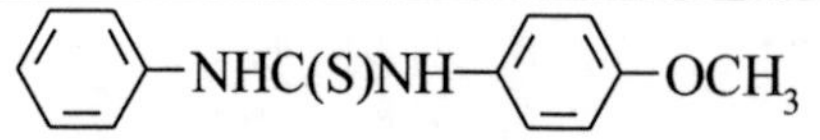	7.7×10^6
–	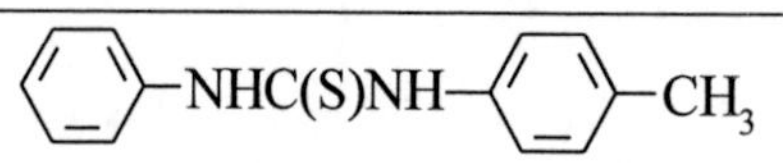	1.2×10^6
Ni	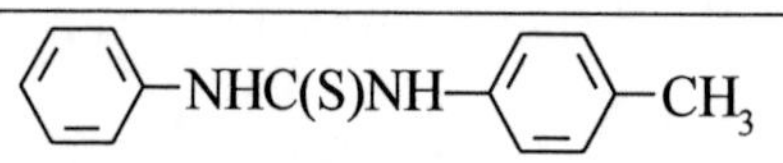	7.9×10^6
Cu	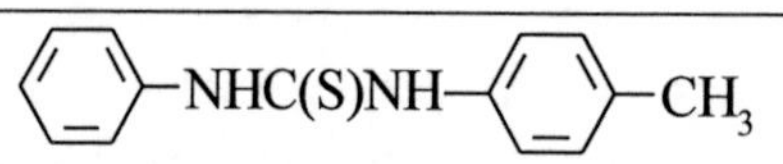	1.0×10^7
Co	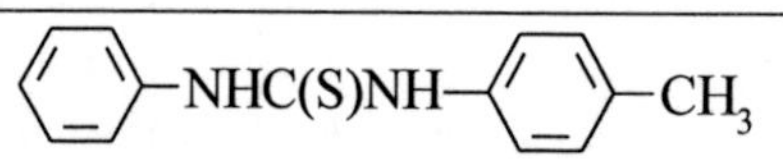	4.9×10^7
–	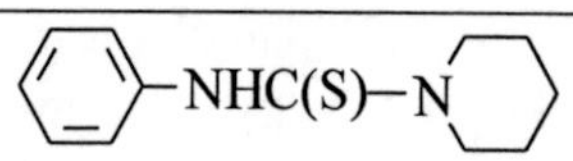	5.0×10^6
Cu	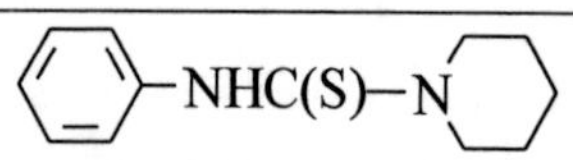	1.2×10^7

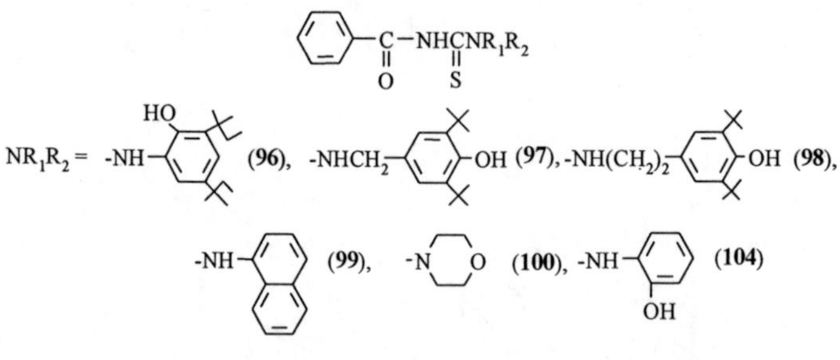

As it is seen from table 2.23 values of inhibition coefficient f for various studied benzoylthiocarbamides during the initiated oxidation of polypropylene change in the sufficiently large range, that is substitutes at a nitrogen atom in a molecule of benzoylthiocarbamides influence considerably the reactivity of investigated compounds in respect of peroxide radicals. In particular, the introduction of sterically hindered phenol fragments (compounds 97 and 98) promotes strengthening of antioxidant activity of benzoylthiocarbamides.

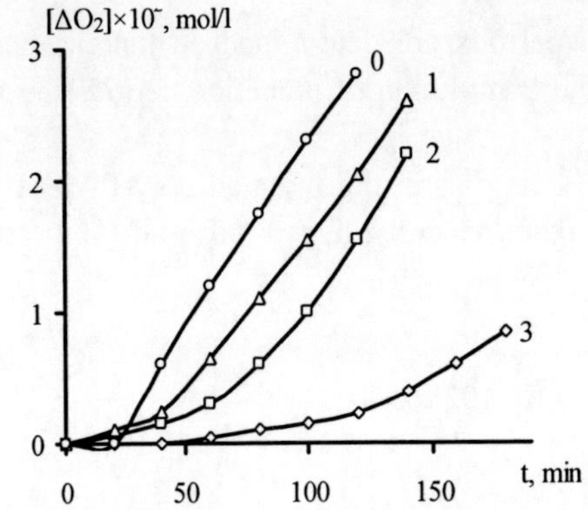

Figure 2.25. Dependence of induction periods of initiated oxidation of polypropylene on the compound concentration (104): 0 – without inhibitor; 1 – 0.5×10^{-4} mole/kg; 2 – 1×10^{-4} mole/kg; 3 – 5×10^{-4} mole/kg; (358 K, oxygen pressure 33.3 kPa).

Table 2.23. Kinetic and stoichiometric parameters of reactions benzoylthiocarbamides $C_6H_5C(O)NHC(S)NRR'$ with peroxide radicals (polypropylene, 358 K, initiator is benzoyl peroxide, [In]=5×10^{-2} mole/kg)

Compound number	f	$k_7\times10^{-2}$ ($l\cdot mole^{-1}\cdot s^{-1}$)	$k_7/k_2\times10^{-2}$
(96)	0.32	4.50	30.00
(97)	1.36	4.70	31.38
(98)	1.40	6.40	42.74
(99)	0.73	4.80	32.02
(100)	1.90	1.25	8.34
(101)	0.23	5.30	35.3
(102)	2.70	9.40	62.74
(103)	1.40	2.35	15.35
(104)	1.93	7.40	49.35
(105)	1.07	12.70	84.71
BHT	1.30	3.10	20.71

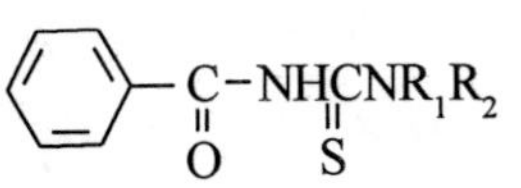

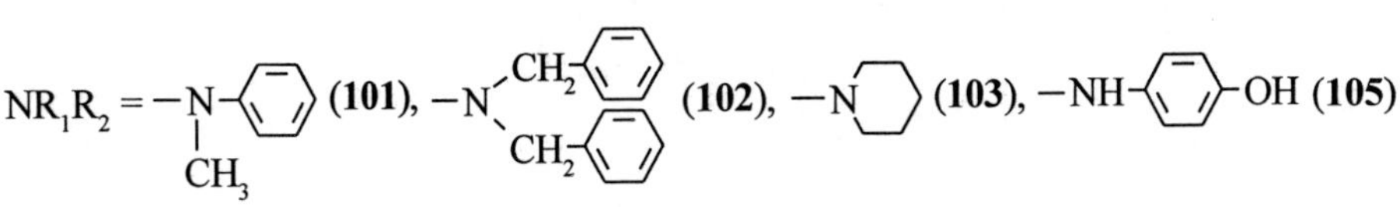

The analysis of kinetic curves for a number of benzoylthiocarbamides (figure 2.26, 2.27) indicates that regularities of polyethylene oxidation are similar to the ones observed during the polypropylene oxidation. The absence of induction periods is characteristic for compounds (96), (101), (102), but nevertheless the introduction of the additive decreases the rate of polymer oxidation.

Considerable induction periods are determined on kinetic curves for compounds (97), (99), (100) and (104), on the completion of induction periods the rate of polymer oxidation starts increasing weakly.

The most stabilizing effects during the initiated polyethylene oxidation are determined for the same benzoylthiocarbamide derivatives (100) and (104), as it was indicated during polypropylene oxidation.

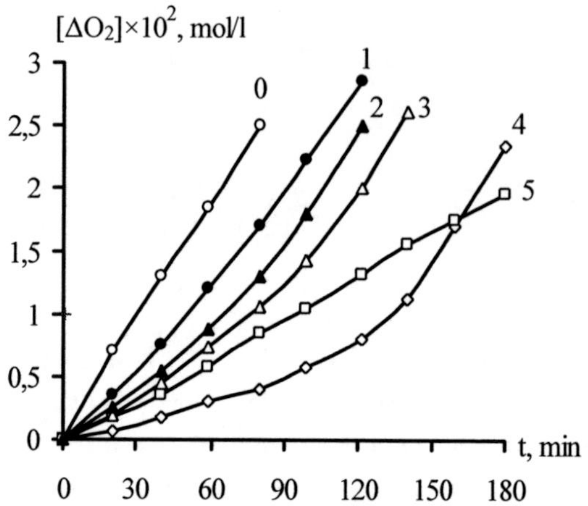

Figure 2.26. Kinetic curves of oxygen uptake by polyethylene in the presence of 1×10^{-4} mole/kg of inhibitor: 0 – without inhibitor; 1 – (96); 2 – (99); 3 – (97); 4 – (100); 5 – (98); (388 K, initiator – benzoyl peroxide).

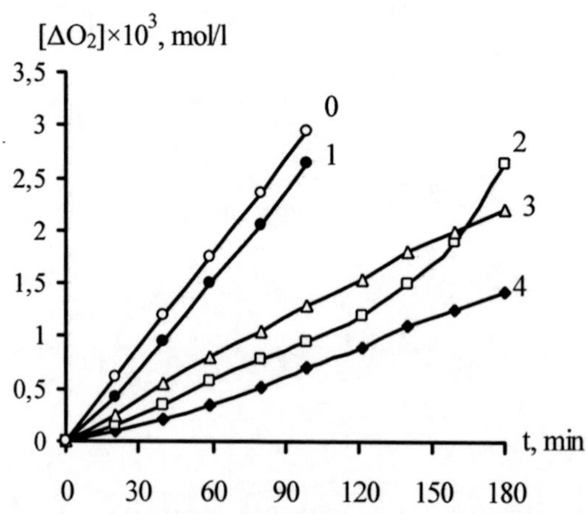

Figure 2.27. Kinetic curve of oxygen uptake by polyethylene in the presence of 1×10^{-4} mole/kg of inhibitor: 0 – without inhibitor; 1 – compound (101); 2 – (104), 3 – (102); 4 – (103), (388 K, initiator – benzoyl peroxide).

Thus, the obtained kinetic data about the inhibiting action of thiocarbamide derivatives as peroxide radical acceptors indicate that benzoylthiocarbamides are the most active in these processes. They are characterized by high values of stoichiometric coefficient (*f*), rate constants of reactions with peroxide radicals (k_7), which exceed similar characteristics of phenol and amine antioxidants in a number of cases.

* * *

On the whole, the characteristic representatives of various classes of chemical compounds (tri- and dialkyl(aryl)phosphites, phosphorus dithioacids (phosphates, phosphonates), thiocarbamides derivatives (aryl- and benzoylthiocarbamides) and their metal complexes, 3,5-di-*tert*-butyl-4-hydroxybenzylamine) are discussed as polyfunctional stabilizers in Chapter 2. The presence of the heteroatoms determining their nucleophilic properties is the important precondition for providing their high reactivity during the stabilization of polymers.

The carried out analysis of the results of modeling chemical processes of inhibited oxidation of polymers has shown that the above described compounds are capable to interact with various electrophilic products of polymer oxidation and to participate both in heterolytic (the interaction with hydroperoxides) and in homolytic (the interaction with alkyl, peroxide and alkoxyl radicals) reactions which are responsible for the stabilization of polymers, displaying thus the polyfunctionality of stabilizing action.

The kinetic methods of quantitative testing of polyfunctional stabilizers allow to determine the general and specific properties of these compounds.

Their high reactivity and heterolytic nature of interaction with hydroperoxides are common for N,S,P-containing polyfunctional stabilizers. It can be the basis for their using as effective nonradical decomposers of hydroperoxides, the main branching agents in the processes of thermal-oxidizing destruction of polymers.

At the same time the specific features of the structure of polyfunctional stabilizers define the distinctions in mechanisms of their reactions with hydroperoxides.

So, the reactions of tri-alkyl(aryl)phosphites are characterized by bimolecular reaction rate constants, the values of constants depend on the nature of substitutes in the molecules. Pentamerous arylene(alkylene)phosphites behave specifically. In reactions with hydroperoxides and with alkoxy- and peroxyradicals they generate a new product, namely, acyclic acidic hydroxyphenylphospate, being a catalyst of hydroperoxide decomposition, on the whole it determines the high stabilizing effect of *ortho*-phenylenephosphorous acid esters.

It was determined for acids of phosphorus that the hydroperoxide decomposition proceeds according to the scheme of proton catalysis. These reactions are characterized baby superstoichiometric coefficients which should determine their high abilities in suppressing degenerated branching of oxidation chains under conditions of polymer oxidation.

Thiocarbamide derivatives (aryl- and benzoylthiocarbamides) display catalytic nature when they interact with hydroperoxides, however the mechanism of hydroperoxide decomposition is caused by the products of thiocarbamide decomposition.

The data about the initiated radical-chain oxidation of polyfunctional stabilizers have shown the possibility of their participation in reactions with $RO_2^{\cdot}$ radicals. The reactivity and mechanisms of such interactions depend on the structure of stabilizers.

So, aromatic phosphites, unlike aliphatic phosphites which are inclined to self-oxidation reactions, are oxidized according to the mechanism of linear termination of chains by phosphite molecules, i.e. the typical mechanisms of action of oxidation inhibitors are accomplished.

In these processes the effectiveness of inhibition by polyfunctional stabilizers depends, first of all, on the nature and ways of the decomposition of formed intermediate kinetically unstable radicals and the possibilities of their changing into stable products. For trialkyl(aryl)phosphites it is a phosphoranyl radical which is capable to be oxidized forming chain-developing alkyl radicals (in a case of aliphatic phosphites), or it is inclined to substitution reactions, releasing thus low-activity aroxyl radical terminating oxidation chains (in case of aromatic phosphites). For hydrophosphorus compounds it is a phosphonradical characterized by dual reactivity, for phosphorus dithioacids it is sulfur-centered radicals etc.

Among thiocarbamides derivatives benzoylthiocarbamides are characterised by the highest values of inhibiting action (superstoichiometric coefficients and significant rate constants of interaction with peroxide radicals), they are comparable with antioxidants of phenolic and amine type as to their effectiveness.

The phenomenon of increasing the efficiency of polyfunctional stabilizers as oxidation inhibitors of hydrocarbon substrata by their linkage into kinetically stable complexes with ions of transition metals is described. So, for trialkyl(aryl)phosphites the basic properties of this phenomenon are determined, allowing to confirm that the presence of strong donor-acceptor interaction between a metal ion and a phosphite-inhibitor in coordination sphere is a necessary condition of strengthening inhibiting properties of phosphites in hydrocarbon substrate oxidation.

The similar facts of strengthening inhibiting actions are indicated for a number of other polyfunctional stabilizers, namely, metal complexes of dithiophosphates and phosphonates, thiocarbamide derivatives, they, as a rule, exceed the corresponding ligands as to their activity.

The increase of stabilizing efficiency of elementorganic compounds can be reached by the introduction of functional groups into their structure, in particular, sterically hindered phenol fragments (for example, 3,5-di-*tert*-butyl-4-hyderoxybenzylated amines, phosphonates, thiocarbamides, etc.). There is a considerable strengthening of their ability to interact with peroxide radicals.

Thus, in Chapter 2 the basic results on the properties of polyfunctional stabilizers defining their possibilities of using as effective stabilizers of polymers are discussed.

REFERENCES

[1] Semenov, N.N. *Tsepnye reaktii (The Chain Reactions)*. Leningrad: *Goskhimizdat;* 1934.

[2] Scott, G. Atmosfernoe okislenie i antioksidanty (Atmospheric Oxidation and Antioxidants). Amsterdam: *Elsevier;* 1965.

[3] Waters U. Mekhanismy okisleniya organicheskikh soedinenii (Mechanism of Oxidation of Organic Compounds). Moscow: *Nauka;* 1966.

[4] Neiman, M.B. Starenie i stabilizatsiya polimerov (Ageing and Stabilization of Polymers). Moscow: *Nauka;* 1964.

[5] Emanuel, N.M., Denisov, E.T., Maizus. Z.K. Tsepnye reaktsii okisleniya v zhidkoi faze (The Chain Reactions of Oxidation in Liquid Phase). Moscow: *Nauka;* 1965.

[6] Voigt, J. Die Stabilisierung der Kuntstoffe gegen Licht und Wärme. Berlin: *Springer-Verlag;* 1966.

[7] Emanuel, N.M., Buchachenko, A.L. Khimicheskaya fizika stareniya i stabilizatsii polimerov (Chemical Physics of Ageing and Stabilization of Polymers). Moscow: *Nauka;* 1982.

[8] Thinius, K. Stabilisierung und Alterung von Plastwerkstoffen. Berlin: *Akademie-Verlag;* 1969.

[9] Grassi, N. and Scott, G. Destruktsiya i stabilizatsiya polimerov. (Destruction and Stabilization of Polymers). Moscow: *Mir;* 1988.

[10] Shlyapnikov, Yu.A. Kiryushkin, S.G., Mar'in, A.P. Antiokislitel'naya stabilizatsiya polimerov (Antioxidative Stabilization of Polymers). Moscow: *Khimiya;* 1986.

[11] Denisov, E.T. Okislenie i destruktsiya karbotsepnykh polimerov (Oxidation and Destruction of Carbochain Polymers). Leningrad: *Khimiya;* 1986.

[12] Piotrovskii, K.B., Tarasova, Z.N. Starenie i stabilizatsiya sinteticheskikh kauchukov i vulkanizatov (Ageing and Stabilization of Synthetic Rubbers and Vulkanizates). Moscow: *Khimiya*; 1980.

[13] Kuz'minskii, A.C. Starenie i stabilizatsiya polimerov (Ageing and Stabilization of Polymers). Leningrad: *Khimiya;* 1989.

[14] Denisov, E.T., Azatyan, V.V. Ingibirovanie tsepnykh reaktsii (Inhibition of the Chain Reactions). *Chernogolovka,* 1997.

[15] Kirpichnikov, P.A., Kolyubakina, N.S., Minsker, K.S. et al. – *Zh. Anal. Khim.,* 1968, vol.23, №10, p. 1532-1583.

[16] Antanovskii, V.L., Buzlanova, M.M. Analiticheskaya khimiya organicheskikh peroksidnykh soedinenii (Analitical Chemistry of Organic Peroxide Compounds). Moscow: *Khimiya;* 1978.

[17] Pobedimskii, D.G., Belyakov, V.A. – *Kinetika Kataliz,* 1969, vol.10, p. 64-68.

[18] Denisov, E.T. Kinetika gomogennykh khimicheskikh reaktsii (Kinetics of Homogeneous Chemical Reactions). Moscow: *Vyssh. Shkola;* 1978.

[19] Kovarskaya, B.M. Blyumenfeld, A.B., Levantovskaya, I.I. Termicheskaya stabil'nost geterotsepnykh polimerov (Thermal Stability of Heterochain Polymers). Moscow: *Nauka;* 1982.

[20] Minsker, K.S., Fedoseeva, G.T. Destruktsiya i stabilizatsiya polivinilkhlorida (Destruction and Stability of Polyvinylchloride). Moscow: *Khimiya;* 1979.

[21] Emanuel, N.M., Buchachenko, A.L. Khimicheskaya fizika molekulyarnogo razrusheniya i stabilizatsii polimerov (Chemical Physics of Molecular Destruction and Stabilization of Polymers). Moscow: *Nauka;* 1988.

[22] Kirpichnikov, P.A., Mukmeneva, N.A. In: *Starenie i stabilizatsiya polimerov (Ageing and Stabilization of Polymers),* Ed. by Kusminskii A.S. Moscow: *Nauka;* 1966; p.165-195.

[23] Kirpichnikov, P.A. Mukmeneva, N.A., Pobedimskii, D.G. – *Usp. Khim.,* 1983, vol.52, №11, p. 1831-1851.

[24] Pobedimskii, D.G., Mukmeneva, N.A., Kirpichnikov, P.A. In: *Developments in Polymer Stabilization*, Ed. by Scott, G. London: *Appl. Sci. Publ.;* 1980, p. 125-184.

[25] Kirby, A.J.,Warren, S.G. *The Organic Chemistry of Phosphorus*. Amsterdam: *Elsevier*; 1967.

[26] Hudson, R.F. *Structure and Mechanism in Organophosphorus Chemistry*. London/New York: Acad. Press.; 1965.

[27] Denney, D., Greenbaum, M. – *J. Amer. Chem. Soc.,* 1957, vol.79, p. 979-984.

[28] Walling, C., Rabinowich, R. – *J. Amer. Chem. Soc.,* 1959, vol.81, p. 1243-1247.

[29] Levin, P.I. – *Zh. Phis. Khim.,* 1964, vol.38, p. 672-675.

[30] Levin, P.I. – *Usp. Khim.*, 1970, vol.39, p. 1687-1706.

[31] Kirpichnikov, P.A., Mukmeneva, N.A., Pudovik, A.N. – Dokl. AN SSSR, 1965, vol. 164, №5, p. 1050-1053.

[32] Pobedimskii, D.G., Chebotareva, E.G., Kadyrova, V.Kh. – *Vysokomol. Soed.*, 1973, vol.B15, №2, p. 90-92.

[33] Chebotareva, E.G., Pobedimskii, D.G., Kirpichnikov, P.A. – *Kinetika Kataliz,* 1973, vol.14, №4, p. 891-895.

[34] Pobedimskii, D.G., Buchachenko, A.L. – *Izv. AN SSSR, Ser. Khim.,* 1968, №6, p. 1181-1186.

[35] Kirpichnikov, P.A., Pobedimskii, D.G., Mukmeneva, N.A. In: *Khimiya i primenenie fosfororganicheskikh soeinenii (Chemistry and Application of Organophosphorus Compounds)*, Ed. by Kabachnic, M.I. Moscow: *Nauka;* 1974. p. 215-220.

[36] Kuliev, F.A. – *Neftekhim.,* 1981, vol.21, №6, p. 898-905.

[37] Mastryukova, T.A., Kabachnik, M.I. – *Usp. Khim.,* 1969, vol.38, №8, p. 1751-1782.

[38] Humphris, K., Scott, G. – *J. Chem. Soc. Perkin II,* 1973, №6, p. 826-831.

[39] Rusavy, D., Slama, Z. – *Chem. Prum.,* 1968, vol.18, №p. 20-22.

[40] Pershin, A.D. Pobedimskii, D.G. Kurbatov, V.A. et al. – *Izv. AN SSSR, Ser. Khim.,* 1975, №3, p. 581-586.

[41] Schwetlick, K., Rüger, C., Noack, R. – *J. Prakt. Chimie,* 1982, Bd.324, H.5, S. 697-705.

[42] Rüger, C., König, T., Schwetlick, K. – *J. Prakt. Chimie,* 1984, Bd.326, H.4, S. 622-632.

[43] König, T., Grossmann, G., Schwetlick, K. – *J. Prakt. Chimie,* 1986, Bd.328, H.5-6, S. 763-771.

[44] König, T., Schwetlick, K. – *Polym. Degrad. and Stab.,* 1986, №15, p. 151-160.

[45] Tkaĉ, A. Rüger, C. Schwetlick, K. – *Collect. Czech. Chem. Comm.*, 1980, vol.45, p. 1182-1196.

[46] Zaichenko, L.P., Babel, V.G., Proskuryakov, V.A. – *Zh. Prikl. Khim.,* 1976, vol.49, №3, p. 465-466.

[47] Humphris, K., Scott, G. – *J. Chem. Soc. Perkin II.,* 1974, №6, p. 617-620.

[48] Haake, P., Westhemier, F. – *J. Amer. Chem. Soc.*, 1961, vol.83, №. p. 1102-1105.

[49] Galkin, V.I., Sayakhov, R.A., Cherkasov, R.A. – *Usp. Khim.,* 1991, vol.60, №8, p. 1617-1641.

[50] Galkin, V.I., Sayakhov, R.A., Cherkasov, R.A. – *Metallorg. Khim.,* 1990, vol.5, №3, p. 986-993.

[51] Cherkasova, O.A., Galkin, V.I., Sayakhov, R.A. et al. – *Zh. Obshch. Khim.,* 1993, vol.63, №6, p. 1234-1239.

[52] Zaichenko, L.P., Nechaev, Yu.D., Babel, V.T. et al. – *Zh Prikl. Khim.,* 1971, vol.44, №9, p. 2137-2138.
[53] Mukmeneva, N.A., Pobedimskii, D.G., Akhmadullina, A.T. et al. – *Vysokomol. Soed.,* 1974, vol.B16, №8, p. 635-637.
[54] Akhmadullina, A.T., Mukmeneva, N.A., Kirpichnikov, P.A. et al. – *Vysokomol. Soed.,* 1974, vol.A16, №2, p. 370-375.
[55] Cherkasova, O.A., Mukmeneva, N.A. – *Zh. Obshch. Khim.,* 1992, vol.61, №10, p.2161-2183.
[56] Schwetlick, K., Pionteck, Z., Winkler, A. – *Polym. Degrad. and Stab.,* 1991, №3, p.219-228.
[57] Cherkasova, O.A., Chebotareva, E.G., Gol'dfarb, E. I. et al. – *Neftekhim.,* 1981, vol.21, №5, p. 728-734.
[58] Nifant'ev, E.E. – *Usp. Khim.,* 1978, vol.47, №9, p.1565-1608.
[59] Ovchinnikov, V.V., Galkin V.I., Cherkasov, R.A. et al. – *Izv. AN SSSR. Ser. Khim.,* 1977, №9, p. 2021-2025.
[60] Kryuchkov, A.A., Komarova, M.R., Kazachenko, A.T. et al. – *Izv. AN SSSR, Ser. Khim.,* 1979, №11, p. 2456-2461.
[61] Ovchinnikov, V.V., Cherkasova, O.A., Verizhnikov, L.V. – *Zh. Obshch. Khim.,* 1982, vol.42, №3, p.174-176.
[62] Antipin, I.S., Gareev, R.F., Ovchinnikov, V.V. et al. – *Dokl. AN SSSR,* 1989, vol.304, №1, p. 135-138.
[63] Ovchinnikov, V.V., Cherkasova, O.A., Yarkova, E.T. et al. – *Izv. AN SSSR, Ser. Khim.,* 1978, №3, p. 689-691.
[64] Pobedimskii, D.G. – *Neftekhim.,* 1978, vol.18, №5, p. 701-707.
[65] Houalla, D., Sanchez, M., Wold, R. et al. – *Tetr. Letters,* 1978, №47, p.4675-4678.
[66] Tsvetkov, E.N., Terekhova, M.I., Petrov, E.S. et al. – *Izv. AN SSSR, Ser. Khim.,* 1978, №9, p. 1981-1983.
[67] Mukmeneva, N.A., Cherkasova, O.A., Cherezova, E.N. – *Zh. Obshch. Khim.,* 1987, vol.57, №12, p. 2696-2698.
[68] Cherkasova, O.A., Mukmeneva, N.A., Chebotareva, E.G, et al. – *Neftekhim,* 1984, vol.24, №1, p. 76-81.
[69] Grossmann, G., Komber, G., Kroshtwitz, G. et al. – *Izv. AN SSSR, Ser. Khim.,* 1989, №1, p. 34-41.
[70] Bridgwater, A.Z., Dever, Z.R., Sexton, M.D. – *Zh. Chem. Soc., Perkin Trans II,* 1978, №7, p. 530-536.
[71] Cherezova, E.N., Sayakhov, P.D., Galkin V.I. et al. – *Zh. Obshch. Khim,* 1990, vol.60, №3, p. 530-532.
[72] Mukmeneva, N.A. et al. The USSR Author Certificate 1404510 (1988); *Bull. Izobret.* (Rus), 1988, №42.
[73] Mukmeneva, N.A. et al. The USSR Author Certificate 1437367 (1988); *Bull. Izobret.* (Rus), 1988, №42.
[74] Cherezova, E.N. et al. The USSR Author Certificate 1437368 (1988); *Bull. Izobret.* (Rus), 1988, №42.
[75] Bridgwater, A.Z., Dever, Z.R., Sexton, M.D. – *Zh. Chem. Soc.,* Perkin Trans II, 1980, №7, p. 1006-1016.

[76] Cherezova, E.N., Cherkasova, O.A., Mukmeneva, N.A. – *Zh. Obshch. Khim.*, 1992, vol.62, №4, p. 781-785.

[77] Ivanov, S., Kateva, I. – *Neftekhim.*, 1978, vol.18, №3, p. 417-425.

[78] Ohms, Y., Grossmann, G., Mukmeneva, N.A. et al. – *J. Prakt. Chemie.* 1989, Bd.33, H.331, S. 361-364.

[79] Grishina, O.N., Bashinova, V.M., Potekhina, M.I. – *Neftekhim.,* 1980, vol.20, №3, p. 451-456.

[80] Howard, J.A., Chenier, Z.H.B. – *Can. J. Chem.*, 1976, vol.54, №3, p. 390-401.

[81] Edilashvili, I.L., Ioseliani, K.B., Zolotova, N.B. et al. – *Neftekhim.,* 1979, vol.19, №6, p. 892-896.

[82] Grishina, O.N., Bashinova, V.M., Potekhina, M.I. – *Neftekhim.*, 1974, vol.14, №1, p. 142-146.

[83] Al-Malaika, S., Scott, G. – *Eur. Polym. J.,* 1983, vol.19, №3, p. 235-240.

[84] Oae, S. Khimiya organicheskikh soedinenii sery (Chemistry of Organic Compounds of Sulfur). Moscow: *Khimiya;* 1975.

[85] Gorbunov, B.N., Gurvich, Ya. A., Maslova, I.P. Khimiya i tekhnologiya stabilizatorov polymernykh materialov (Chemistry and Technology of Stabilizers for Polymeric Materials). Moscow: *Khimiya;* 1981.

[86] Mukmeneva, N.A., Cherezova, E.N., Tuzova, N.G. – *Zh. Prikl. Khim.,* 1994, vol.67, №4, p. 633-635.

[87] Mukmeneva, N.A., Cherezova, E.N., Cherkasova, O.A. et al. – *Zh. Obshch. Khim.*, 1994, vol.64, №6, p. 1025-1027.

[88] Cherezova, E.N., Zhukova, R.S., Tatarintseva, T.B. et al. – *Zh. Obshch. Khim.,* 1997, vol.67, №8, p. 1356-1360.

[89] Krivenko, L.V., Cherezova, E.N., Mukmeneva, N.A. – *Zh. Prikl. Khim.*, 2000, vol.73, №7, p. 1193-1197.

[90] Sibirskaya, V.V., Kukushkin, Y.N. – *Koordin. Khim.,* 1978, vol.4, №7, p. 963-991.

[91] Voss, W.W. – *Chem. Amides,* 1971, Bd.743, H4, S. 383-475.

[92] Pobedimskii, D.G., Mukmeneva, N.A., Kirpichnikov, P.A. – *Usp. Khim.,* 1972, vol.41, №7, p. 1242-1259.

[93] Bentrude, W., Fu I., Rogers, P. – *J. Amer. Chem. Soc.*, 1973, vol.95, p. 3625-3629.

[94] Waters, G., Ingold, K. – *J. Amer. Chem. Soc.*, 1972, vol.94, p. 2528-2533.

[95] Davis, A., Griller, D., Roberts, B. – *J. Chem. Soc., Perkin Trans II,* 1972, №15, p. 2224-2228.

[96] Humphris, K., Scott, G. – *Pure and Appl. Chem.*, 1973, vol.36, p. 163-167.

[97] Kirpichnikov, P.A., Mukmeneva, N.A.,Pudovik, A.N. et al. – *Zh. Obshch. Khim.,* 1964, vol.34, №3, p. 1683-1684.

[98] Kirpichnikov, P.A. et al. The USSR Author Certificate 162958 (1964); *Bull. Izobret.* (Rus), 1964, №11.

[99] Pobedimskii, D.G., Levin, P.I., Chelnokova, Z.B. – *Izv. AN SSSR, Ser. Khim.,* 1969, №9, p. 2066-2068.

[100] Schwetlick, K., König, T., Rüger, C. et al. – *Zeitshrift für Chemie,* 1986, H.26, S. 360-366.

[101] Schwetlick, K., König, T., Rüger, C. et al. – *Polym. Degrad. and Stab.*, 1986, vol.15, №2, p. 97-108.

[102] Schwetlick, K., Pionteck, J., König, T. et al. – *Eur. Polym. J.,* 1987, vol.23, №5, p. 383-388.

[103] Rüger, C., König, T., Schwetlick, K. et al. – *Acta Polymerica,* 1986, vol.37, p. 435-441.

[104] Levin, Ya.A., Il'yasov, A.V., Pobedimskii, D.G.,– *Izv. AN SSSR, Ser. Khim.*, 1970, №7, p. 1680-1684.

[105] Pobedimskii, D.G., Kirpichnikov, P.A., Samitov, Yu.Yu. et al. – *Org. Magn. Resonance*, 1973, vol.5, №10, p. 503-506.

[106] Buchachenko, A.L., Pobedimskii, D.G., Kurbatov, V.A. – *Izv. AN SSSR, Ser. Khim.,* 1973, №11, p. 2454-2458.

[107] Pobedimskii, D.G., Kurbatov, V.A., Temyachev, I.D. et al. – *Teoret. Eksperim. Khim.,* 1974, vol.10, №4, p.492-499.

[108] Pobedimskii, D.G., Nasybullin, Sh. A., Kadyrova, V.Kh. et al. – *Dokl. AN SSSR,* 1976, vol.226, №3, p.634-636.

[109] Pobedimskii, D.G., Nasybullin, Sh. A., Svitich, R.B. et al. – *Org. Magn. Resonance,* 1977, vol.9, №2, p. 61-65.

[110] Nasybullin, Sh. A., Valeev, N.A., Pobedimskii, D.G. et al. – *React. Kinet. and Catal. Lett.,* 1977, vol.7, №1, p. 69-74.

[111] Kurashov, V.I., Pobedimskii, D.G., Kirpichnikov, P.A. – *Dokl. AN SSSR*, 1978, vol.28, №6, p. 1407-1410.

[112] Kurashov, V.I., Pobedimskii, D.G., Krivileva L.G. – *Zh. Phys. Khim.,* 1979, vol.53, №2, p. 501-502.

[113] Mukmeneva, N.A., Akhmadullina, A.G., Sabirova, L.N. et al – *Vysokomol. Soed.,* 1976, vol.B18, №2, p. 108-111.

[114] Levin, Ya.A., Vorkunova, E.I. Gomoliticheskaya khimiya fosfora (The Homolytical Chemistry of Phosphorus). Moscow: *Nauka;* 1978.

[115] Buchachenko, A.L., Vasserman, A.M. *Stabil'nye radikaly (Stable Radicals).* Moscow: *Khimiya;* 1973.

[116] Kirpichnikov, P.A., Kolyubakina, N.S., Mukmeneva, N.A. et al –*Plast. massy.* – 1971. – №4. – pp. 45-47.

[117] Mukmeneva, N.A., Akhmadullina, A.G., Pobedimskii, D.G.et al. – *Vysokomol. Soed.,* 1974, vol.B16, №8, p. 635-637.

[118] Levin, Ya.A., Il'yasov, A.V., Goldfarb, E.I. et al. – *Org. Magn. Resonance,* 1973, vol.5, №10, p. 487-495.

[119] Levin, Ya.A., Il'yasov, A.V., Gol'dfarb, E.I. et al. Primenenie metodov spectroscopii v issledovaniyakh kinetiki i mekhanizma khimicheskikh reaktsii (The Use of Spectral Methods in Studies of Kinetiks and Mechanism of Chemical Reactions). Kiev: *Znanie;* 1972.

[120] Mukmeneva, N.A., Cherezova, E.N., Mar'in, A.P. et al. – *Zh. Prikl. Khim.,* 1994, vol.67, №8, p. 1335-1338.

[121] Gross, H., Seibt, H., Keitl, T. – *J. Prakt. Chemie,* 1975, Bd.317, H.6, S. 890-896.

[122] Cherezova, E.N., Kadyrova, V.Kh., Mukmeneva, N.A. – *Bashkirskii Khim. Zh.,* 1996, vol.3, №1-2, p. 133-143.

[123] Strigun, M.M., Vartanyan, L.V., Emanuel, N.M. – *Usp. Khim.,* 1968, vol.39, №6, p. 998-1024.

[124] Pudovik, A.N., Gur'yanova, I.V., Ishmaeva, E.A. In: *Reaktsii i metody issledovaniya organicheskikh soedineni (Reactions and Research Methods of Organic Compounds)*, Ed. by Kasanskii, B.A. Moscow: *Khimiya;* 1968, vol. 19, p. 7-848.

[125] Denisov, E.T. – *Usp. Khim.*, 1978, vol.47, №6, p. 1109-1118.

[126] Howard, J.R., Ohkatsu, G., Chenier, I.H.B. et al. – *Can. J. Chem.*, 1973, №51, p. 1543-1553.

[127] Wasson, J.R., Woltermann, G.M., Stoklova, H.I. – *Fortschr. Chem. Forschung,* 1973, Bd.35, S. 65-129.

[128] Beresnev, V.V., Kirpichnikov, P.A. – *Vysokomol. Soed.,* 1967, vol.B9, №11, p. 809-812.

[129] Ivanov, S.K., Kateva, V.D. – *Dokl. Bulgar. AN,* 1972, vol.25, p. 645-648.

[130] Kateva, V.D., Ivanov, S.K. – *Dokl. Bulgar. AN,* 1975, vol.28, p. 83-85.

[131] Dzhambekov, N.F. – *Neftekhim,* 1979, vol.19, №6, p. 892-896.

[132] Pobedimskii, D.G., Nasybullin, A.Sh. et al. – *Vysokomol. Soed.,* 1977, vol.19A, №2, p. 352-356.

[133] Nugumanova, G.N., Cherezova, E.N., Sofronova, O.V. et al. – *Kauch. Resina,* 1998, №2, p. 34-36.

[134] Nugumanova, G.N., Zhukova, R.S., Cherezova, E.N., Mukmeneva, N.A. et al. – *Zh. Prikl. Khim.,* 1998, №9, p. 1520-1524.

[135] Mukmeneva, N.A., Cherezova, E.N., Tatarintseva, T.B. et al. – *Russian Polym. News,* 2000, vol.5, №2, p. 1-6.

[136] Mukmeneva, N.A., Cherezova, E.N., Rusina, I.F. et al. – *Inter. J. Polym. Mater.,* 2000, vol.47, p. 317-318.

[137] Mukmeneva, N.A., Shanina, E.L., Zaikov, G.E. In: *Ageing of Polymers, Polymer Blends and Polymer Composites – 2*, Ed. by Zaikov, G.E., Buchachenko, A.L., Ivanov, V.B. New-York: Nova Science Publisher, 2002, №5, p.205-209.

[138] Karpukhina, G.V., Maizus, Z.K., Emanuel, N.M. – *Dokl. AN SSSR,* 1968, vol.182, №4, p. 870-873.

[139] Karpukhina, G.V., Maizus, Z.K., Meskina, M.Ya. – *Kinetika i kataliz,* 1968, vol.9, №2, p. 245-249.

[140] Karpukhina, G.V., Maizus, Z.K., Meskina, M.Ya. – *Izv. AN SSSR, Ser. Khim.,* 1977, №2, p. 313-316.

[141] Karpukhina, G.V., Emanuel, N.M. – *Dokl. AN SSSR,* 1984, vol.276, №5, p. 1163-1167.

[142] Mukmeneva, N.A., Kadyrova, V.Kh., Cherezova, E.N. – *Vysocomol. Soed.,* 1988, vol.30B, №5, p. 378-380.

Chapter 3

STABILIZATION OF POLYMER COLOR

The aim of retaining natural color of polymer products (color stabilization) is a special one in the complicated problem of polymer stabilization. The coloring of polymers limits considerably and often excludes the possibility of practical using.

As a rule, polymers during their processing and service performance contact with air being exposed to thermal oxidizing destruction which is of special importance among other types of destruction including the issues of polymer coloring, as the action of oxygen promotes the coloring of most polymers.

The intensive accumulation of products of polymer oxidation takes place already during the induction period. As many of the primary products of oxidation are usually more oxidized than initial polymers, with the conversion of the latter the diversity of decomposition products increases. The formation of volatile and high-molecular alcohols, ketones, aldehydes, acids, unsaturated compounds, conjugate systems is observed during the processes of thermal oxidation of unstabilized polymers as a result of the decomposition of their polymeric chains that is the degree of their coloring depends directly on the intensity and depth of oxidizing conversions in polymer resulting in the accumulation of chromophore and auxochromous groups [1-12].

On the whole, the change of polymeric material color is the result of various chemical processes, among them two main directions should be pointed out, that is chemical conversions occurring directly in macromolecules and conversions of additives and associated admixtures in the polymer composition. The structure of polymers defines mainly the way of coloring.

For some polymers coloring starts at initial stages of destruction as a result of forming chromophores of visible region, namely, conjugate systems (polyene sequences in polyvinylchloride, polyvinylacetate, conjugate cyclic structures in polyacrylonitrile etc.) even at relatively low temperatures and without considerable changing physical and mechanical properties of polymers. These are polymers containing functional groups in the side chain, these groups are capable of eliminating (polyvinylchloride etc.) [13-16].

Polymers of other type (polyolefines etc.) are relatively chemically stable. Though the formation of active centers initiating the decomposition of such polymers can take place during the whole destructive process, their intensive coloring is seen at sufficiently deep stages of destruction. For these polymers the primary reason of coloring is considerably quicker conversion of attendant additives (stabilizers and others) into chromophores [17-19] and catalytic influence of admixtures, for example, compounds of variable valency metals etc., without considerable changing their physical and mechanical properties [20-23]. Thus, coloring of polymers may proceed according to various mechanisms at all stages of polymer ageing.

It is important to point out, that the processes of coloring and thermal-oxidative degradation of polymers can have common regularities, revealing in the occurrence of similar centers. Alongside with it, in case of inhibiting the polymer coloring there can be other, specific mechanisms, uncharacteristic of the processes of chain inhibition (deactivation of promoters of polymer destruction, destruction of anomalous fragments, initiating decomposition of polymers, reactions with polyene structures etc.). It is in this connection that the discussion of color stabilization is given in a separate chapter.

Alongside with chemical (according to functional groups) and technical (according to the application) classifications of stabilizers they are also classified as coloring and noncoloring [24].

Coloring stabilizers include aromatic amines and their derivatives, a number of phenol compounds of specific structure etc., which are efficient antioxidants. Alongside with it, during the process of oxidation inhibited by them stabilizers change into colored products imparting coloring to polymers. It is in this connection that these stabilizers can be used only for the production of dark materials.

Noncoloring stabilizers don't influence the change of polymer color, as the products of their conversions are colorless. This type of stabilizers includes the compounds of various chemical nature, namely, organophosphorous compounds and, first of all, phosphorous acid esters, and also thiocarbamides, dialkyldithiocarbamides, thio(dialkyl)propionates, phenol compounds, etc.

Phosphorous acid esters are the most studied and used as noncoloring stabilizers among the above listed. They are widely used for producing colorless, transparent and nontoxic polymeric materials, in combinations with other stabilizers (metal stearates, epoxies, phenols, etc.) as well. Stabilizing efficiency of phosphorous acid esters should be connected with their high reactivity in respect of various active products in the composition of polymers, catalyzing their decomposition and coloring.

But at present the review information on stabilizing properties of phosphorous acid esters is very limited [25, 26].

3.1. Inhibition of Polymer Coloring by Means of Neutralization of Active Products of Macromolecule Conversions by Phosphorous Acid Esters

The results of investigations making the foundation for searching and choosing highly efficient color stabilizers on the basis of phosphorous acid esters are systematized in this section. The regularities of inhibition of polymer coloring by phosphites are considered, polymers of two types, polyvinylchloride (PVC) and polyethylene taken as typical examples.

3.1.1. Inhibition of Polyvinylchloride Coloring

Coloring (chromophor effect) of PVC starts at initial stages of destruction (already when the elimination of 0.1–0.2 wt% HCl takes place) and is the most symptomatic effect of dehydrochlorination of polymer [27].

PVC, like other linear polymers, macromolecules of which don't have double bonds, should be transparent for visible light and ultraviolet light. However, PVC absorbs a number of waves of ultraviolet light spectrum owing to the presence of chromophores, namely, unsaturated groups of atoms C=C, C=O, C(O)-C=CCl in various combinations, in the composition of macromolecules. These groups are able to initiate the destruction of PVC. During the process of polymer destruction long polyconjugated polyene systems appear, as a result, coloring occurs [28-30]. The active centers of destruction and coloring in PVC practically coincide.

Phenomenological investigations of processes of the PVC destruction in the presence of phosphorous acid esters allowed to determine a number of specific effects [31-34].

Firstly, the distinctive feature of organic phosphites is their ability to maintain the initial color and transparency of PVC during its destruction depending on the nature of a substitute in the ester group of phosphite (tables 3.1-3.3).

Table 3.1. Influence of cyclic phosphites on the properties of PVC- composite [a]

Phosphite	Content of phosphite, wt %	Thickness of the film is 3 mm		Thickness of the film is 0.5 mm	
		Transparency, %		τ_C [b], min. (448 K)	τ_T [c], min (448 K)
		420 nm	520 nm		
(1)	1.0	25.7	72.8	50-65	51
	1.5	20	70		68
(2)	1.0	0.7	39	15-30	50
	1.5	0	23		48

[a] composite formulation (parts by weight per 100 parts by weight of PVC): barium stearate– 0.6, cadmium stearate– 1.4, epoxidated soy oil – 2.0; [b] color resistance – time before color changing of PVC; [c] thermal stability – time before the beginning of dehydrochlorination of PVC.

Table 3.2. Influence of phosphites on the thermal stability of PVC-composite [a] (448 K)

Phosphite	τ_T, min	τ_C, min	Transparency of initial samples, %	
			420 nm	520 nm
(1)	254	210	80 [b] (47.0) [c]	83.2 (66.1)
(3)	167	150-180	80 (42.5)	86 (63.0)

[a] composite formulation (parts by weight per 100 parts by weight of PVC): dioctylphthalate – 40, barium stearate – 1.4, cadmium stearate – 1.6, epoxidated soy oil – 10, phosphite – 0.5;
[b] thickness of the film is 0.5 mm; [c] thickness of the film is 3 mm.

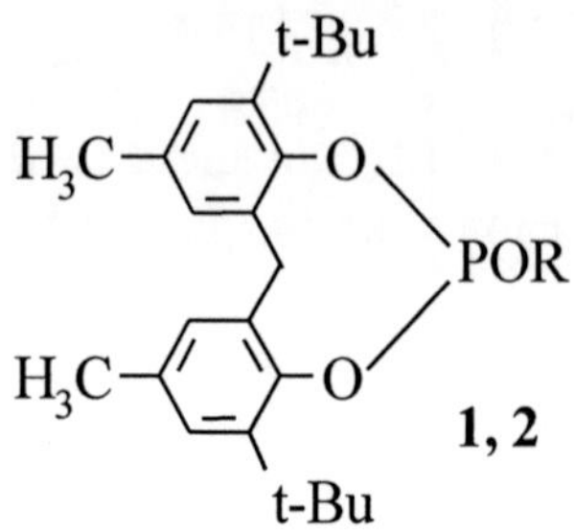

R = C_2H_5 (**1**), C_6H_5 (**2**); $(C_6H_5O)_2P(OC_8H_{17}\text{-}i)$

1, 2 **3**

Table 3.3. Influence of diphosphites and oligophosphites on the properties of PVC-composite [a]

Phosphite	Content of phosphite, mass parts	Initial color [b]	Transparency [c], %		τ_C, min. (448 K)	τ_T, min (448 K)
			420nm	520nm		
–	–	Light yellow	75	82	30-35	28
(4)	0.75	Light grey	78	84	30-35	53.5
	1.0	Yellow	–	–	50-55	12
	1.5	Yellow	–	–	50-55	15
(5)	0.75	Light grey	81	85	45-50	47.5
	1.0	Colorless	85.5	87.8	70-80	63
	1.5	Colorless	84	87	60-70	54
(6)	0.75	Light grey	79	85	36-40	36.5
	1.0	Colorless	85	86.5	60-70	51
	1.5	Colorless	86	87.5	65-70	45
(7)	0.75	Light yellow	77	84	50-55	30
	1.0	Light yellow	78	84.5	40-45	25
(8)	0.75	Colorless	85	87	70-80	70
	1.0	Colorless	86	88	75-80	70

[a] composite formulation (part by weight per 100 parts by weight of PVC): barium stearate– 0.6, cadmium stearate – 1.4, octylepoxystearate – 1; [b] thickness of the film is 3 mm; [c] thickness of the film is 0.5 mm.

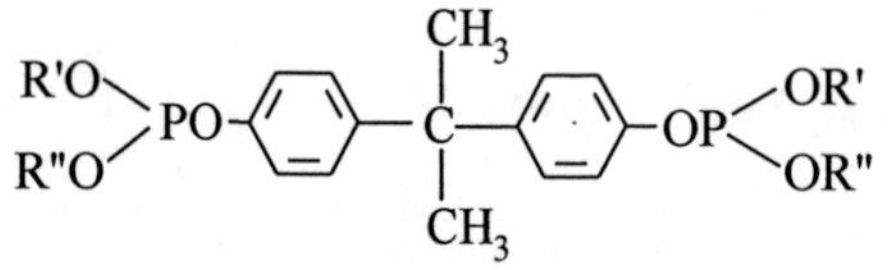

4-6

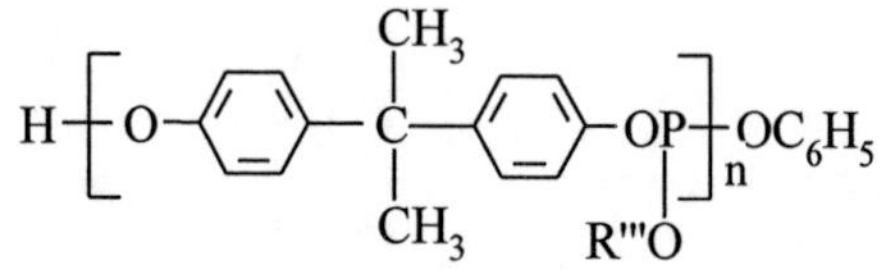

7-8

R' = C_6H_5 (**4,5**), i-C_8H_{17} (**6**); R'' = C_6H_5 (4), i-C_8H_{17} (5,6); R''' = C_6H_5 (7), i-C_8H_{17} (8)

Another important feature of phosphites is that they decrease the rate of dehydrochlorination both in inert gas and air (table 3.4).

Table 3.4. Rate constants of PVC dehydrochlorination in the presence of phosphites (1 part by weight per 100 parts by weight of PVC, 448 K)

Phosphite	$k \times 10^7$, s^{-1}		Phosphite	$k \times 10^7$, s^{-1}	
	In air	In nitrogen		In air	In nitrogen
–	9.3	6.8	11	3.4	2.7
(1)	1.9	1.5	12	2.3	1.6
(2)	2.2	1.7	13	2.5	2.3
(3)	2.0	1.2	14	2.7	2.2
(5)	2.1	1.3	15	3.2	2.5
(8)	1.4	1.3	16	2.4	1.6
(9)	3.3	2.5	17	2.5	1.8
(10)	3.6	3.0	18	2.2	1.5

$(RO)_3P$ R = C_6H_5 **(9)**, i-C_8H_{17} **(10)**, 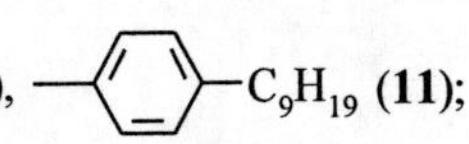 **(11)**; 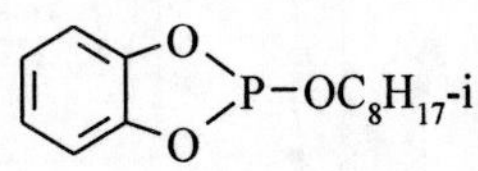 **(14)**;

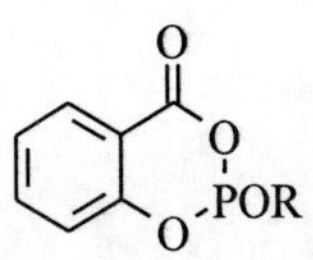 R = i-C_5H_{11} **(12)**, 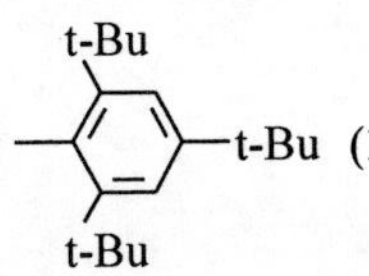 **(13)**; 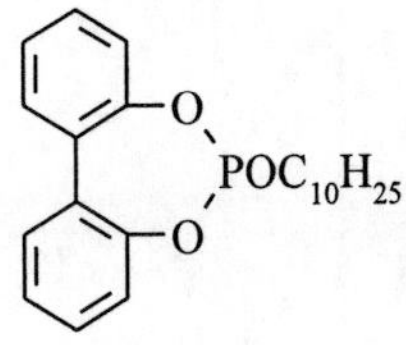 **(15)**;

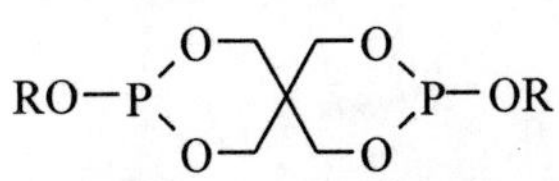

R = $C_{12}H_{25}$ **(16)**, C_6H_5 **(17)**;

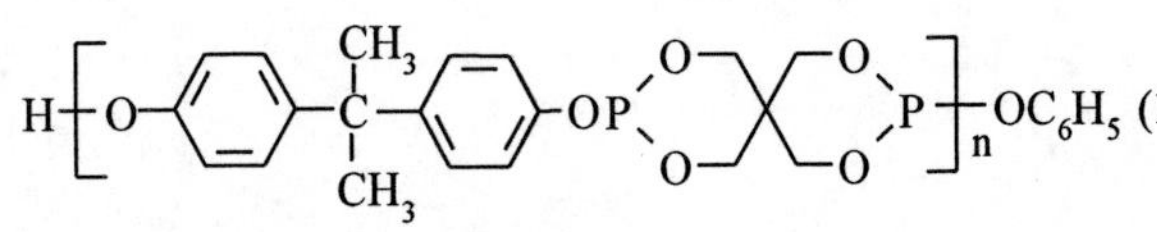 **(18)**

Finally, the essential advantage of phosphites is their ability to show distinct synergistic effect when they are jointly used with some other stabilizers. Simultaneously the thermal stability index is practically dependent directly on the color resistance index: their maximums are situated in the same region of phosphite content in the composition (figure 3.1) [31, 32].

Functionalized derivatives of phosphorous acid esters are known as stabilizers of PVC, particularly, polyfunctional stabilizers, containing oxiran (epoxy) groups (further they are called epoxyphosphites) [35-37].

This group of stabilizers is of interest as their structure models mix composition (phosphite with epoxy compound), being a part of widely used formulation for uncolored and transparent PVC materials. The presence of three-coordinated phosphorus and oxiran cycle (HCl acceptor) in a molecule of epoxyphosphites creates preconditions for displaying intramolecular synergism by the former. Thus, if we consider the efficiency of inhibition of PVC thermal destruction the epoxyphosphites are comparable with equimolar mixtures consisting epoxy compounds and phosphites or surpass them (table 3.5) [36].

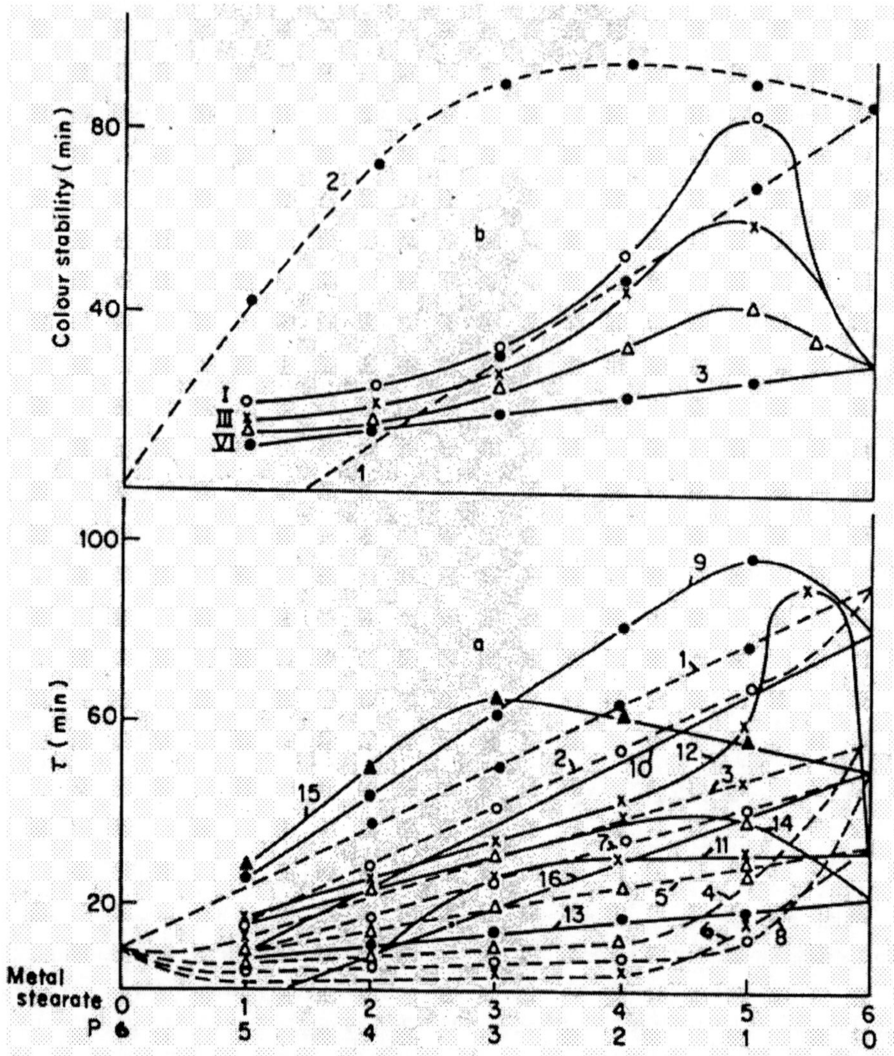

Figure 3.1. The dependence of τ (a) and color (b) at 448 K in the presence of metal stearates upon the tri(n-nonylphenyl)phosphite (P) content (P + metal stearate; total 6g/100g PVC)/ (a) 1 and 9, Ba stearate; 2 and 10, the same + P; 3 and 11, Cd stearate; 4 and 12, the same + P; 5 and 13, Ca stearate; 6 and 14, the same + P; 7 and 16, Zn stearate (413 K); 8 and 15, the same + P (413 K); 1–8 in powder; 10–16 in film. (b) Zn stearate (413 K); 2, Zn stearate + P; 3, Ba stearate; I, Cd stearate + P; III, Cd stearate + $(C_6H_5O)_2P(OC_{10}H_{21}\text{-}i)$; VI, Cd stearate + $(C_6H_5O)_2P(OC_8H_{17}\text{-}i)$.

Table 3.5. Rate of PVC dehydrochlorination (448 K, 10^{-2} Pa, HCl acceptor is barium stearate 2.5×10^{-3} mole/mole PVC) in the presence of epoxyphosphites and mixtures of organic phosphites with epoxy compounds

Stabilizer	$C_0\times10^3$, mole/mole PVC [a]	$V_{HCl}\times10^6$, mole HCl/mole PVC [b]
–	–	0.68
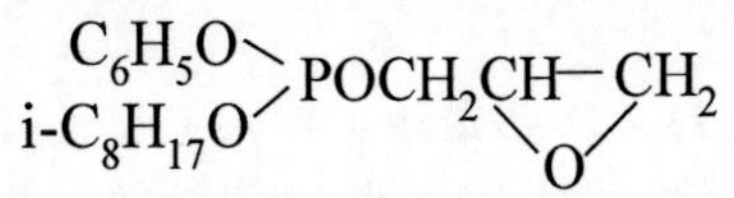	1 5 10	0.31 0.30 0.30
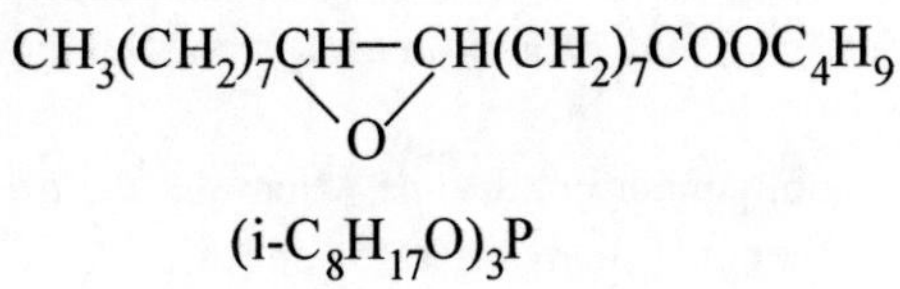	5/5 10/10	0.33 0.31
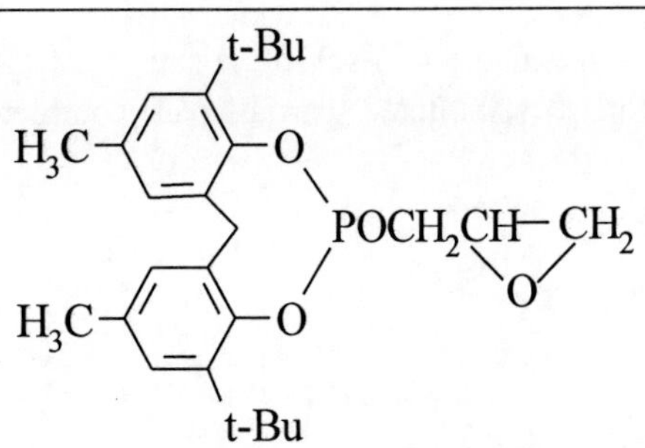	1 5 10	0.54 0.42 0.36
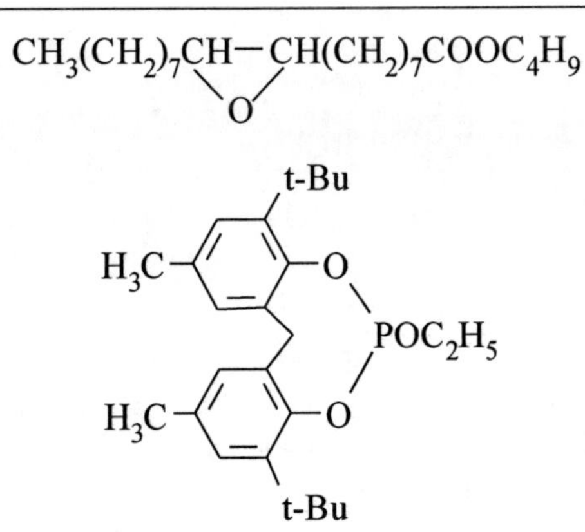	1/1 5/5 10/10	0.46 0.33 0.33
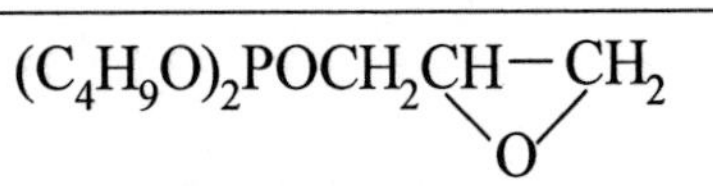	1 5 10	0.41 0.32 0.25
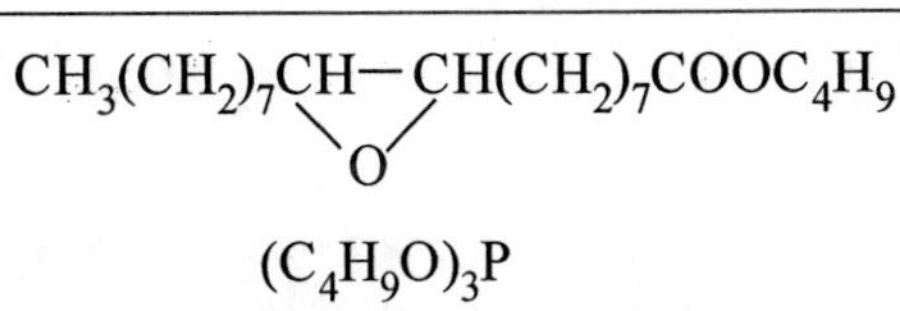	0.5/0.5 2.7/2.7 5/5	0.26 0.16 0.14

[a] phosphite content in PVC; [b] rate of PVC dehydrochlorination.

As a rule, in a number of certain groups of phosphites (mono-, di- and oligophosphites) there is the same tendency regarding the increase of color resistance of PVC-composites, that is mixed alkylaromatic phosphites (diphenyl-*i*-octylphosphite, ethyl ester of 2,2'-methylene-bis-(4-methyl-6-*tert*-butylphenyl)phosphorous acid, alkyl esters of *i*-propylidenediphenylphosphorous acid, polyalkylarylenephosphites) and corresponding epoxyphosphites are more efficient than phosphorous acid esters containing only aromatic or aliphatic substitutes [25, 26].

3.1.1.1. Reactions with the Products of Polyvinylchloride Destruction

As a result of investigations it was established that phosphorous acid esters can inhibit destruction and coloring of PVC owing to the neutralization of active agents (hydrogen chloride, variable valency metals, hydroperoxides, etc.) and anomalous fragments containing or forming in PVC and catalyzing its distruction.

The interaction of phosphites with HCl is possible under conditions when the rate of HCl elimination is very high and a metal-containing acceptor (metal stearate) doesn't provide its quick and complete binding.

Full esters of phosphorous acids depending on the nature of a substitute can interact with HCl with the formation of dealkylation products (substituted partially or completely) according to the reactions [38]:

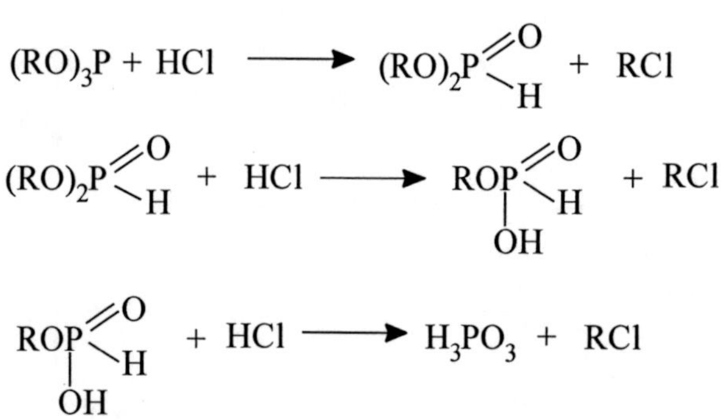

The degree of phosphites dealkylation depends on their structure.

Thus, for $(AlkO)_3P$ all three stages are characteristic; two first stages are characteristic of $(AlkO)_2POAr$; only the first stage is characteristic of $(ArO)_2POAlk$; $(ArO)_3P$ is not dealkylated practically under the influence of HCl. As the accumulation of acid products (monoalkylphosphites and H_3PO_3) results in the acceleration of the process of polyvinylchloride dehydrochlorination, the most efficient thermal and color stabilizers are alkylarylphosphites.

The presence of epoxy groups in molecules of epoxyphosphites predetermines their additional participation in binding hydrogen chloride and the increase of the time of thermal stability of polymeric compositions [35].

The influence of chemical nature of substitutes at a phosphorus atom can be observed, cyclic and acyclic epoxyphosphites taken as an example (table 3.6). Thus, the interaction of 2,2'-methylene-bis(4-methyl-6-*tert*-butylphenyl)- (20) and 2,2'-methylene-bis(methyl-α-methyl-cyclohexylphenyl)epoxyphosphite (21) with HCl results in nearly total consumption of epoxy groups. The primary direction of the interaction of phosphites having such structure with HCl is the opening of the epoxy cycle with the formation of β-chlorohydrin. The content of P^{III} doesn't change practically.

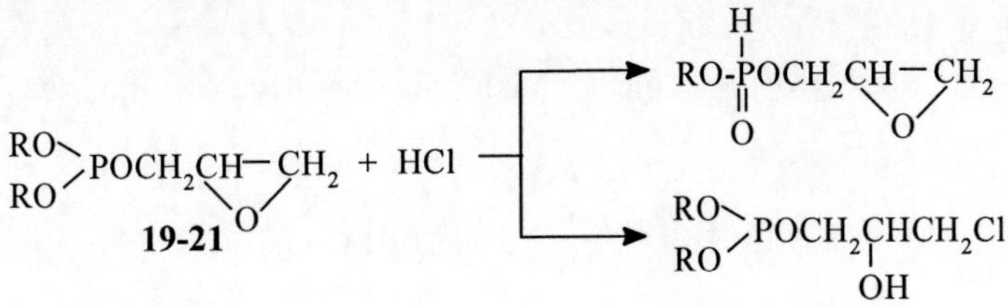

R = C_4H_9 (**19**), 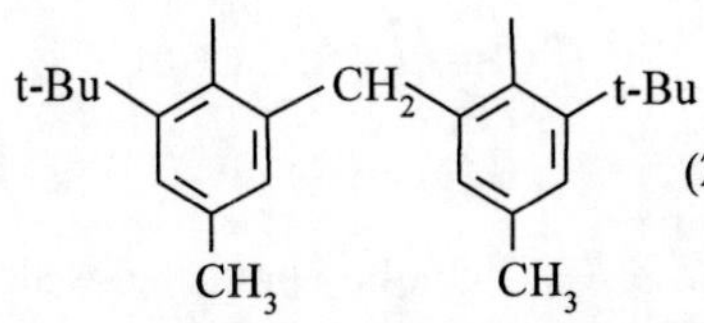 (**20**), 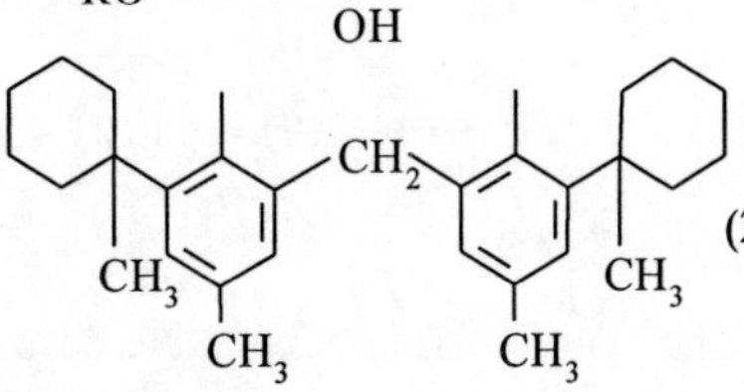 (**21**)

Table 3.6. The interaction of epoxyphosphites with hydrogen chloride (293 K, acetone)

Phosphite	Initial content		Content after the reaction with c HCl	
	Epoxy groups [a], %	P^{III}, % [b]	Epoxy groups, %	P^{III}, %
(19)	16.60	12.15	1.57	следы
(20)	9.44	7.01	0.85	6.90
(21)	7.20	5.65	1.7	5.50

[a] it is determined by the method of hydrobromination [39];
[b] it is determined by the method of polarography [40].

At the same time for dialkylepoxyphosphites, particularly, for dibutylepoxyphosphite (**19**), the reaction with HCl results in the formation of both β-chlorohydrin, and dealkylation of dibutylepoxyphosphite.

It can be expected that the reaction similar to the interaction of phosphites with HCl will take place in case with other protondonor agents, for example, with carboxylic acids, formed in the reaction of thermal stabilizers with HCl [38]:

$$(RO)_3P + RCOOH \rightarrow (RO)_2P(O)H + RCOOR$$

Thermal-oxidative destruction of PVC, similar to carbochain polymers (for example, polyolefines), can be presented in the form of the following stages:

- oxidation of macromolecules with the formation of alkyl radicals [27, 41-43]:

$$\sim CH_2\text{-}CHCl\text{-}CH_2\text{-}CHCl\sim \xrightarrow{O_2} \sim CH_2\text{-}\dot{C}Cl\text{-}CH_2\text{-}CHCl\sim + HOO^{\bullet}$$

- interaction of alkyl radicals with oxygen and the formation of peroxide radicals:

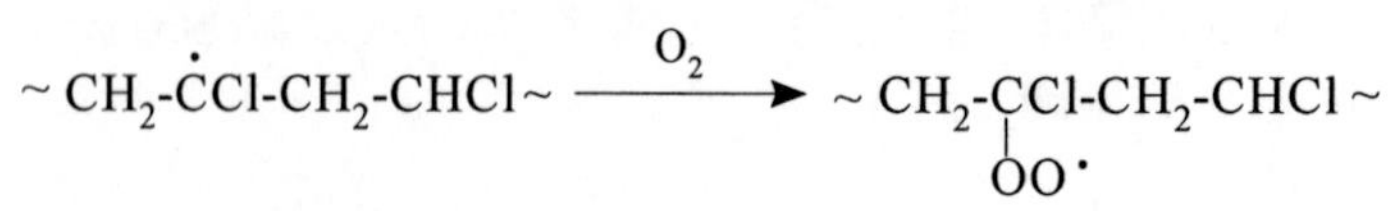

- reactions of peroxide radicals with macromolecules and the formation of hydroperoxides:

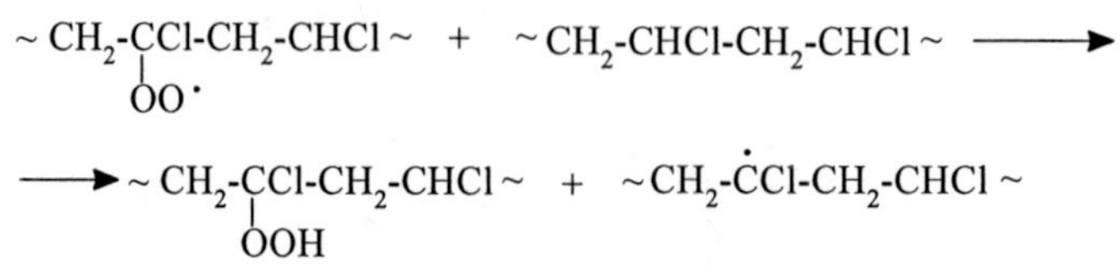

The peculiarities of oxidative destruction of PVC can be connected with heterolytic acid-catalyzed destruction of hydroperoxides at the moment of their formation with HCl elimination, for example:

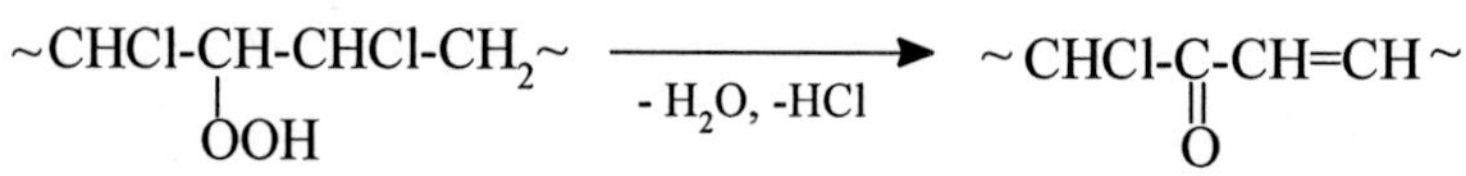

During this process, active ketochlorallyl groups are formed, they decrease thermal and color stability of PVC. In this connection, the interaction of phosphites with hydroperoxide takes on special meaning, as a result of these reactions the forming of initiating anomalous fragments is prevented.

It is known, trialkyl(aryl)phosphites interact with hydroperoxides without radical formation, that leads to forming stable colorless products of the reaction, namely phosphates and corresponding alcohols (see Section 2.2.1.1) [25, 26].

Epoxyphosphites react with cumene hydroperoxide similarly. The reactions have the first order regarding to hydroperoxide and epoxyphosphite [44]:

The values of kinetic constants of their interaction are given in table 3.7.

Table 3.7. Kinetic parameters of the reaction of cumene hydroperoxide (CHP) with epoxyphosphites (EPP) (293-323 K, chlorbenzene, [CHP]=[EPP]=5×10^{-2} mole/l)

Phosphite	k×10^2, l/(mole·s)				E_a, kJ/mole	lgA_0
	293 K	303 K	313 K	323 K		
(20)	0.4	0.7	1.9	3.5	54	35.1
(21)	0.2	0.3	0.7	1.2	45	28.1
(22)	0.5	1.8	3.3	6.9	43	7.9
(23)	1.6	2.5	6.2	8.3	61	5.0
(24)	0.5	0.7	2.1	3.8	65	7.2
(25)	2.5	4.2	6.9	9.3	34	3.5

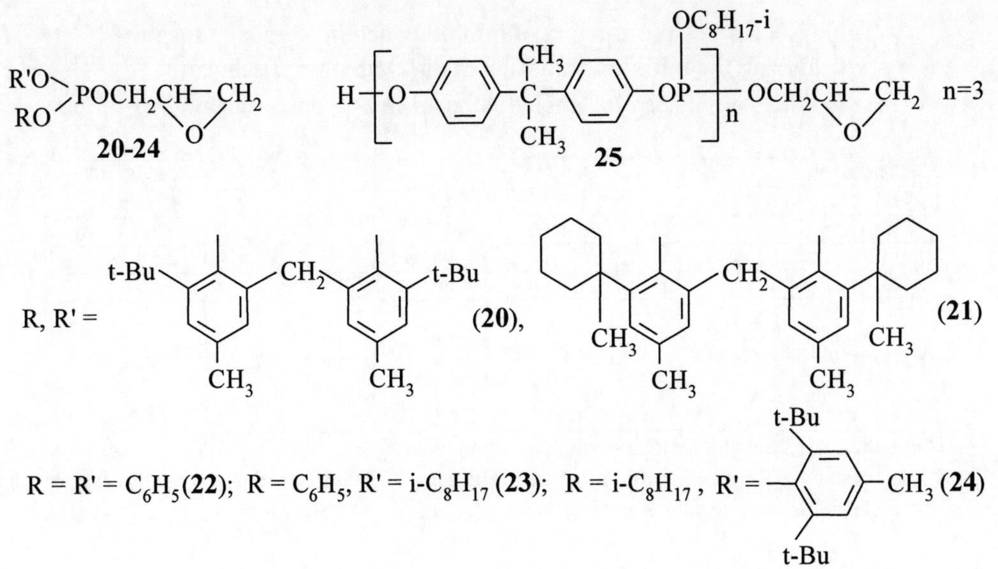

As it is seen from table 3.7, the parameters of Arrhenius equation for the investigated epoxyphosphites, defined in the temperature range of 293-323 K (E_a = 43–65 kJ/mole; lgA_0 = 3.5–7.9), are very close to the values known for monofunctional full esters of phosphorous acid having similar structure (E_a = 25–55 kJ/mole, lgA_0 = 4.5–6.4).

Based on the assumption about the identity of mechanisms of the oxidation of carbon-chain polymers and PVC it is reasonable to analyze the possibility of participation of phosphites in the processes of chain inhibition of PVC by termination of oxidation kinetic chains during the interaction with peroxy radicals. It is even more important that the necessary components of plasticized PVC-composites are esteric plasticizers and their oxidative stability defines substantially thermal-oxidative stability of PVC-plasticate [27]. Inhibiting properties of trialkyl(aryl)phosphites under conditions of thermal oxidation of polymers are investigated and described reasonably well (see Section 2.2.2.1), that's why there is no necessity to discuss them in detail in the given Section.

The quantitative assessment of antioxidative activity of epoxyphosphites is defined during the initiated oxidation of model sebacic acid 2-ethylhexyl ester which is the easiest oxidized plasticizer. Epoxyphosphites are characterized by the high rate of oxidation chain termination ($k_7/k_2=10^4$) and commensurable with the commercial antioxidant diphenylolpropane regarding its activity (table 3.8) [44].

In whole, the described reactions of phosphorous acid esters with HCl, hydroperoxides and peroxide radicals can develop under certain conditions and that's why they are considered as accompanying, being of secondary importance during inhibition of the destruction and coloring of PVC.

Table 3.8. Kinetic parameters of inhibiting action of epoxyphosphites during the initiated oxidation of di-(2-ethylhexyl)sebacate (338 K, [a]W_i=20.55×10^{-8} mole/(l·s), pressure O_2 650 atmosphere)

Compound	f^b	$k_7/k_2{}^c \times 10^{-4}$	$f \cdot k_7/k_2 \times 10^{-4}$
(22)	0.3	3.2	0.96
(23)	0.3	4.3	1.3
(24)	1.5	1.7	2.5
(25)	1.5	1.8	2.7
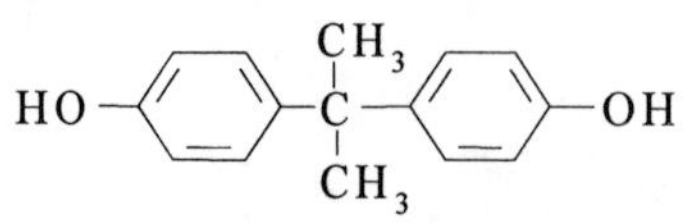	2.0	3.4	5.6

[a] W_i is initiation rate; [b] stoichiometric coefficient of inhibition;

[c] k_7 is rate constant of the reaction of an inhibitor with radicals $RO_2^{\cdot}$; k_2 is rate constant of the reaction of transferring a chain to substrate.

3.1.1.2. Reactions with Low-Molecular Models of Anomalous Groups of Polyvinylchloride

It follows from considering the reasons of developing coloring polymers during ageing that the presence of chains of anomalous structure is of essential significance alongside with the main structure of the chain defining the structure of polymer and complex of chemical and physical properties inherent in it. It is along labile structural anomalies (primary active centers of macromolecules) that the initiation of polymer destruction usually occurs [27]. If these processes lead to the formation of chromophor and auxochromous structures in the composition of a macromolecule, polymer is colored intensively.

PVC dehydrochlorination is a complex process, consisting of parallel-serial reactions [45-49]:

- HCl elimination in random manner with the formation of individual, double bonds (>C=C<). During this process there is statistical dehydrochlorination of PVC (V_S) and forming chlorallyl groups (CAG):

~CH_2-CHCl-CH_2-CHCl-CH_2CHCl~ → ~CH_2CH=CH-CHCl-CH_2-CHCl~
CAG

- growth of polyconjugated sequences (PCS) due to the activation of HCl elimination by adjoining >C=C<-bond (chlorallyl activation) – V_P:

~CH_2-CH=CH-CHCl-CH_2-CHCl~ → CH_2-CH=CH-CH=CH-CHCl~
PCS

- forming conjugated ketochlorallyl groups (KAG) in the processes of obtaining and storing PVC in the presence of oxygen, i.e. under real-life environment:

~CH_2-CH=CH-CHCl-CH_2~ → ~C(O)-CH=CH-CHCl-CH_2~
KAG

As it follows from the given reactions during PVC dehydrochlorination anomalous structures are formed which are capable of initiating the following PVC destruction with the formation of conjugated systems responsible for the development of intensive coloring of PVC.

The differentiated approach to the investigation of the influence of phosphites on the forming of individual double bonds in random manner (V_S) and the formation of polyene sequences (V_P) provides more complete data about the process of PVC dehydrochlorination [50].

The change of average viscosity molecular weight during ozonization and subsequent oxidation-reduction bond cleavages in a polymer product gives the notion of the amount of inner statistical double bonds in macromolecules.

Alongside with it the number of breakages γ or inner double bonds (which is the same as the former one) in macromolecules resulting in the decrease of molecular weight of polymer is defined by the ratio:

$$\overline{\gamma} = \frac{62.5 \cdot 1.86}{\overline{M}_{\eta}^{0}} \{([\eta]_0 / [\eta])^{1/0.725} - 1\}$$

where $\overline{M}_{\eta}^{0}$ is average viscosity moleular weight of initial PVC before ozonization;

$[\eta]_0$, $[\eta]$ are characteristic viscosities of polymer solution in cyclohexanone before and after ozonization with the following oxidative destruction, respectively

As the forming of double bonds in random manner is the basis of statistical dehydrochlorination, the change of γ is nothing more than the rate of statistical dehydrochlorination (V_S). As it is known, PVC already contains certain amount of inner double bonds in the macromolecule composition.

The study of the interaction of PVC with organic phosphites indicated that the latter changed the rate of statistical PVC dehydrochlorination (V_S) when the overall rate of dehydrochlorination (V_{HCl}) was decreased. The influence of phosphorous acid esters on the rate of forming polyene sequences ($V_P = V_{HCl} - V_S$) is also observed: trialkyl(aryl)phosphites decrease it substantially or increase it slightly; for epoxyphosphites V_P is considerably lower (table 3.9, 3.10) [51].

20, 23

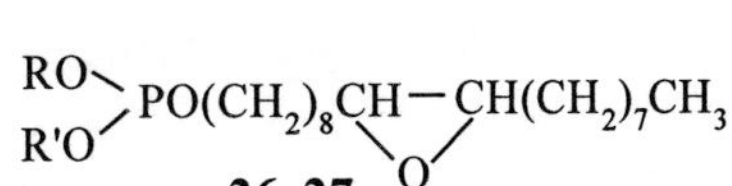

26, 27

R,R' = 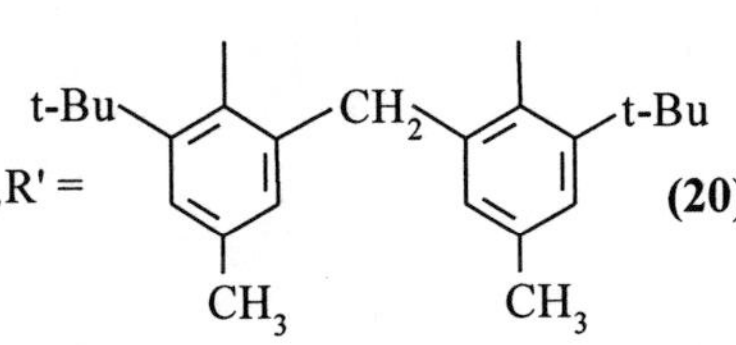 **(20)**, R = C_6H_5, R' = i-C_8H_{17} **(23)**, R=R' = C_4H_9 **(26)**, C_6H_5 **(27)**

Table 3.9. Kinetic parameters of the process of PVC dehydrochlorination (448 K, 10^{-2} Pa) in the presence of epoxy phosphites

Phosphite	$C_0\times10^3$, mole/mole PVC	$V_{HCl}\times10^6$, mole HCl/(mole PVC·s)		
		V_{HCl}	V_S	V_P
PVC without additives (γ_0=1.20×10^{-4} mole/mole PVC)	–	0.98	0.08	0.90
(20)	1.0	0.86	0.09	0.77
	5.0	0.75	0.19	0.56
	10.0	0.73	0.11	0.62
(23)	1.0	0.57	0.10	0.47
	5.0	0.55	0.14	0.41
	10.0	0.54	0.19	0.35
(26)	2.7	0.62	0.10	0.47
	5.5	0.55	0.20	0.41
	10.0	0.44	0.27	0.35
(27)	2.9	0.90	0.09	0.81
	5.8	0.78	0.11	0.67
	10.0	0.54	0.14	0.40

Table 3.10. Kinetic parameters of the process of PVC dehydrochlorination (448 K, 10^{-2} Pa) in the presence of organic phosphites and epoxy compounds

Phosphite	$C_0\times10^3$, mole/mole PVC	$V_{HCl}\times10^6$, moleHCl/(mole PVC·s)		
		V_{HCl}	V_S	V_P
PVC without additives (γ_0=0.76×10^{-4} mole/mole PVC)	–	0.68	0.08	0.60
$(C_4H_9O)_3P$	1.0	0.44	0.08	0.36
	5.0	0.42	0.08	0.34
	10.0	0.42	0.08	0.34
$(i\text{-}C_8H_{17}O)_3P$	1.0	0.51	0.08	0.43
	10.0	0.48	0.08	0.40
$(C_6H_5O)P(OC_8H_{17}\text{-}i)_2$	1.0	0.40	0.08	0.32
	5.0	0.23	0.08	0.15
	10.0	0.24	0.08	0.16
$(C_6H_5O)_2P(OC_8H_{17}\text{-}i)$	1.0	0.38	0.08	0.30
	5.0	0.32	0.08	0.24
	10.0	0.30	0.08	0.22
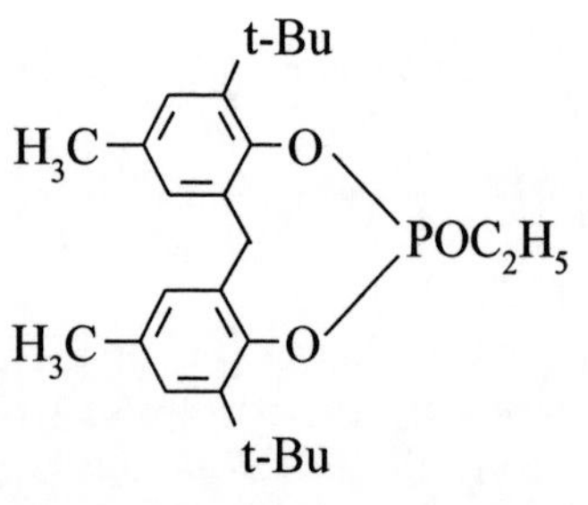	1.0	0.51	0.08	0.43
	10.0	0.36	0.08	0.26
PVC without additives (γ_0=0.93×10^{-4} mole/mole PVC)	–	0.78	0.08	0.70

Phosphite	$C_0\times10^3$, mole/mole PVC	$V_{HCl}\times10^6$, moleHCl/(mole PVC·s)		
		V_{HCl}	V_S	V_P
$CH_3(CH_2)_7CH-CH(CH_2)_7COOC_4H_9$ (epoxy, O bridging)	5.0 10.0 20.0	0.52 0.46 0.47	0.08 0.08 0.08	0.44 0.38 0.39
$CH_3(CH_2)_7CH-CH(CH_2)_7COOC_8H_{17}$-i (epoxy, O bridging)	5.0 10.0 20.0	0.56 0.54 0.40	0.08 0.08 0.08	0.48 0.46 0.32

As all vinylchloride chains in a PVC macromolecule can participate equally in the process of HCl elimination in random manner, the observed decrease of V_S (figure 3.2) is conditioned by the chemical interaction of phosphites with individual C=C – bonds being available in PVC [52].

Indeed, there is quantitative "closing" of inner C=C – bonds in polymer chains under thermal PVC exposure (353-373 K) when full phosphites are used. It particularly reflects in maintaining the initial average viscosity molecular weight of PVC after ozonization of a macromolecule and the subsequent oxidative hydrolysis (table 3.11) [52].

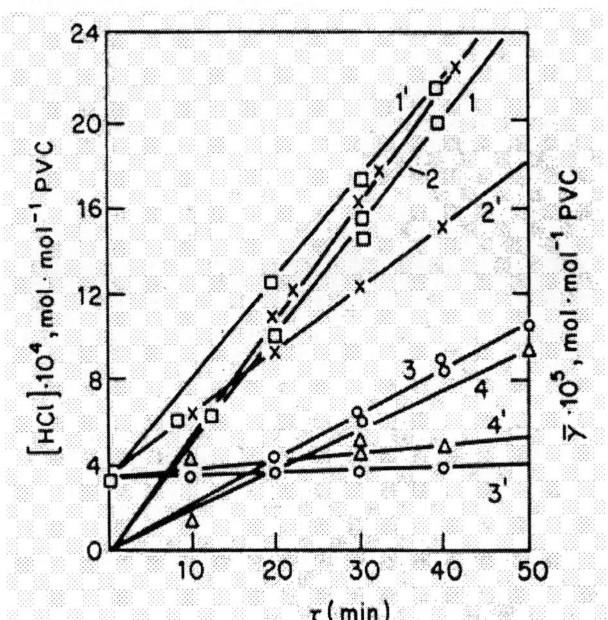

Figure 3.2. Overall (1–4) and statistic (1'–4') PVC dehydrochlorination in the presence of 0.02 mole of phosphite per mole of PVC (448 K, the residual presser 10^{-4} torr; HCl acceptor is Ba stearate, 0.02 g.equiv./g.equiv. PVC, 1 and 1', pure PVC; 2 and 2', with phosphite (28); 3 and 3', with phosphite (29); 4 and 4', with phosphite (30).

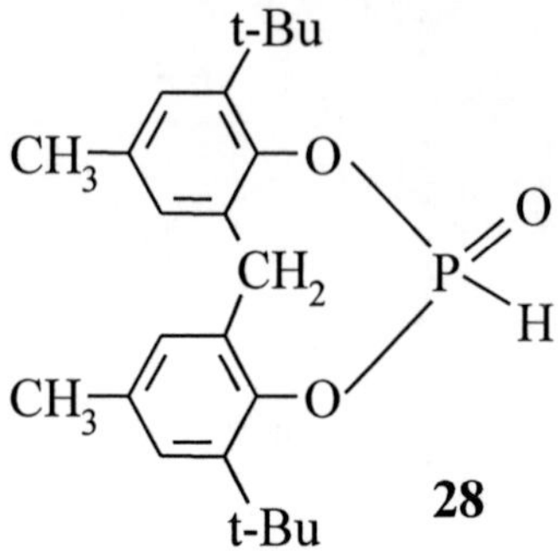

28

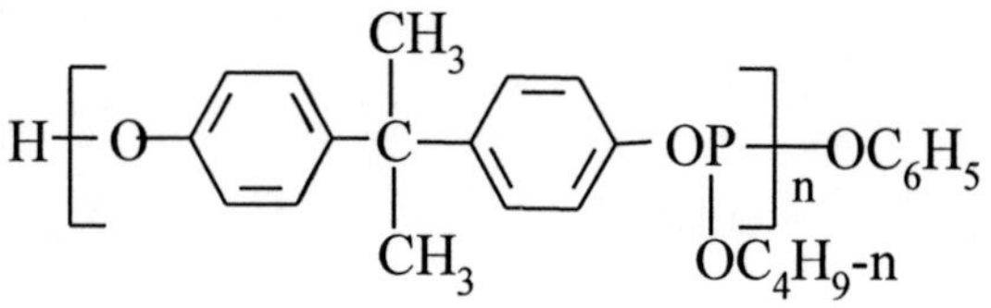

29

$C_6H_{13}OP(OC_{10}H_7\text{-}\alpha)_2$

30

Table 3.11. Changing the content of inner C=C – bonds in PVC macromolecules during thermal exposure of the mixture of polymer with phosphite (373 K, 1.33 Pa)

Composition	Time of thermal treatment, hours	$[\eta]_0$, decilitre/g	$[\eta]$, decilitre/g	$\gamma_0\times10^4$, mole/mole PVC
PVC $M^0_\eta = 127500$	–	1.23	1.19	0.35
PVC + phosphite (8)[a]	2	1.23	1.23	0
PVC + phosphite (30)[a]	8	1.23	1.22	0
PVC $M^0_\eta = 99000$	–	1.025	0.92	1.44
PVC + phosphite (8)[b]	11	1.025	1.03	0

$[\eta]_0$ and $[\eta]$ are characteristic viscosities of polymer solutions in cyclohexanone before and after ozonization with the subsequent oxidation-reduction eliminating; M^0_η is average viscosity molecular weight of initial PVC before ozonization; [a] content of phosphite is 0.02 mole/mole PVC; [b] content of phosphite is 0.015 mole/mole PVC.

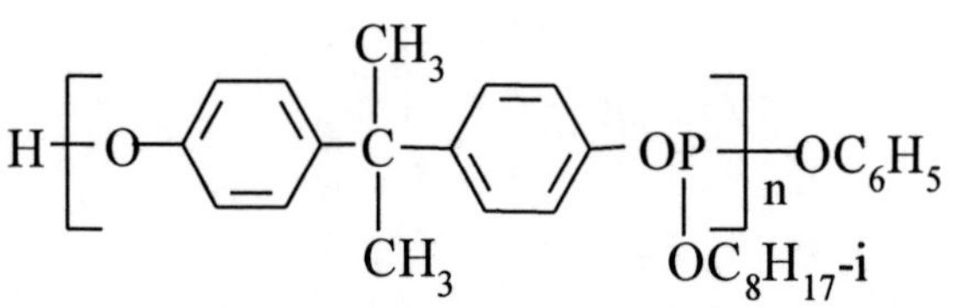

8

$C_6H_{13}OP(OC_{10}H_7\text{-}\alpha)_2$

30

The results of increasing the PVC stability were usually connected with the possibility of interaction of phosphites with β-chlorallyl group, which was considered by many authors as the main one in the general process of inhibition of PVC destruction [53-56].

Two alternative ways of this interaction were discussed: either a reaction with chlorine, being in β-position towards C=C bond, or, taking into account the fact of exhaustion of inner C=C-bonds – chemical addition of phosphites to double bond β-chlorallyl fragment:

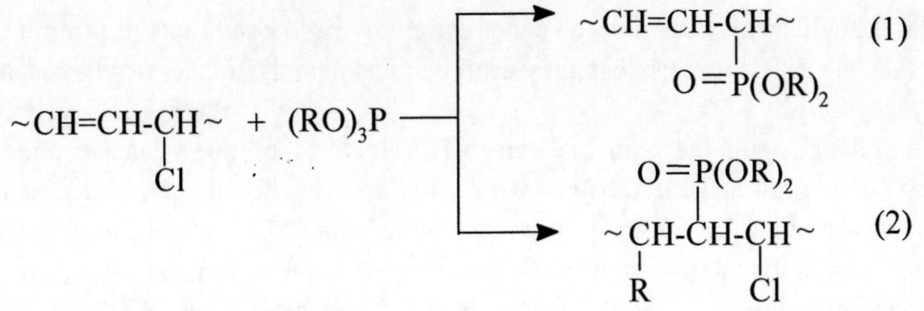

Scheme 3.1.

However, as a result of heating (353 K, for over 100 hours) of equimolar amounts of tributylphosphite with 4-chlorpentene-2 (low-molecular model of β-chlorallyl groups in PVC), the inertness of phosphites concerning 4-chlorpentene-2 is determined (the products of their interaction are not detected) [57].

At the same time according to the data of NMR ^{31}P and infrared spectroscopy, epoxyphosphites are capable of reacting with 4-chlorpentene-2 in the following directions [36]:

- dehydrochlorination of 4-chlorpentene-2 (occurrence of conjugated double bond ν = 1660 sm^{-1}) (Reaction 3);
- interaction of epoxy groups with releasing HCl, it leads to the formation of phosphite containing chlorohydride group (δ_p = +135 ppm., ν_{OH} = 3400 sm^{-1}) (Reaction 4);
- dealkylation of phosphites by releasing HCl, it leads to the formation of alkyl(aryl)phosphorous acid (Reaction 5).

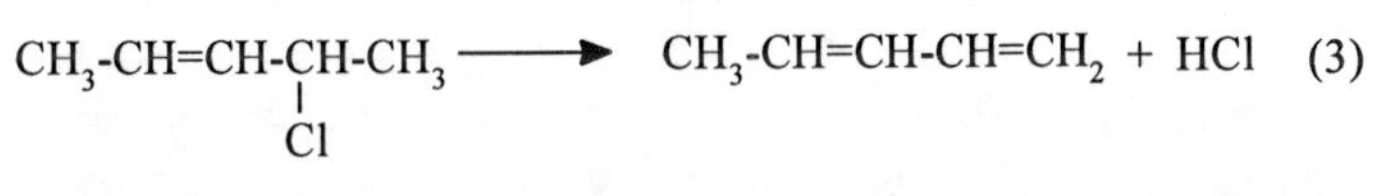

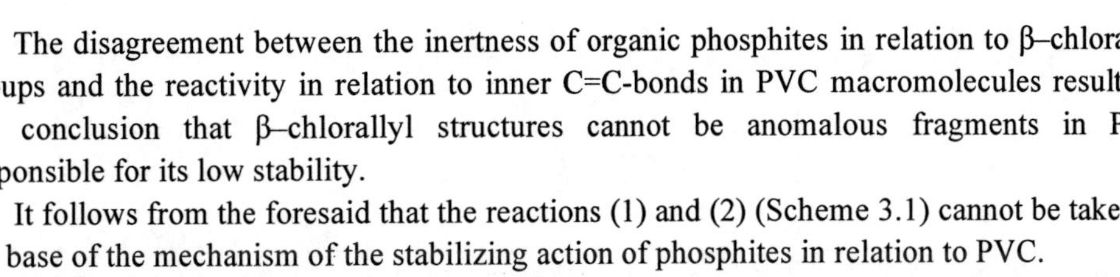

The disagreement between the inertness of organic phosphites in relation to β–chlorallyl groups and the reactivity in relation to inner C=C-bonds in PVC macromolecules results in the conclusion that β–chlorallyl structures cannot be anomalous fragments in PVC responsible for its low stability.

It follows from the foresaid that the reactions (1) and (2) (Scheme 3.1) cannot be taken as the base of the mechanism of the stabilizing action of phosphites in relation to PVC.

At the same time PVC treated by phosphite is characterized by the low initial rate of HCl elimination. Phosphorilation of PVC can be connected with the interaction of phosphorous acid esters with a number of other reactive groups, formed in the process of PVC destruction. Polyenes are attributed to a number of such centers, the formation of which is possible already at the early stages of destruction. At the same time it is known [58] that reactions of phosphites with diene systems proceed under extremely rigid conditions. It is more realistic to suppose that the products of phosphites dealkylation, namely, dialkyl(aryl)phosphites react with polyenes [59].

In compliance with the high reactivity of derivatives of three-valence phosphorus in relation to α,β-unsaturated ketones [60-63] the addition reaction of phosphites to a carbonyl fragment of ketochlorallyl group is more probable when PVC stabilization takes. In this connection a model reaction of tributylphosphite with methylvinylketone (low-molecular analog of anomalous carbonylallyl fragment in PVC) was conducted. The kinetic parameters (k_{ef}, E_{act}) of the reaction of phosphites with methylvinylketone indicate higher activity of trialkyl- and aryl(alkyl)phosphites in these reactions in comparison with triarylphosphites (table 3.12).

Table 3.12. Kinetic parameters of reactions of phosphites with methylvinylketone

Phosphite	$k_{ef}\times10^4$, l/(mole·s)			E_{act}, kJ/ mole
	293 K	308 K	323 K	
$(C_4H_9O)_3P$	0.47	0.89	1.66	30.1
$(i\text{-}C_8H_{17}O)_2POC_6H_5$	0.30	0.60	0.92	28.5
$i\text{-}C_8H_{17}OP(OC_6H_5)_2$	0.14	0.32	0.61	36.1
$(C_6H_5O)_3P$	0.10	0.25	0.44	37.3
$C_6H_{13}OP(OC_{10}H_7\text{-}\alpha)_2$	–	–	0.4	–

To confirm such mechanism in polyvinylchloride the model system was used, it was the mixture of methylvinylketone, 4-chlorpentene-2 and tributylphosphite. At 353 K phosphites react selectively and easily with methylvinylketone, and 4-chlorpentene-2 stays changeless after the reaction, without considering a certain amount of products of its dehydrochlorination. Dibutyl-3-oxobutylphosphonate was identified as the main product of the reaction (65-70%) [64, 65].

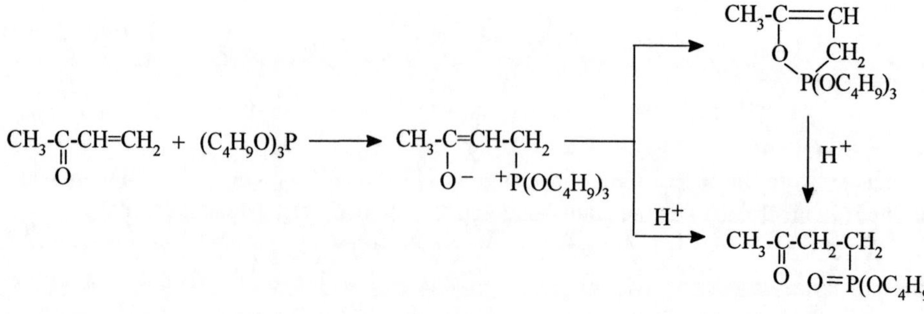

This model allows to verify once more the competitive ability of phosphites in relation to ketoallyl (methylvinylketone) and β–chlorallyl (4-chlorpentene-2) fragments of PVC.

The interaction of phosphites, particularly, tri-*i*-octylphosphite, with ketochlorallyl groups contained in PVC leads to the decrease of their content in polymer macromolecules. The forming of saturated structures and, as a result, the decrease of depression of limit value of viscosity of ozonized polymer, proceeds with the evident rate even at room temperature [66].

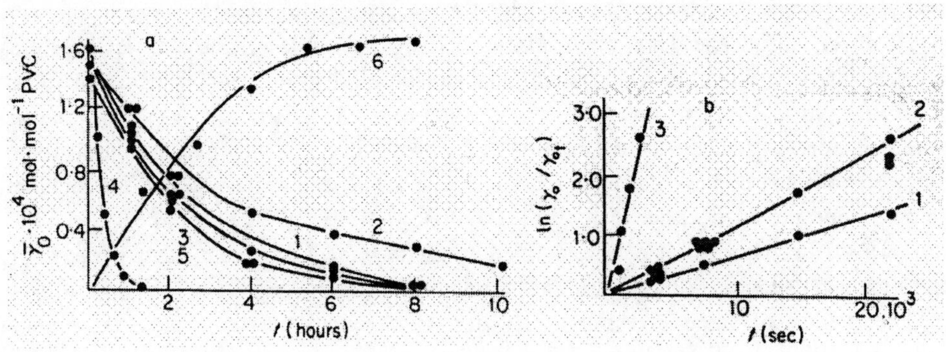

Figure 3.3. Kinetic curves for the interaction of ketoallylchloride groups (KAG) with $(i\text{-}C_8H_{17}O)_3P$ (a) and their semilogarithmic anamorphose (b). The initial KAG content, $\gamma_0 \times 10^4$ mole/mole PVC. (a) 1 and 6 – 1.52; 2 and 4 – 1.62; 5 –1.42; temp. (°C): 2 – 16; 1, 3 and 5 – 25; 4 – 40. (b) temp. (°C): 1 –16; 2 – 25; 3 – 40.

As phosphite was introduced into polymer to large excess in relation to ketochlorallyl groups (it meets the actual conditions of PVH stabilization) the reaction rate is defined only by the content $(RO)_3P$. The decrease of the content of ketochlorallyl groups results in the increase of thermal stability of polymer. As a result, the initial rate of dehydrochlorination of PVC, modified by the phosphite treatment with the subsequent removal of phosphite residues from polymer by ether extraction, decreases (figure 3.4).

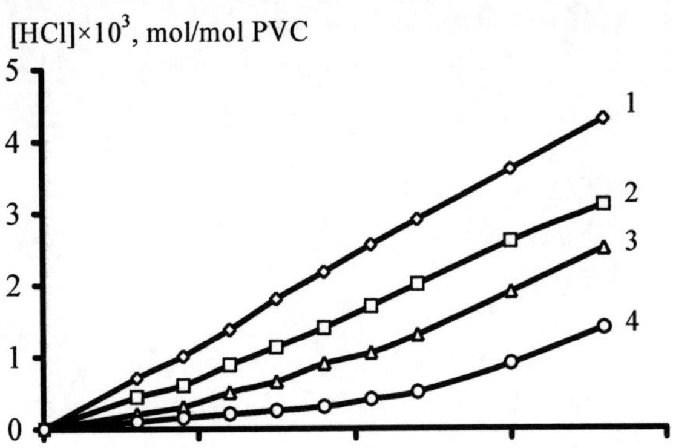

Figure 3.4. Kinetic curves of HCl elimination from PVC (γ_0=1.52×10^{-4} mole/mole PVC): (1) containing no phosphite and pretreated at 298 K by phosphite $(C_8H_{17}O)_3P$ for 1 hour (2), for 2 hours (3), for 6 hours (4).

Processing the curve of the accumulation of ketochlorallyl groups according to the first-order equation results in the rate constant of ketophosphonate groups decomposition ($k_{decomp.}$), which is equal to $0.93\times10^{-3}s^{-1}$ at 448 K. At the temperature of the polymer destruction the value of the ratio of the rate constant of forming ketophosphonate groups and the rate constant of their decomposition ($k_{form.}/k_{decomp.}$) makes more than 1000 (it is calculated according to the activation energy).

However, during the destruction of modified PVC the rate of releasing HCl restores to the value of pure polymer. The content of ketochlorallyl groups increases again to initial value γ^0 (figure 3.3b). According to the observed experimental fact it is evident that under certain conditions of PVC destruction the possibility of thermal decomposition of ketophosphonate groups is not excluded and, as a result, the regeneration of ketochlorallyl groups is not excluded either. Thus, the stabilizing efficiency of phosphites in relation to thermal PVC decomposition will be defined by the ratio of the rate constants of reactions of formation and decomposition of saturated ketophosphonate groups.

The results of later investigations when new models were used being optimally appropriate to ketochlorallyl groups (regarding their structure) in PVC are convincing confirmation of the interaction of phosphites with ketochlorallyl groups [67, 68]. Such compounds are oxochloralkenes, i.e. 2-oxo-3-methyl-5-chlorhexene-3 and 2-oxo-5-chlorheptene-3, their structures contain simultaneously all reactive centers (C=O, C=C, C-Cl), inherent to ketochlorallyl groups. It develops preconditions to prove directly polymer analogous reactions of phosphites along one or several centers of this group.

The reactions of these models with tributylphosphite (31) and epoxyphosphite (23) are investigated. The interaction of phosphite (31) with 2-oxo-3-methyl-5-chlorhexene-3 (32) proceeds through the intermediate forming of phosphorane (33) which converts into stable ketophosphonate (34) under thermal exposure, that was established by NMR ^{31}P spectroscopy.

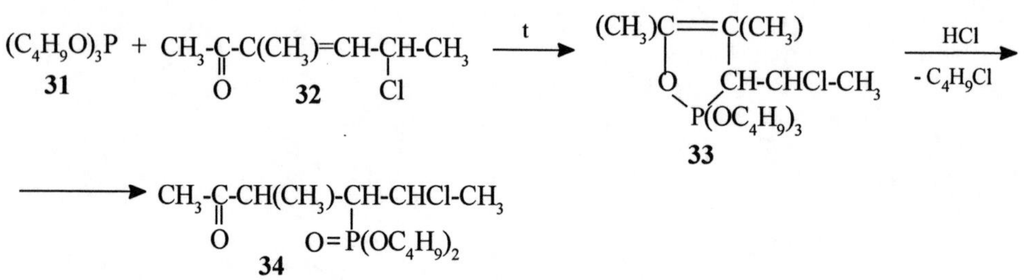

When the reactions of phosphite (**23**) with 2-oxo-5-chlorheptene-3 (**35**) were investigated in the reaction mixture under the same experimental conditions by method of NMR ^{31}P spectroscopy, the presence of phosphorane structure was not registered. In this case the formation of O-phenyl-2-ethylhexyl-(2-oxo-5-chlorheptyl-4)-phosphonate (**36**) can be presented by the following scheme:

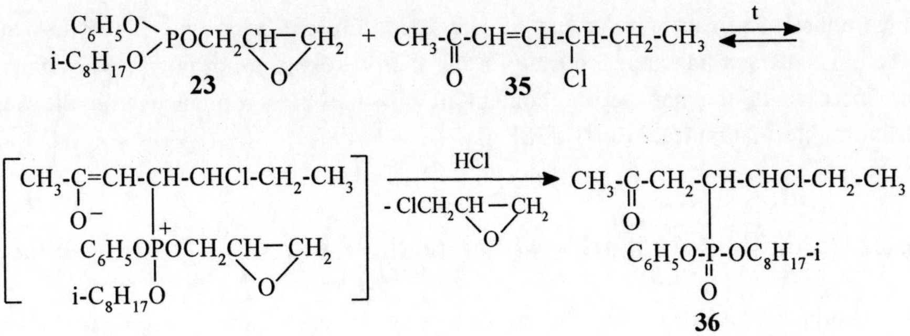

Thus, at the stage of the initiation of PVH destruction the reactions of phosphites with polyenes and β–chlorallyl fragments have no fundamental importance during the inhibition of PVH decomposition and, as a result, of its coloring.

The main reaction of organic phosphites responsible for the stabilization of PVC and retaining its initial color should be considered the interaction of organic phosphites with ketochlorallyl fragments ~CH(Cl)-CH=CH-C(O)~. This reaction is followed by the removal of inner double bonds being in ketochlorallyl groups. During this process the interruption of conjugation in groups invariably results in sharp decrease of the rate of PVC dehydrochlorination. It is the main reason of the inhibition of forming polyene sequences and other colored compounds by phosphites [69].

In the developed PVC destruction process the deactivation of the formed reaction centers, particularly, -C(O)H, -C(O)OH, C=O, C=C(O), C=CH-CH=C etc. (responsible for PVC coloring) by phosphites is possible. The decrease of dehydrochlorination rate and weakening of the coloring of dehydrochlorinated PVC in the presence of phosphites is an experimental proof of that (table 3.13).

Table 3.13. Phosphorilation of dehydrochlorinated PVC (concentration of phosphite is 7×10^{-4} (g·mole)/g PVC, 413 K, 90 min)

Phosphite	Polymer color	Amount of released HCl, mg/g	Content of phosphorus in polymer, %
–	Dark brown-	0.45	
$(i\text{-}C_8H_{17}O)_2POC_6H_5$	Beige	–	0.26
$(C_6H_5O)_3P$	Dark brown	2.28	0.22
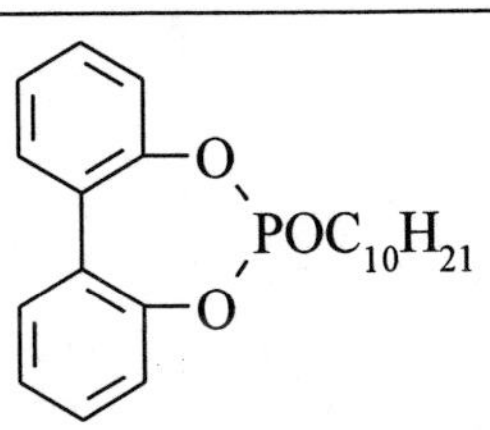	Beige	–	0.25

When dehydrochlorinated PVC is heated phosphite joins quickly at the beginning of the process, then its amount in polymer increases steadily. It is evident that at the initial reaction stage phosphites react with groups already being present in polymer, then they interact with reaction centers which appear during further thermal exposure of polymer products. It leads to the increase of the phosphorus content in polymer in compliance with the degree of its dehydrochlorination (table 3.13) [38].

3.1.2. Inhibition of Coloring of Polyolefines and Some Other Polymers

The possibilities of deactivation of reactive centers of a macromolecule ("weak zones") by phosphites are considered, these reactive centers initiate and develop the processes of destruction and coloring (processes at the stage of initiation included) in polyolefines.

Such reactive centers in polyolefines can be double bonds, branchings of polymer chains with labile hydrogen atoms, carbonyl containing groups [70]. When polyethylene containing various functional groups (C=O, OH, C=C) was investigated it turned out that phosphites are inert in relation to hydroxyl groups and double bonds (table 3.14).

In the given section the results on selective interaction of phosphites with carbonyl groups, which are contained in the initial polyethylene and formed in the result of its thermal-oxidative destruction, are given. They can be formed in the structure of polymer as anomalous fragments at synthesis process owing to the copolymerization of carbon oxide (as an admixture) with monomer and at the initial stages of oxidation [1].

Table 3.14. Composition of polyfunctional groups of HDPE in the presence of $(i\text{-}C_8H_{17}O)_3P$

Sample of HDPE	Content of groups per 1000 hydrogen atoms					
	Initial			After the introduction of phosphite		
	C=O	OH	C=C	C=O	OH	C=C
1	0.3	–	–	0.1	–	–
2	0.39	–	–	0.2	–	–
3	–	–	0.23	–	–	0.23
4	–	–	0.49	–	–	0.49
5	–	0.26	–	–	0.26	–

The model introduction of phosphorous acid esters into preoxidized polyethylene and hexadecane (at 473 K, oxygen pressure is $2 \cdot 10^{-4}$ Pa) results in the considerable decrease of intensity of absorption band of C=O-group (1720 sm^{-1}) in infrared spectra of polymer samples. The phosphorylation of polymer takes place during that process, in infrared spectra of polyethylene reprecipitated after oxidation there are absorbtion frequencies of phosphoryl groups (ν=1250 sm^{-1}) [71, 72].

The general regularities of phosphites behavior under thermal oxidation of polyolefines are studied, polyethylene of high pressure (LDPE) and polyethylene of low pressure (HDPE) taken as examples [73].

It is established that aromatic esters, alkylated into the aromatic ring by bulky substitutes, for example cyclic esters on the basis of 2,2'-methylene-bis(4-methyl-6-*tert*-butylphenol), as well as oligophosphites are the most efficient at the inhibition of thermal-oxidative HDPE destruction (table 3.15). When assessing color stabilizing properties of phosphites it is established that HDPE films containing these phosphites preserve natural color when exposed to air at 473 K for 2 hours. The samples stabilized by trialkyl-, alkylaryl- and arylphosphites acquired yellow color during the same period of time.

The comparison of the efficiency of phosphites as antioxidants and color stabilizers of LDPE testifies that dependence between color stabilizing and antioxidative properties is defined by the structure of phosphite. In compliance with it, phosphorous acid esters can be divided into two types. Some of them are strong inhibitors of free radical oxidation of polymer owing to accepting peroxide radicals (chain termination) and develop color stabilizing properties at the same time. For the other type of phosphites the nonchain reactions of inhibition are characteristic. They interact with molecular products of polymer destruction, chromophor and auxochrome (ROOH, fragments containing C=O, C=C, etc. groups) included, thereby preventing coloring polymers.

Table 3.15. Influence of phosphites on thermal stability of HDPE during milling at 433 K

Phosphite (5 wt%)	τ_{br}[a], h	Phosphite (5 wt %)	τ_{br}[a], h
–	1-2	$(C_8H_{17}O)_3P$	2
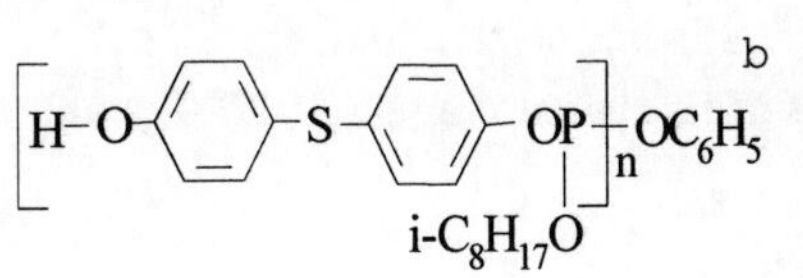	11	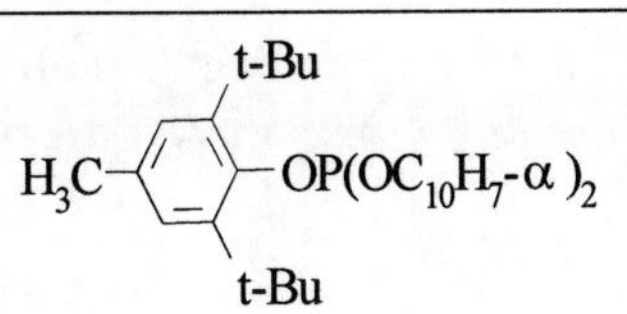	6
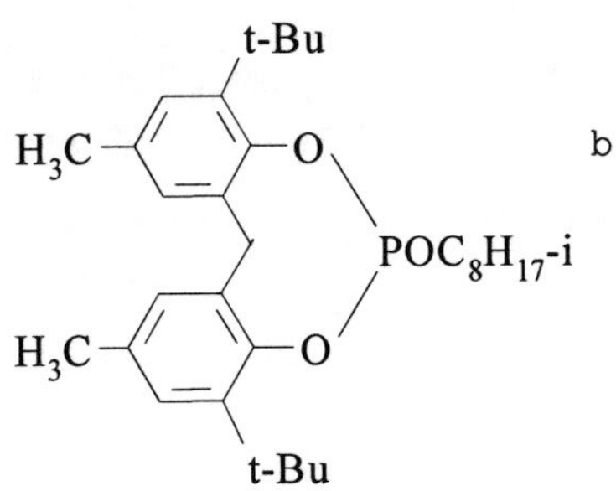	13	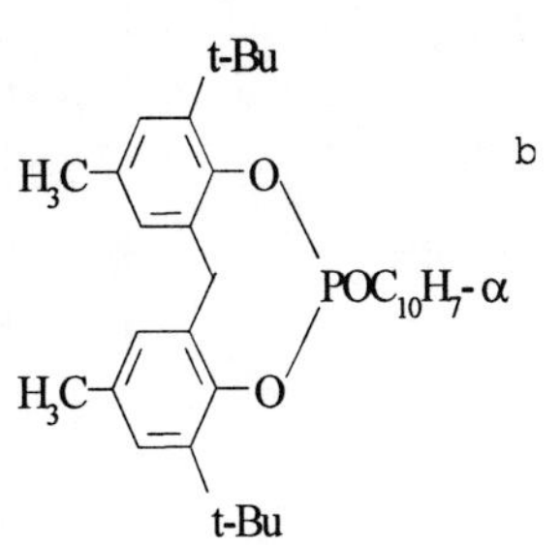	12

Table 3.15. (Continued).

Phosphite (5 wt%)	τbr a, h	Phosphite (5 wt %)	τbr a, h
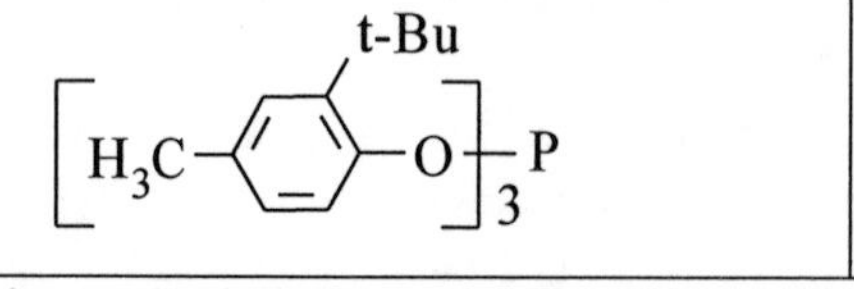	8	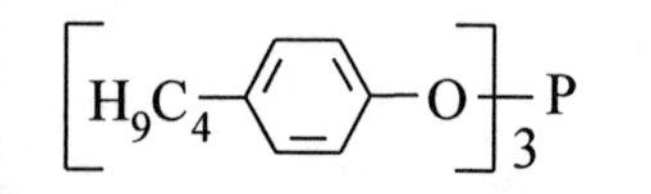	3

[a] time period before coming of brittleness of polyethylene; [b] the samples have natural color.

In most cases coloring, physical and mechanical properties and molecular weight of polymers during the induction period change weakly. The sharp worsening of their properties is observed during the autocatalytic stage of oxidation. At that the oxidation of macrochains (during processing and application) is accompanied by intensive forming of C=O–groups practically at all stages of thermo-oxidative destruction of polymer, generally, during the decomposition of hydroperoxides. When accumulated in polymer C=O–groups can participate in the forming of chromophore conjugated systems combining with each other and other structures (auxochromes), coloring intensively polymer and worsening considerably its performance characteristics [1-12]. Apparently, it is high reactivity of phosphites in relation to these structures [58-60, 63, 71, 72] that predetermines their efficiency under conditions of autocatalytic polymer oxidation (table 3.16). The efficient stabilization of polymer color under conditions of developed destructive processes when phosphites are used in combinations with other additives, in the first place, with inhibitors of free radical oxidation of polymers, should be explained by the increase of the contribution of the abovementioned reactions [74].

Table 3.16. Changing of content of C=O-groups in polyethylene stabilized by phosphites (0.05 wt%) during the process of its thermal ageing (493 K, air)

Phosphite [a]	Content of C=O – groups (D, 1720 sm^{-1})				
	0 hour	5 hours	10 hours	15 hours	20 hours
–	0.23	2.01	2.59	2.61	2.69
(1)	0.22	0.93	1.62	1.88	2.08
(2)	0.23	0.53	0.95	1.54	1.80

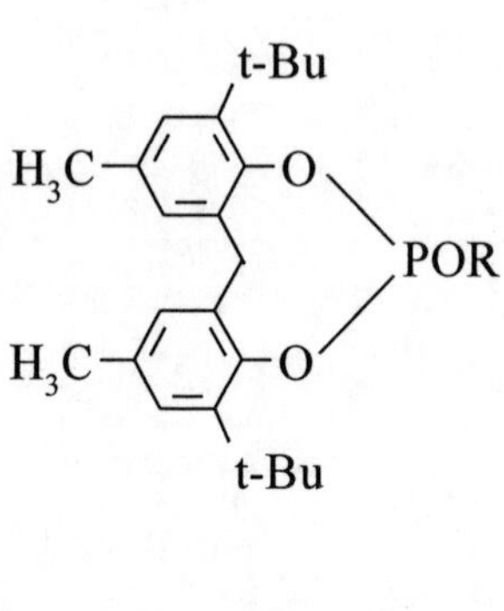
R = C_2H_5 (**1**), C_6H_5 (**2**);

In the presence of inhibitors of free radical oxidation the consumption of phosphites in reactions with peroxide radicals should decrease, as a result, the function of phosphites as nonradical destructors of hydroperoxides increase considerably, for example. The confirmation is the preserving of considerable synergic effect according to the value of induction period of oxidation and color index when in the composition with phenol antioxidant 2,2'-methylene-bis(4-methyl-6-*tert*-butylphenol) a typical destructor of hydroperoxides, namely, dilauryl thiodipropionate is replaced by trinaphthylphosphite [74].

When the mechanism of action of organic phosphites during the inhibition of polymer coloring in the developed process of oxidation is discussed the possibility of participation of dialkyl(aryl)phosphorous acids, formed *in situ,* in the processes of stabilization should be also taken into consideration. The forming of these acids under conditions of autocatalysis is simplified by the presence of strong dealkylating agents (polymer decomposition products and introduced additives), such as HCl, H_2O, RCOOH etc, in the system [75].

It is even more probable that dialkyl(aryl)phosphorous acids in relation to some structures formed during polymer decomposition can be more reactive than full phosphites. Thus, the investigation of dependence of induction periods of LDPE oxidation in the presence of acid esters of phosphorous acid indicated that the inhibiting ability of their aromatic derivatives exceeded the inhibiting ability of full phosphites [76].

Diarylphosphites containing *tert*-butyl groups in *o*-position of phenol rings (28, 37) and oligophosphites (38) turned out to be the most efficient. Acids (39) and (40) containing aliphatic and nonalkylated aromatic radicals at a phosphorus atom are considerably inferior to phosphites (28), (37), (38) according to their inhibiting properties.

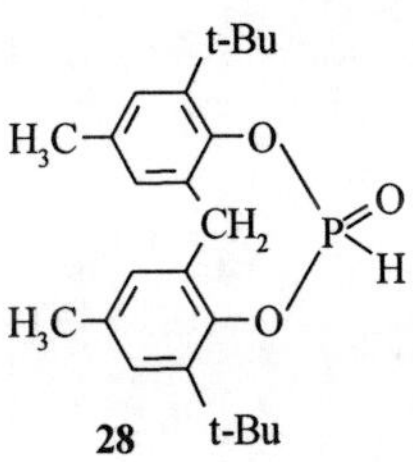

28

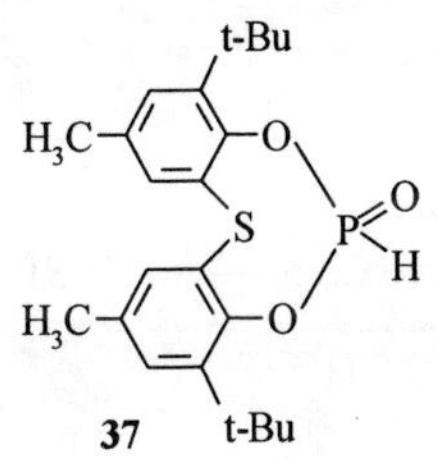

37

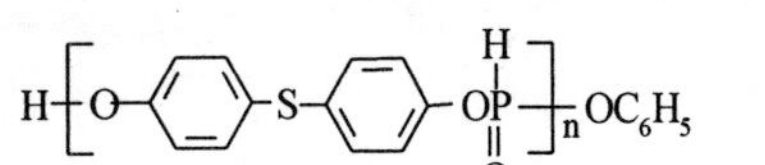

38

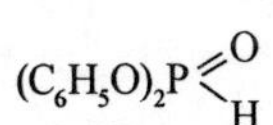

39

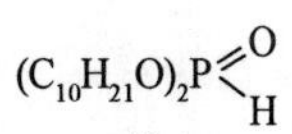

40

The activity of dialkyl(aryl) phosphorous acids when polyethylene is stabilized coincides with reactivity in relation to stable free radicals, such as α,α-diphenyl-β-picrylhydrazyl, nitroxyl radical, that indicates the possibility of their participation in the inhibition of free radical oxidation [76].

The reaction rate constants of acids (28) и (39) with nitroxyl radicals were determined by method of EPR, they turned out to be equal to 1.36×10^{-2} и 4.2×10^{-4} l/(mole·s), respectively, at 328 K in toluene. For acid (28) in the range of 328-375 K the value of activation energy (61.74 kJ/mol) and preexponential factor (6.6×10^{6} l/(mole·s) are defined. The influence of a substitute at a phosphorus atom preserves its tendency in case of reactions with nitroxyl radical [76].

On the other hand, dialkyl(aryl)phosphorous acid can react with active products of polymer decomposition (C=O-groups, etc.), similar to full esters of phosphorous acid [58, 59, 71]. Thus, under equal conditions di-*i*-octylphosphite decreases the concentration of C=O-groups by 45%, 2,2'-methylene-bis(4-methyl-6-*tert*-butylphenyl)phosphite does it by 28%, tri-*i*-octylphosphite does it by 20.5% [25].

The abovementioned examples testify that under certain conditions (especially in the presence of dealkylated reagents in the system) full esters of phosphorous acid can act as "depot" of products (particularly, dialkyl(aryl)phosphorous acids), which, acting according to the principle *in situ*, provide high stabilizing result at the stage of the developed process of polymer destruction.

Thus, the investigation of the mechanism of stabilization of polymer color by phosphites in developed processes of destruction allows to implement efficient means of secondary stabilization, i.e. the inhibition of just those processes which result in substantial changing of polymer properties (reactions with ROOH, C=O groups, etc.).

To identify products formed during the interaction of phosphorous acid esters with carbonyl groups of polymer the infrared spectra of polymer, purified from excess of phosphorus additive by reprecipitation, were taken. The spectra of such polyethylene is characterized by the decrease of intensity of carbonyl absorption band (C=O, 1720 sm^{-1}), by the appearing of wide intensive peak in the range of 1020-990 sm^{-1} (P-O-C) and absorption band in the range of 1280-1250 sm^{-1} (P=O) [72], that testifies to phosphorylation of polyethylene. Infrared spectra of modified polymer and the results of model reactions phosphites with low-molecular analogs of oxidized polyethylene (methylpropylketone and phosphites at 448 K without a catalyst and in the presence of aluminium alcoholate taken as samples) allows to examine the course of the reaction in compliance with the scheme [71, 72]:

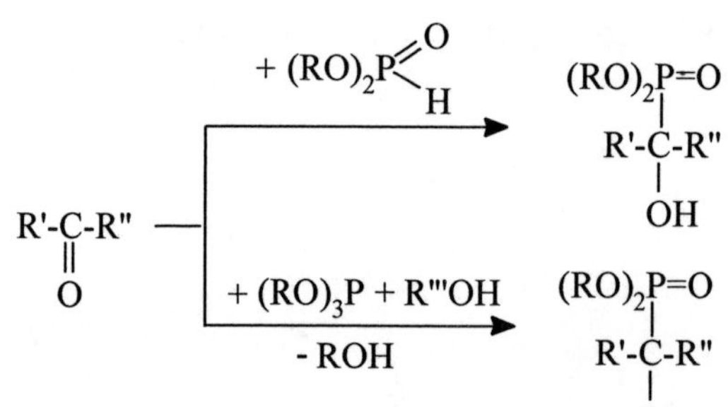

The ability of phosphites to interact with groups C=O is interconnected with performance characteristics of polymer, such as tangent dielectric loss angle (tg δ), polymer color (ρ) (table 3.17), and physical and mechanical properties (table 3.18) [72].

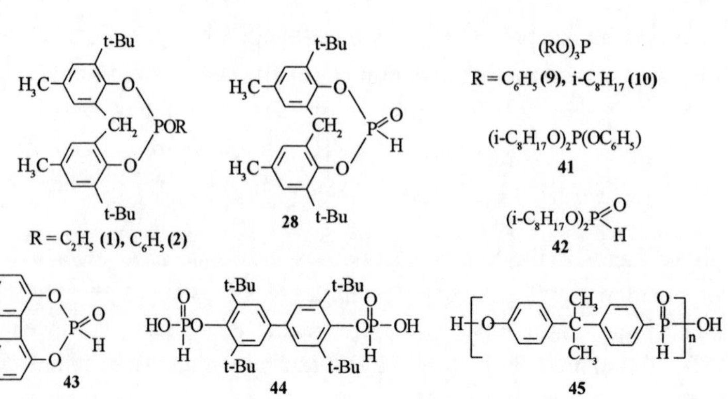

Table 3.17. Changing of tangent dielectric loss angle (at 10^6 Hz) and HDPE color during the process of thermal-oxidative ageing (383 K)

Phosphite (0.5 wt%)	tg δ×10⁴	P[a], %	tg δ×10⁴	ρ, %	tg δ×10⁴	ρ, %	tg δ×10⁴	ρ, %
	Time of ageing, hours							
	0		70		140		210	
–	3.7	83.0	4.3	80.0	4.7	78.0	9.1	74.0
(1)	5.9	81.7	8.7	80.4	6.3	80.0	6.3	78.5
(2)	4.0	80.1	3.4	79.6	4.5	79.1	4.6	78.2
(9)	5.7	80.6	6.2	77.3	6.3	74.8	7.3	73.1
(10)	3.2	82.3	4.4	79.5	3.8	75.2	5.6	74.8
(28)	2.7	82.0	3.0	81.3	4.7	80.1	4.6	79.0
(41)	4.4	81.5	5.3	80.6	5.9	76.4	6.8	74.1
(42)	2.7	92.2	2.5	90.7	3.0	88.3	3.3	86.8
(43)	4.5	89.2	4.5	88.0	4.5	83.6	7.8	81.1
(44)	2.9	85.4	3.0	85.0	3.2	84.5	3.3	83.8
(45)	2.6	86.8	2.7	87.0	3.0	86.0	3.4	85.1

[a] ratio of luminous flux reflected by a sample to luminous flux of the compared sample.

Table 3.18. Changing of physical and mechanical properties of HDPE stabilized by phosphite (42) during the process of thermal-oxidative ageing at 383 K

Polymer	Time of ageing, h	Melt flow index, g/10 min	Elongation at break, %	Yield strength, MPa	Tensile strength, MPa
HDPE	0	0.39	926	27	27
	95	2.06	20	28	29
	120	brittle	brittle	brittle	brittle
HDPE + 5 wt% (42)	0	0.4	910	26	25
	95	0.56	300	27	22
	120	0.3	300	27	27
HDPE + 0.3 wt% (42)	95	0.9	330	27	26

As it is seen from the experimental data the values tg δ being quantity measure of carbonyl groups in the oxidizing polymer don't change practically in the presence of phosphites. The data of table 3.17 testifies to the ability of phosphites to improve the initial color of polyethylene and to decelerate its coloring in the process of ageing. It should be connected with the ability of phosphorous acid esters to interact with carbonyl groups being in the structure of conjugated systems, formed in the process of polymer oxidation and conditioning the occurrence of its coloring.

Thus, the deactivation of active centers, formed at the stage of development and branching of destructive processes (thermal oxidation of polyethylene taken as an example) and defining the decomposition and coloring of polymers, by organic phosphites is one of the ways in the stabilization of polymer color, it is of specific importance in the commercial processing and service performance of polymers. At the same time the chemical modification of macromolecules takes place owing to polymeranalogous reactions, it leads to the increase of polymer's own stability [77].

The issues connected with the stability to color changing in the processes of thermal oxidation are typical for many polymers. Determined approaches to color stabilization by phosphites, polyethylene being an example, can be applied to other types of polymers (on conditions that similar products of polymer decomposition are formed). The submitted below results should be considered as illustrations confirming that commercial use of phosphites for thermal and color stabilization of a number of carbon-chain and heterogeneous-chain polymers which are easily exposed to coloring owing to peculiarities of their structure is promising.

So, the presence of branching in the structure of poly-4-methylpentene-1 (monomeric unit contains *i*-butyl group including two tertiary carbon atoms, one being in the main chain, the other being in the side chain) promotes the decrease of stability of this polymer towards any type of ageing, coloring included [78]. Regarding its stability to oxidation poly-4-methylpentene-1 is considerably inferior to polyethylene, however, the regularities during its stabilization, particularly when phosphorous acid esters are used, are similar. Cyclic esters on the basis of 2,2'-methylenebis(4-methyl-6-*tert*-butylphenol) (48), (49) and oligophosphites (18) and (50), as well as mix compositions of phosphites with phenol antioxidants (table 3.19, 3.20) are the most efficient in polyolefines, they preserve the initial color of polymer at most [78].

Table 3.19. The influence of phosphites on the thermal stability of poly-4-methylpentene-1

Phosphite (0.01 mole/kg)	Melt flow index, g/10 min		Time before brittleness beginning (milling, 443K), h
	After extrusion	After jet molding	
–	11	23	2
(10)	5.5	9	2.5
(17)	3	4	9
(18)	1.3	2.3	10
(46)	2	3	4
(47)	2	2.5	6
(48)	2	2	16
(49)	1.3	1.5	10
(50)	1.3	1.5	20

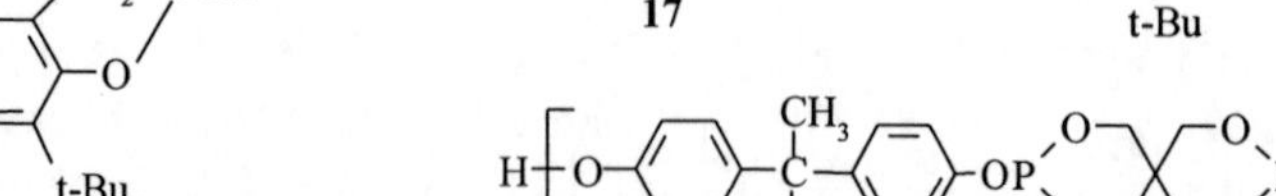

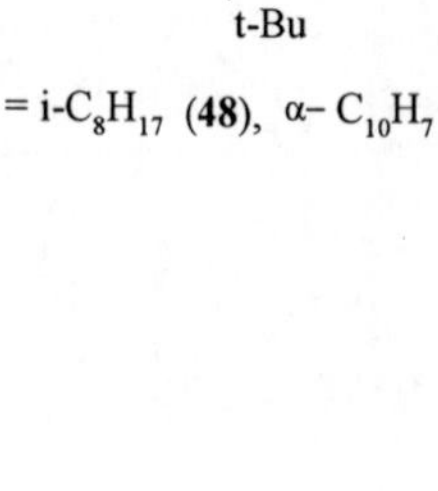

R = i-C_8H_{17} (**48**), α- $C_{10}H_7$ (**49**)

17

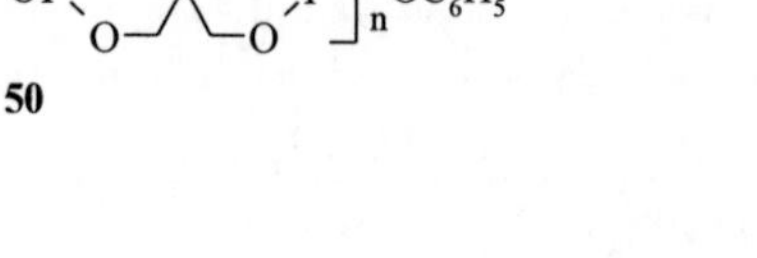

18

50

Table 3.20. Characteristics of poly-4-methylpentene-1, stabilized by phosphites and Irganox 1010

Composition	Content of stabilizer, wt%	Transparency, % light transmission	Melt flow index, g/10min		Time of starting of polymer destruction (473 K), hour	Impact elasticity, kJ/m^2
			After extrusion	After jet molding		
Irganox 1010	0.5	90	15.6	31.2	30	12
Phosphite (49)	0.5	90.5	9.0	16.5	5	34
Irganox 1010 Phosphite (49)	0.3 0.2	90	7.5	9.75	40	28
Irganox 1010 Phosphite (49)	0.2 0.3	92	6.4	7.1	38	32
Irganox 1010 Phosphite (2)	0.2 0.3	91	6.5	7.1	38	32
Irganox 1010 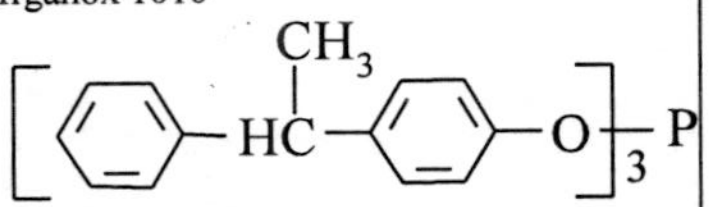	0.2 0.3	89	12	19.8	14.5	12
Irganox 1010 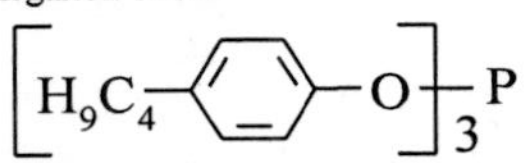	0.2 0.3	90	10	24	14	12.5
Irganox 1010 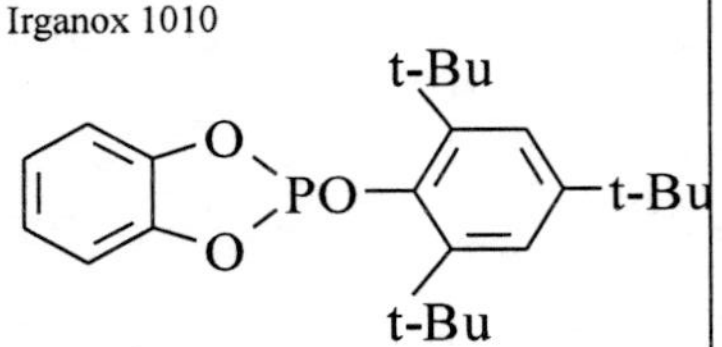	0.2 0.3	91	10	13.5	10	11

The processes of oxidation of carbon-chain rubbers (isoprene, butadiene-styrene, chloroprene rubbers, etc.) are characterized by the regularities which are common for carbon-chain polymers in whole. Therefore, it should be expected that the same peculiarities of stabilizing action of organic phosphites would occur as in case of polyolefines, for example. So, it was established that oligophosphites, sterically hindered esters on the basis of 2,2'-methylene-bis(4-methyl-6-*tert*-butylphenyl)phosphorous acid are highly efficient during ageing of a number of rubbers, they surpass commercial tri(*p*-nonylphenyl)phosphite in stabilizing properties. Oligophosphites on the basis of alkylated thiobisphenols provide light color and high induction periods (τ) of rubber oxidation (table 3.21, 3.22), these values are comparable with the similar characteristics for commercial antioxidants.

The data on the stabilization of chloroprene rubber given in table 3.23 define the efficiency of stabilizing action of phosphites under conditions of high temperature, ultraviolet- and γ-irradiation, determined according to the value of the induction period of oxidation (τ) and the rate of oxidation process [79].

Table 3.21. The stability of isoprene rubber stabilized by oligophosphites and amine antioxidants (oxidation in air, 393 K)

Stabilizer	Content of stabilizer, wt%	τ, h	[η] [a]	
			Before ageing	After ageing (393 K, 3 hours)
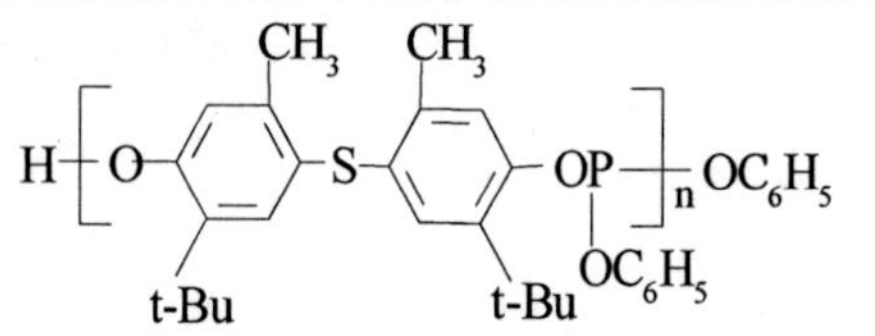	0.65 1.00	7.0 10.5	2.87 –	2.72 –
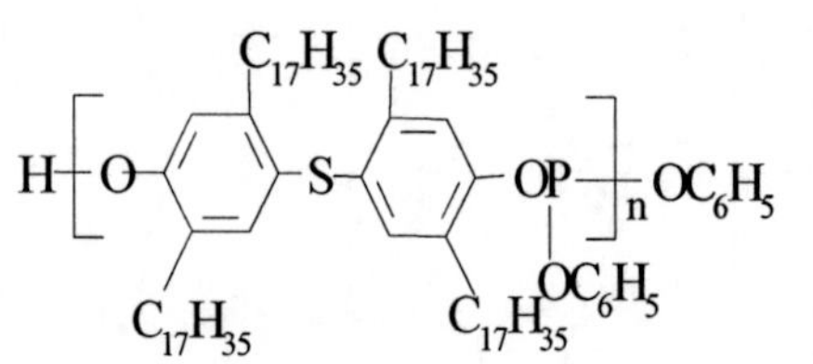	0.65 1.00	9.0 14.0	2.88 –	2.76 –
N,N'-diphenyl-*p*-phenylenediamine N-phenylnaphthylamine-2	0.25 0.40	6.5	2.90	2.78
N,N'-diphenyl-*p*-phenylenediamine N-phenylnaphthylamine-2	0.35 0.65	10.0	2.91	2.80

[a] intrinsic viscosity.

Table 3.22. The stability of butadiene-styrene rubber stabilized by phosphites (oxidation in air, 393 K)

Phosphite [a]	$[\eta]_0$	τ, min	Δ[η]	k_τ [b] $\times 10^4$
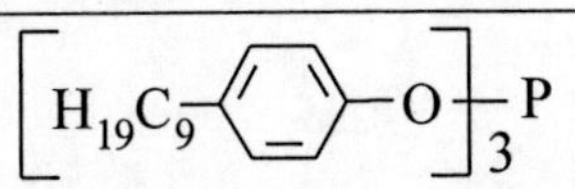	3.07	147	0.82	56
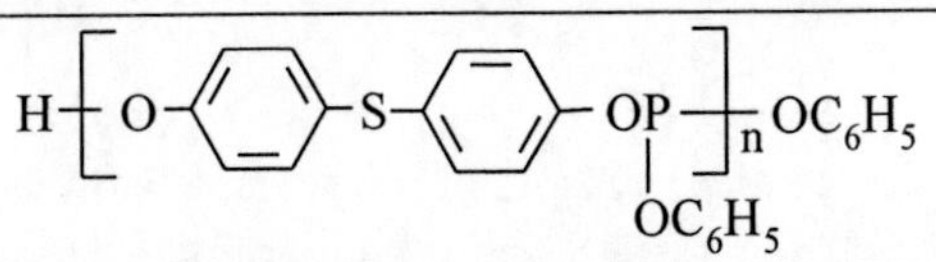	2.94	308	0.26	9
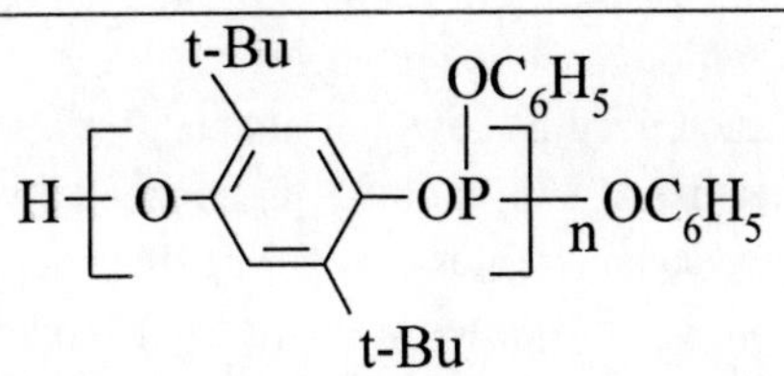	3.06	308	0.11	4
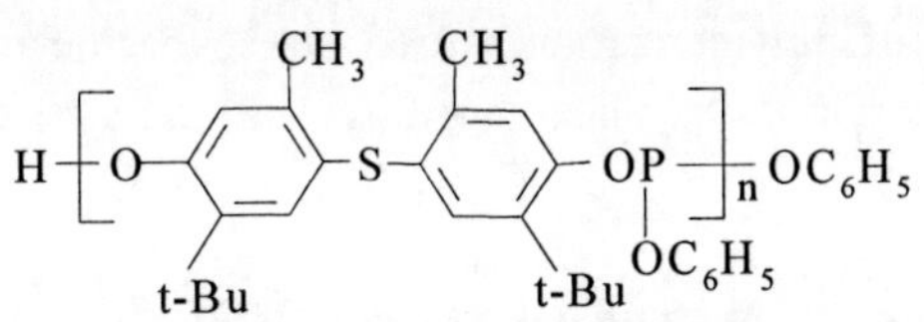	2.96	350	0.34	10

[a] content of stabilizer is 1.5 parts by weight per 100 parts of rubber;
[b] the rate of changing of intrinsic viscosity of rubber per time unit in the induction period of oxidation.

Table 3.23. Characteristics of stabilized chloroprene rubber

Stabilizer	Content of stabilizer, wt%	Thermal ageing (373 K, air, 373 K)		Ultraviolet irradiation		γ-irradiation (Co^{60})	
		τ [a], h	W [b]	τ [a], h	W [b]	τ [b], h	W [b]
(49)	0.5 1.0 2.0	12 40 90	– 8 –	25 70 110	– 6 –	2 4 8	– 16 –
(51)	1.0	2.2	7.9	30	5	2.5	14.5
(49) + (51) [c]	0.5+0.5 1.0 + 1.0	25 86	– 1.6	50 160	– 0.2	7.5 10.3	– 1.3

[a] time before the formation of carbonyl and carboxyl groups in polymer; [b] W=tgα is angle tangent of kinetic curve slope of the dependence of optical density of absorption C=O groups (D_{1720}) on time;
[c] a sample is color stabilized.

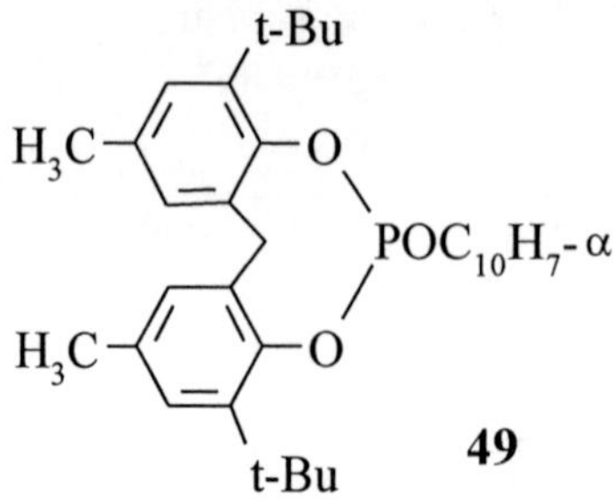

49

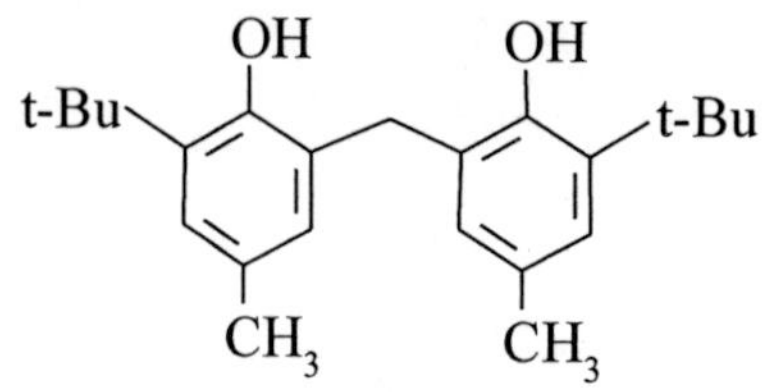

51

In whole, when rubbers are stabilized for studied phosphites there is a correlation between their antioxidizing and color stabilizing characteristics, that is similar to the action of phosphites of similar structure during the thermal oxidation of polyethylene.

Copolymers of styrene with acrylonitrile are characterized by low thermal stability. Short heating even at 423 and 453 K (to say nothing of heating at 473 and 493 K) connected with the processing of material into products leads to considerable change of their color, copolymer turns yellow and even darkens. Organic phosphites can prevent coloring copolymers of acrylonitrile with styrene at 473-493 K in the air for a long time. Cyclic esters of phosphorous acid introduced at the stage of polymerization turned out to be the most efficient (table 3.24), especially in combination with phenol antioxidant 2,4,6-tri-*tert*-butylphenol [80].

Table 3.24. Influence of phosphites on preserving color of copolymers of styrene and acrylonitrile (thermal exposure in air, 493 K)

Stabilizer	Content of stabilizer, wt%	Color stability [a], min
–	–	10
2,4,6-tri-*tert*-butylphenol	0.2	20
2,6-di-*tert*-butyl-4-methylphenol	1.0	20
Phosphite (49)	0.1	45
Phosphite (2)	0.25	30
Phosphite (49) 2,4,6-tri-*tert*-butylphenol	0.1 0.2	180
Phosphite (2) 2,4,6-tri-*tert*-butylphenol	0.25 0.2	150
Tri(*p*-nonylphenyl)phosphite	1.0	8

[a] time before the change of copolymer color.

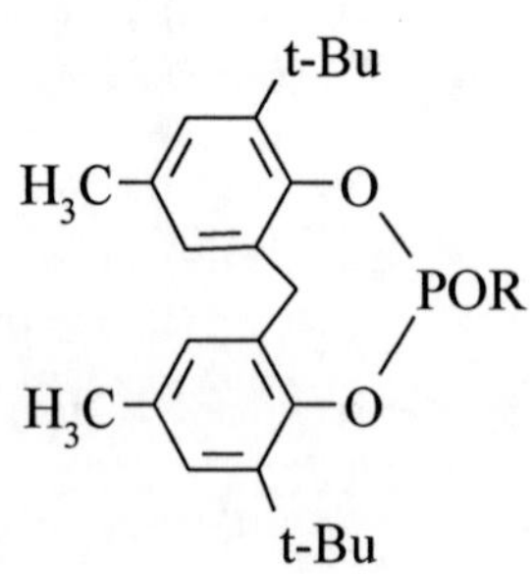

$R = C_6H_5$ (**2**), $C_{10}H_7$-α (**49**)

Organic phosphites inhibit efficiently the development of destructive processes of many heterogeneous-chain polymers, polyethers included, such as polycarbonate [81], polyethyleneterephthalate [82], phenol ether resin [83], polyarylenethersulfone [84], etc. under the influence of high temperatures and oxygen. Since, as a rule, such polymers are processed at high temperatures (523-573 K), oligomeric and other esters of phosphorous acid characterized by thermal stability and low volatility are used for their stabilization.

The main characteristic of phosphites in the process of thermal-oxidative destruction of polycarbonate is their ability to weaken polymer color during their ageing, it changes in the range from yellow to dark brown in the absence of such stabilizers (table 3.25).

Table 3.25. Characteristics of stabilized polycarbonate

Phosphite	Content of stabilizer, wt%	Optical transmission at λ=425 [a], %	
		solutions	disks
–	–	52	23
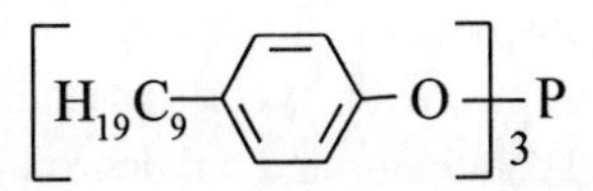	0.5	65	45
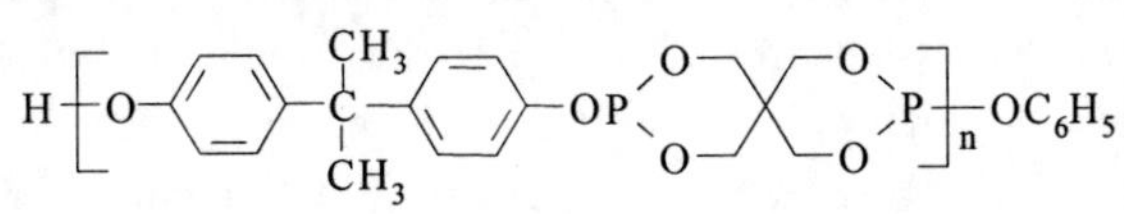	0.1 0.2 0.5	68 69 70	67 70 70
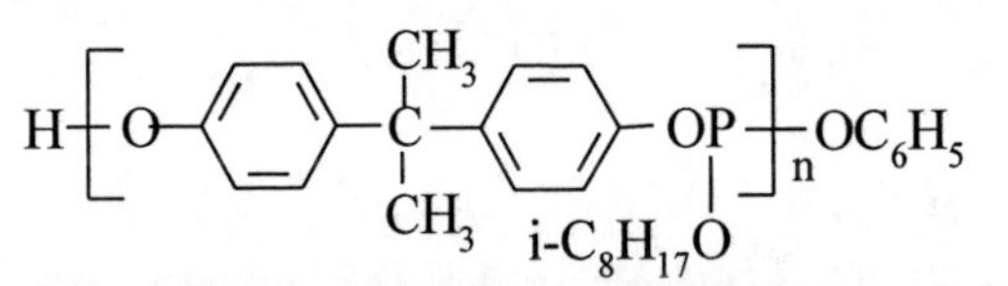	0.1 0.5	67 69	53 62

[a] per cent of optical transmission of solutions of polycarbonate before ageing is 87%.

The decrease of intensity of polycarbonate yellowing depends on the chemical structure of phosphites. As it is seen from table 3.25, oligophosphite on the basis of diphenylolpropane and pentaerythritol are the most efficient for preserving polymer color. However, phosphites, as a rule, are weak thermal stabilizers of polycarbonates.

When naphthyl (49) and phenyl (2) esters of 2,2'-methylenebis(4-methyl-6-*tert*-butyl phenyl)phosphorous acid are introduced into phenol ether resin destructive processes and coloring of polymer decelerate considerably [83]. So, unstabilized samples preserve 50% of tensile strength during heating at 423 K for 20 days, and samples with phosphite (49) preserve it for 60 days. High inhibiting properties of phosphite determine also the possibility of continuous performance (at high temperatures) of products made from phenol ether resin stabilized by organic phosphites [83].

Oligophosphites on the basis of diphenylolpropane and pentaerythritol are efficient stabilizers for polyarylenethersulfone (table 3.26) [84].

Table 3.26. Characteristics of stabilized polyarylenethersulfone

Stabilizer	Content of stabilizer, wt%	MFI, g/10 min	Optical transmission [a], %
–	–	28	30
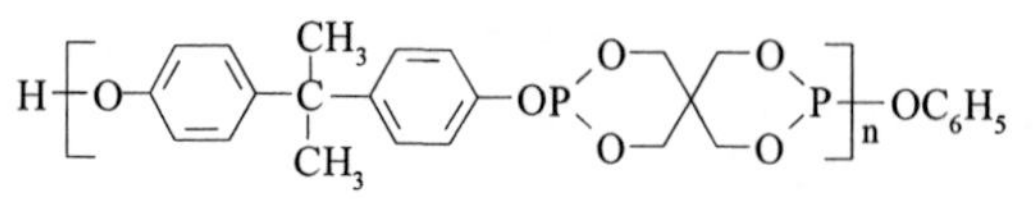	0.2 0.5 1.0	3.9 5.1 8.0	50 52 52
$(C_6H_5O)_3P$	0.5	3.0	40

[a] optical transmission of 1%-solution of polymer in DMF after heating within 30 min at 593 K.

In whole, color stabilization of various types of polymers by phosphites during the process of thermal oxidation accompanied by macromolecules' own transformations is implemented as the complex of reactions. The reactions of phosphites with products which are potentially active in relation to coloring polymer, for example, with hydroperoxides or with products promoting directly the coloring of polymer, i.e. chromophor and auxochromous groups contribute greatly to the forming of positive color effects in polymers. Phosphites which are capable of such reactions, unlike classical antioxidants, not always inhibit oxidation processes, however polymer coloring goes down considerably in their presence. Because of weak inhibiting properties of a number of phosphites the mechanism of interaction with peroxide radicals (the reactions of chain termination) in the general process of color stabilization of polymer materials is not always implemented in full.

3.2. Inhibition of Polymer Coloring by Chemical Deactivation of Admixtures of Compounds of Variable-Valency Metals and Products of Stabilizer Conversions

It is established that phosphorous acid esters have weak antioxidant properties when they are used individually in comparison with typical inhibitors of free radical reactions. Therefore, in the general case, phosphites are used as unique synergists in mixtures with strong antioxidants (sterically hindered phenols, etc.), providing thus obtaining not only color-stable, but high-thermostable polymer compositions. The explanation of the mechanism of forming synergetic thermo- and color-effects, when compositions of phenol antioxidants with phosphites are used, is important for the development of noncoloring mix stabilizers of polymers. At that it is reasonable to consider the role of each component of such stabilizing mixtures in the processes of polymer oxidation and coloring inhibited by them.

3.2.1. The Correlation of the Structure of Sterically Hindered Phenols with Their Coloring Properties

Sterically hindered phenol compounds being strong inhibitors of free radical oxidation can simultaneously color polymers as a result of their own transformations.

The primary reason of coloring of chemically stable polymers, polyolefines in the first place, is the formation of chromophor products of oxidation of phenol stabilizers. It is understandable that tendency to the formation of such products is defined by the structure of these stabilizers [85].

There are data in publications that the tendency to coloring is conditioned by methylene bridge of di(hydroxyphenyl)methane stabilizers [86, 87], as when they are oxidized colored methylenequinones are formed [88].

So, for example, long-wave absorption maximum of hydro-galvinoxyl (52), the product of oxidation of bis-(3,5-di-tert-butyl-4-hydroxyphenyl) of methane (53), is at 398 nm and determines the yellow color of this methylenequinone.

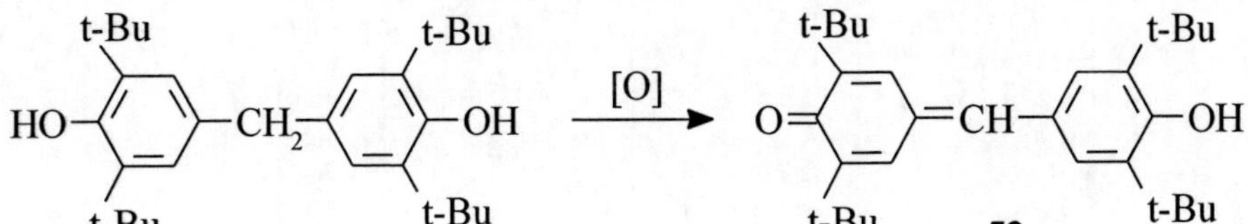

53 52

As it is seen from table 3.27, methylenequinones having substitutes at α-carbon atom which are less electronodonor in comparison with 4-hydroxy-3,5-di-tert-butylphenyl substitute, absorb only in ultraviolet region.

Table 3.27 Long-wave absorption maximums in electronic spectra of α-substituted 2,6-di-tert-butylmethylenequinones

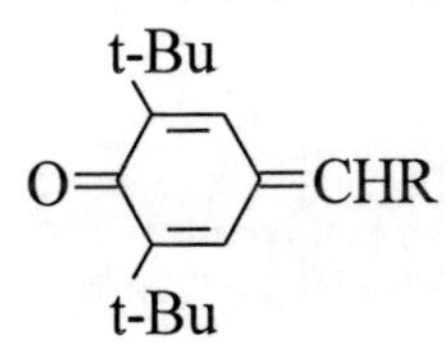

R	H	CH_3	OCH_3	Ph	$N(CH_3)_2$	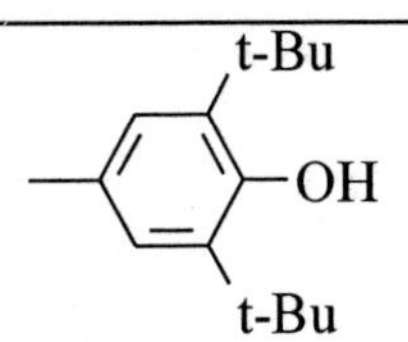
α_{max}, nm	285	300	325	344	380	398
lg ε	–	4.29	4.12	4.48	4.39	4.54

In addition, in some cases di(hydroxyphenyl)methane stabilizers can give rise to deeper violet coloring of polymers [89, 90].

The reasons of coloring of various solutions of hydrogalvinoxyl (52) are investigated in the publication [91]. Compound (52), unlike many sterically hindered phenols has sufficiently evident acid properties and forms salts easily. Negative logarithm of dissociation constant of compound (52) defined according to the data of pH-metric titration in aqueous methanol solution (80 vol%) is equal to 10.38±0.1. For the comparison, pk_a of 2,6-di-*tert*-butylphenol dissociation in methanol makes 17.08 [92]. Therefore, it can be concluded that intensive violet coloring in alkaline solutions of hydrogalvinoxyl (52) is likely to be connected with the formation of salts.

Table 3.28. Color of hydro-galvinoxyl salts solution in в various solvents

Cation	Solvent	Concentration, mole/litre	Color
Li	Acetone	$1{\cdot}10^{-1}$ - $1{\cdot}10^{-5}$	Violet
Li	Benzene	$1{\cdot}10^{-2}$ - $1{\cdot}10^{-5}$	Yellow
Li+18-crown-6	Benzene	$1{\cdot}10^{-2}$	Red [a]
K	Acetone	$1{\cdot}10^{-1}$ - $1{\cdot}10^{-5}$	Violet
K	Benzene	$1{\cdot}10^{-2}$ - $1{\cdot}10^{-5}$	Yellow
K+18-crown-6	Benzene	$1{\cdot}10^{-2}$ - $1{\cdot}10^{-4}$	Violet
Cs+2·18-crown-6	Benzene	$1{\cdot}10^{-2}$ - $1{\cdot}10^{-4}$	Violet
$N(C_2H_5)_4$	Acetone	$1{\cdot}10^{-1}$ - $1{\cdot}10^{-5}$	Violet
$N(C_2H_5)_4$	Benzene	$1{\cdot}10^{-4}$	Violet
$N(C_2H_5)_4$	Chloroform	$1{\cdot}10^{-2}$	Violet

[a] in electronic spectrum there are both maximums discussed below.

As it is seen from table 3.28, solutions of compound (52) salts can have different color depending on polarity of a solvent, the size of a cation, the absence or presence of crown ether [91]. Solutions of violet color have absorption maximum at 550-583 nm (ε = 80000 l·mole^{-1}·sm^{-1}), solutions of yellow color have absorption maximum at 390-400 nm, which coincides with the absorption maximum of hydro-galvinoxyl solution (λ=398 nm; ε = 35000 l·mole^{-1}·sm^{-1}) (figure 3.5). Unlike the latter the yellow color of solutions of hydro-galvinoxyl salts can be changed into violet by addition of a more polar solvent, for example, acetone.

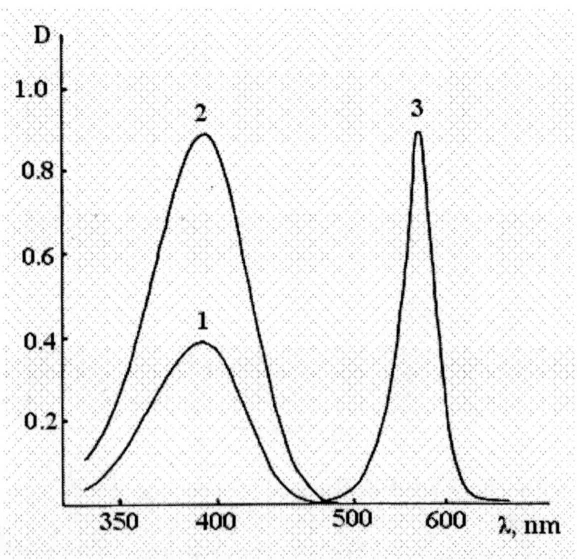

Figure 3.5. Electronic spectra of hydro-galvinoxyl solutions and its salts. 1 – compound (52) in acetone (concentration $2.5{\cdot}10^{-4}$ mole/l); 2 – potassium salt of hydro-galvinoxyl in benzene (concentration $1.2{\cdot}10^{-5}$ mole/l); 3 – tetraethylammonium salt of hydro-galvinoxyl in acetone (concentration $1.1{\cdot}10^{-4}$ mole/l).

The dependence of phenolate absorption on cation size, solution polarity and a number of other factors can be conditioned by the change of the state of an ion pair, i.e. the energy of electrostatic interaction between ions [93, Chapter 3]. In the ground state of a contact ion pair a cation of less radius interacts more efficiently with an anion, as it is situated closer to it. The redistribution of anion charge takes place at light excitation. As, according to Franck-Condon principle, at that a cation doesn't change its position, the excited state is less stabilized by a counterion. More high-frequency transition corresponds to phenolate with a cation of lesser size which has lesser ground state energy.

The interaction of an anion with a cation decreases not only with the increase of cation size, but also with the increase of solvation degree or as a result of forming the complex of a metal cation with crown ether. As a rule, the indicated factors give rise to small shift of absorption maximum of contact ion pairs [93]. Large bathochromic shifts are observed when contact ion pairs are changed into solvent-separated. In the latter a cation is practically surrounded by solvent molecules and is appreciably separated from an anion.

The large difference in the position of long-wave absorption maximum of hydro-galvinoxyl salts solutions (table 3.28) allows to conclude that there are two different types of ion pairs in these solutions. In solutions of violet color ($\lambda_{max} \approx 580$ nm) there are solvent-separated ion pairs (54) or free ions, in solutions of yellow color ($\lambda_{max} \approx 400$ nm) there are contact ion pairs (55).

$$\underset{\mathbf{55}}{A^{-}\,Kat^{+}} \rightleftharpoons \underset{\mathbf{54}}{A^{-}//\,Kat^{+}}$$,where A^{-} – hydro-galvinoxyl anion (**52**)

The qualitative difference in the state of ion pairs of yellow and violet solutions of hydrogalvinoxyl salts can be observed in spectra NMR ^{1}H. As it is seen from the comparison of data from table 3.28 and 3.29, in solutions of violet color ring protons and protons of *tert*-butyl groups, belonging to phenoxyl and cyclohexadiene rings, are indistinguishable, that can be the result of the delocalization of the negative charge of an anion along conjugation system in solvent-separated ion pairs. At the same time in solutions of yellow color in contact ion pairs the charges are compensated and nonequivalence of protons of cyclohexadiene and phenoxyl rings in spectra NMR ^{1}H is preserved.

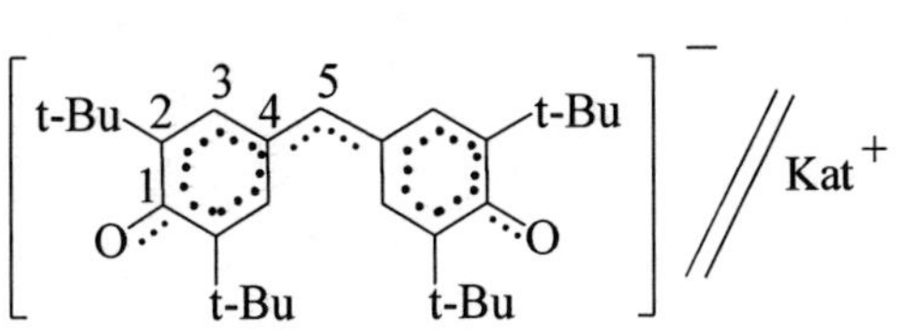

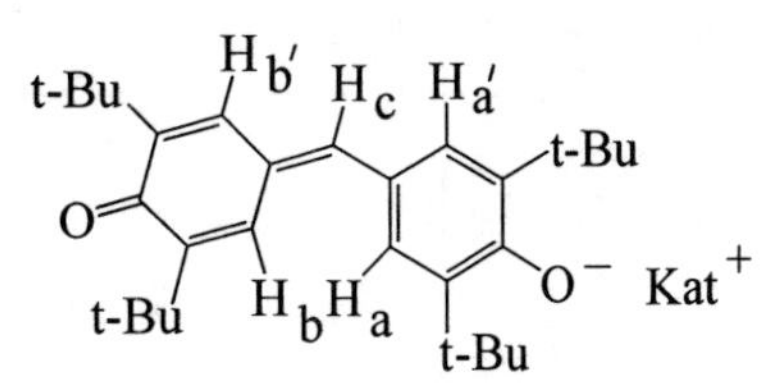

The equivalence of phenoxyl and cyclohexadiene rings of hydro-galvinoxyl salts in solutions of violet color is observed also in spectra NMR ^{13}C, where there are only 5 signals of unsaturated hydrocarbon atoms (table 3.30). There are 11 such signals in the NMR ^{13}C spectrum of hydro-galvinoxyl solution.

Table 3.29. Parameters of NMR ^{1}H spectra of hydro-galvinoxyl and its salts in various solvents

№ [a]	Cation	Solvent	Concentration, mole/l	$\delta C(CH_3)_3$, ppm	δH_a and $\delta H_{a'}$ ppm	δH_b or $\delta H_{b'}$, ppm	δH_c, ppm	$^{4}J_{H\text{-}H}$, Hz [d]
1	$N(C_2H_5)_4$	$CDCL_3$	$3 \cdot 10^{-2}$	1.37	7.23		7.00	
2	Li	acetone-d_6	0.1	1.36	7.45		7.13	
3[b]	Li	benzene-d_6	$2 \cdot 10^{-3}$	1.36; 1.46; 1.53	7.40	6.99; 7.76	6.89	2.4
4[c]	Li	benzene-d_6 + acetone-d_6 (1:1)	$1 \cdot 10^{-3}$	1.37	7.42		7.16	
5	H	benzene-d_6	$5 \cdot 10^{-3}$	1.35; 1.50; 1.57	7.42	7.01; 7.81	6.88	2.4

[a] solvents №1,2,4 have violet color, №3,5 have yellow color;
[b] the solution is obtained by dilution of solution №2 by benzene-d_6;
[c] the solution is obtained by adding acetone-d_6 to solvent №3;
[d] the constant of spin-spin interaction between protons b and b′.

Table 3.30. Parameters of NMR ^{13}C spectra of hydro-galvinoxyl salts in solutions of violet color (salt concentration is 0.1 mole/l)

Cation	Solvent	δC_1, ppm	$^{3}J_{C1\text{-}H}$, Hz	δC_2, ppm	δC_3, ppm	$^{1}J_{C3\text{-}H}$, Hz	δC_4, ppm	δC_5, ppm	$^{1}J_{C5\text{-}H}$, Hz
$N(C_2H_5)_4$	$CDCL_3$	180.0(t)	14	118.7(s)	131.3(d)	151	141.8(s)	148.6(d)	146
Li	acetone-d_6	178.1		117.1	129.7		139.4	148.2	

It should be noted that the conjugation between rings in molecules of hydro-galvinoxyl salts, conditioning their coloring (the presence of long-wave absorption maximums in electronic spectra), expects coplanar arrangement of these rings. At the same time the equivalence of aromatic protons H_a and $H_{a'}$ in NMR ^1H spectra of nonlinear molecules of hydro-galvinoxyl salts and hydro-galvinoxyl (52) itself can be explained only by the rotation of a phenoxyl ring around the bond C_{Ar} – C_{vinyl}. It is true, since the equivalence of the indicated protons disappears at temperature minus 353 K. This contradiction (between coloring hydro-galvinoxyl salts and rotation of a phenoxyl ring interrupting conjugation) is seeming, since frequencies of rotatory and vibration transitions are several orders lower than frequencies of electron transitions [94]. Concerning electron transitions the rotation is decelerated, i.e. at the time of electron transition molecules have fixed coplanar arrangement of rings, that conditions the occurrence of long-wave absorption maximums in electronic spectra and coloring of solutions of hydro-galvinoxyl salts.

X-ray structure analysis of tetraethylammonium salt of hydro-galvinoxyl (56) showed [95] that in the intensively colored crystal of this compound the bulky tetraethylammonium cations are situated in the area between the anions of hydro-galvinoxyl and don't form contact ion pairs with them (figure 3.6). It leads to flattening of anions in comparison with a neutral molecule of hydro-galvinoxyl (the angle between the planes of the rings in these compounds is equal to 14 and 19° respectively), it promotes the increase of conjugation degree between rings.

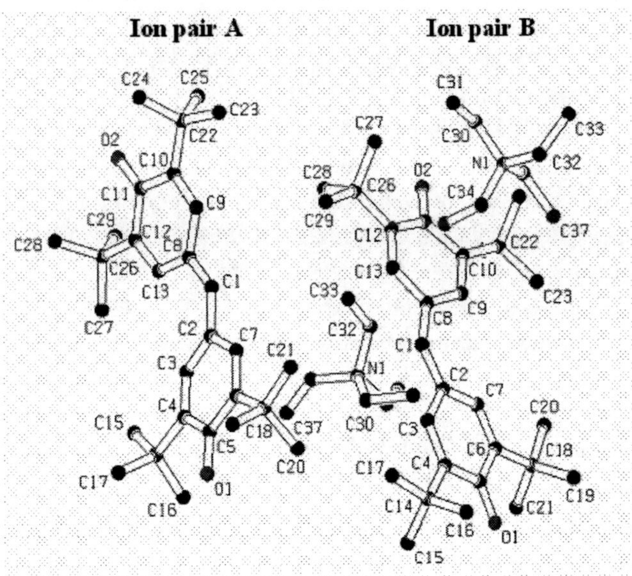

Figure 3.6. Geometry of the independent part of crystal structure of hydro-galvinoxyl tetraethylammonium salt (56).

The crystals of the complex of 18-crown-6 with potassium salt of hydro-galvinoxyl (57) have also intensive violet color. As it is seen from figure 3.7, a potassium cation in a crystal of this compound is situated opposite an oxygen atom of an anion [96]. However, seemingly crystal packaging doesn't allow to implement reasonably efficient electrostatic interaction of charges.

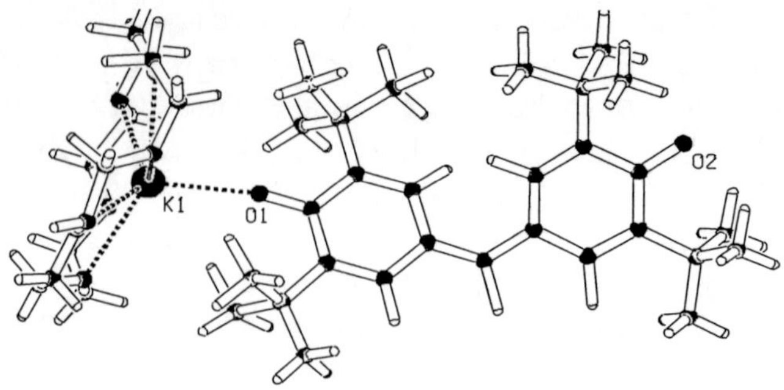

Figure 3.7. Geometry of complex of 18-crown-6 with potassium salt of hydro-galvinoxyl (57) in crystal.

Solid potassium salt of hydro-galvinoxyl has the same intensive color as the complex of this salt with crown ether that indicates the identical state of ion pairs in crystals of these compounds.

Thus, it can be stated that the occurrence of intensive violet coloring hydro-galvinoxyl solutions in the presence of bases is connected with salt formation and is conditioned by the occurrence of solvent-separated ion pairs. The factors influencing electrostatic interaction between ions, such as nature of a solvent, volume of a cation, define the color of compound (52) salt solutions. The same reason, i.e. the decrease of electrostatic interaction between ions, conditions the intensive color of solid salts of hydro-galvinoxyl. Therefore, the formation of mesomer anions leads to the occurrence of violet coloing both in solutions and in solid phase. It is important to indicate that unlike sterically hindered phenol not having electron-accepting substitutes in *para*-position, acid dissociation of phenols, which phenol ring is in conjugation with cyclohexadiene one, particularly hydro-galvinoxyl, proceeds easily and is observed already in the presence of salts of weak acids, such as sodium carbonate, sodium sulfide, sodium acetate and in base solutions. Solvents acquire intensive violet color, disappearing when acids are added.

In polymer compositions the ionization of α-hydroxyphenyl substituted methylenequinones, formed during oxidation of di(hydroxyphenyl)methane antioxidants, takes place, apparently, under the action of components of basic nature contained in these compositions which are used, specifically, in the production of rubbers [97]. Thus, for example, during the production of butyl rubber and isoprene rubber calcium stearate is used as antiagglomerator, it is introduced together with the stabilizer bis(2-hydroxy-3-*tert*-butyl-5-methylphenyl)methane (51) at the stage of degassing as alkaline suspension. The dependence of coloring of isoprene rubber, stabilized by compound (51) on the content of iron stearate formed from calcium stearate and iron salts contained in polymer is given in the publication [98].

In polyethylene compositions calcium stearate is used as a nontoxic stabilizer having lubricating properties.

The formation of methylenequinone compounds takes place in the course of inhibited oxidative process with the participation of di(hydroxyphenyl)methane stabilizers, as well as in the result of their oxidation and oxidative dehydrogenation by quinones, oxidative dehydrogenation proceeds especially easy in the presence of bases.

The same reasons condition similar coloring of other di(hydroxyphenyl)methane stabilizers, for example, 3,5-di-(3′,5′-di-*tert*-butyl-4′-hydroxybenzyl)-2,4-dihydroxybenzophenone (58).

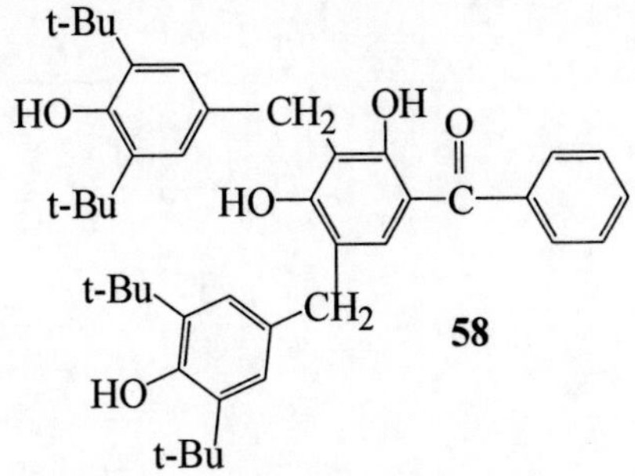

58

When it is oxidized various colored products can arise, part of them is given in Scheme 3.2. In addition, in the presence of bases colored phenolates can be formed from the compound (58) itself.

As it is seen from figure 3.8, the addition of alkali into solutions of 2,4-dihydroxybenzophenone (59) and stabilizer (58) leads to the occurrence of more intensive and longer-wave absorption maximums in their electronic spectra. The addition of acid restores the initial view of spectra, thus, these maximums can be referred to phenolates of compounds (59) and (58).

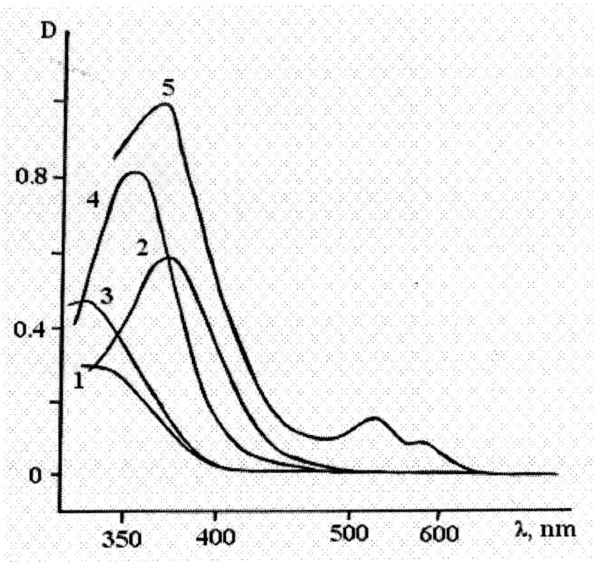

Figure 3.8. Electronic spectra of acetone solutions: 1 – benzophenone (58), ε=4650; 2 – benzophenone (58) in the presence of KOH, ε = 8380; 3 – 2,4-dihydroxybenzophenone (59); 4 – 2,4-dihydroxybenzophenone (59) in the presence of KOH; 5 – solution 2 in two days.

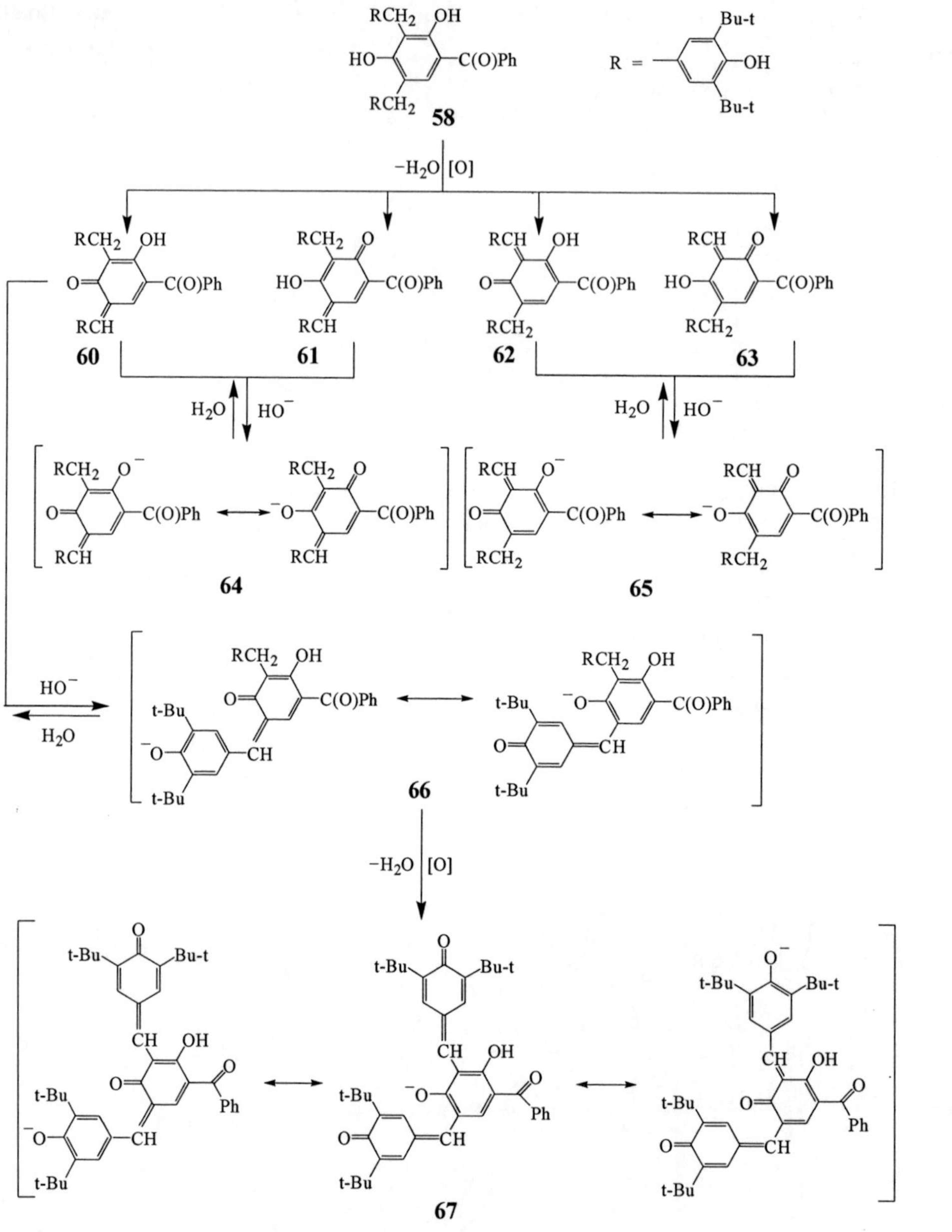

Scheme 3.2.

When compound (58) was oxidized by lead dioxide a product was obtained, the molecular weight of which conforms to compounds (60-63) according to data of chromatography–mass analysis. As it is seen from figure 3.8 and 3.9, the absorption maximums of unionized methylenequinone (60-63) in the visible region of electronic spectrum is shifted only a little into long-wave region in comparison with the corresponding maximum of phenol (58). The differences in molar coefficients of absorption of these compounds in the given region are also small. It is different when alkali is added into the solution of methylenequinone (60-63). The solution acquires dark red color and intensive absorption maximums at λ_{max}= 375 и 520 nm arise in the electronic spectrum. The absorption band at λ_{max}= 520 nm has a shoulder in the region of 580 nm. The intensity of absorption in the region of the shoulder increases when alkaline solution is kept in the air. The acidification of the solution leads to the disappearance of long-wave maximums and restoration of spectrum (1) at figure 3.9.

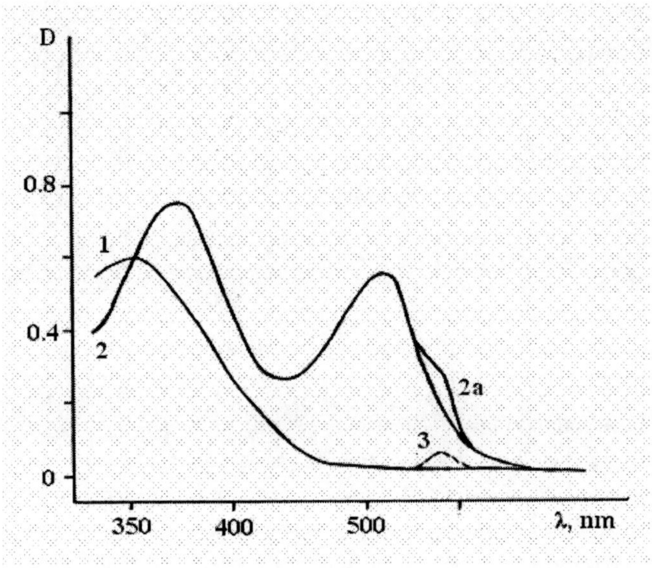

Figure 3.9. Electronic spectra of acetone solutions:1 – methylenequinones (60-63), ε = 8690; 2 – methylenequinones (60-63) in the presence of KOH, ε^{375} = 22100, ε^{520} = 16200; 2a – solution 2 after two days of keeping in the air; 3 – absorption maximum obtained as a result of subtraction of spectrum 2 from spectrum 2a.

Thus, spectral behavior of the mixture of model methylenequinones (60-63) conforms well with the conversions given in Scheme 3.2. The band at λ_{max}= 375 nm should be related to the absorption of less conjugated anions (64), (65). The intensity of absorption at 580 nm increases when alkaline solutions (58) and (60-63) are kept in the air, which is seemingly connected with the deepening of oxidative process. In connection with it should be related to the absorption of the most oxidated and the most conjugated mesomer anion (67). In this case the band at λ_{max}= 520 nm relates to the absorption of a mesomer anion (66).

Thus, coloring of chemical-resistant polymers in the presence of stabilizer (58), as in case of using antioxidants (51) and (53), arises in the result of oxidative processes proceeding first of all with the participation of these compounds. Initial products of oxidation, i.e. α-(hydroxyphenyl)substituted methylenequinones (60-63), as well as hydro-galvinoxyl (52), have relatively short-wave and low intensive absorption maximums in the visible region and, therefore, can be the reason of only weak yellow coloring. The main contribution to color generation is made by mesomer anions, in the considered case these are structures (66) and (67), formed from the oxidation products (60-63) even in the presence of weak bases, for example, calcium stearate contained in polyethylene compositions.

It should be noted, that such phenol antioxidants, the structure of which doesn't contain di(hydroxyphenyl)methane fragments, capable to form colored mesomer anions as a result of oxidative conversions in the presence of bases, are more appropriate as noncoloring stabilizers. So, methylenequinone (68), formed when polyphenol stabilizer (69) is oxidized and absorbing in ultravilolet region (table 3.27), cannot generate similar anion considering the absence of conjugation between phenol and cyclohexadiene rings.

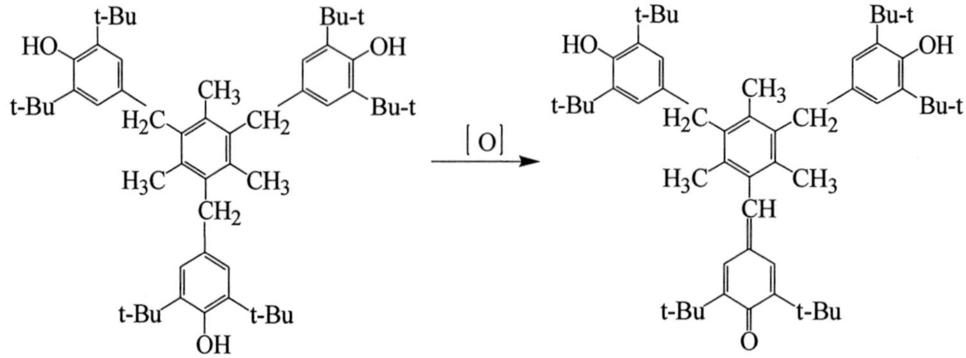

69 **68**

It is easy to be convinced that 2,4,6-tris(3′,5′-di-*tert*-butyl-4′-hydroxybenzyl)resorcin (70) is inferior to stabilizer (69) in this connection, as cyclohexadiene fragments of oxidation product of compound (70) are in conjugation with the central resorcin ring, that conditions the possibility of forming colored mesomer anions.

It should be taken into consideration that some sterically hindered phenol antioxidants can generate di(hydroxyphenyl)methane compounds as a result of oxidative transformations. In the publication [99] it is indicated that the formation of a mesomer anion is the reason of violet coloring of 2,6-di-*tert*-butyl-4-methylphenol when the latter is heated continuously in air in alkaline medium.

Thus, to solve the problem of polymer coloring, stabilized by di(hydroxyphenyl)methane antioxidants, it is necessary to prevent the accumulation of α-hydroxy phenylmethylen equinoide products in the system. Firstly, it permits to avoid yellow coloring, characteristic for these products, secondly, it excludes the possibility of forming colored mesomer anions.

3.2.2. Inhibition Reactions of Coloring in Polymer Systems Containing Compounds of Variable-Valency Metals, Phenol Antioxidants and Phosphorous Acid Esters

In practice metal compounds are contained virtually in all polymers (either as specially introduced additives or getting from the outside). In compliance with this it is necessary to consider their possible influence on the process of polymer coloring and its inhibition [100-106].

In whole, derivatives of polyvalent metals can efficiently influence on conversions of macromolecules at all stages of polymer oxidation, and, thereby, promote the formation of various products of polymers destruction [20, 23]:

$$RH + M^{n+}L_x \rightarrow M^{(n-1)+}L_{x-1} + R^{\cdot} + HL.$$

In the developing oxidation process the interaction of metals with hydroperoxides is a decisive force:

$$M^{n+1} + ROOH \rightarrow M^{n+} + RO_2^{\cdot} + H^+$$

$$M^{n+} + ROOH \rightarrow M^{n+1} + RO^{\cdot} + HO^-$$

Alternation of reactions provides the consistency of the ratio of valence forms of a catalyst in the course of the process and conditions the efficiency of catalysis.

In the result of destructive processes chromophores and auxochromes are formed, they lead to polymer coloring.

Another important consequence of the presence of polyvalent metals in the composition of polymers is the possibility of their interaction with the components of stabilizing mixture leading to the formation of products coloring polymer [107-110].

3.2.2.1. Reactions of Phenol Antioxidants with Compounds of Variable-Valency Metals

Coloring of polymer systems, containing ions of polyvalent metals, arises just after the introduction of phenol antioxidants. According to the data from table 3.31, color index of polymer (ρ) changes from 82% (natural color) for nonstabilized polyethylene, to 65% (light yellow) and 48% (light brown) for polyethylene, stabilized by 2,2'-methylene-bis(4-methyl-6-*tert*-butylphenol) (51) and 2,2'-thio-bis(4-methyl-6-α-methylbenzylphenol) (71) respectively, physical and mechanical properties retaining invariable [22].

The intensity of polymer coloring depends considerably on the nature and concentration of metals derivatives. For example, the color of polyethylene in the presence of phenol (51) changes from natural to brown (table 3.32) in the range of concentration change of metal chlorides (from 0.005 to 0.05wt%).

Table 3.31. Properties of HDPE containing phenol antioxidants (0.1wt%)

Phenol	Elongation at break, %	Yield strength, MPa	$tg\delta^{a} \cdot 10^4$	ρ^{b}, %
–	830	23	2.0	82
(51)	800	26	2.3	65
(71)	720	25	3.5	48

[a] tangent dielectric loss angle; [b] ratio of luminous flux reflected by the sample to luminous flux of the compared sample.

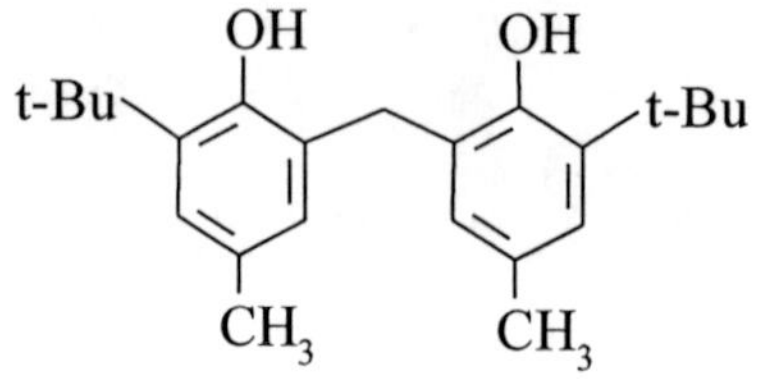

51

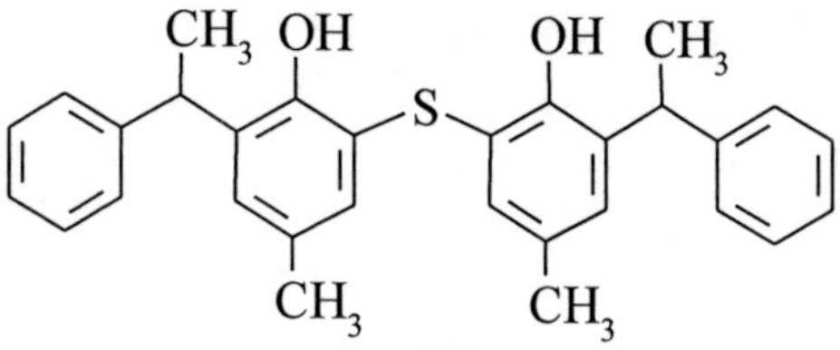

71

Table 3.32. Color of LDPE containing metal chlorides and 2,2'-methylene-bis(4-methyl-6-tert-butylphenol) (51)

Metal chloride	Content of metal chloride, wt%	Color of polyethylene			
		Without phenol (51)		In the presence of phenol (51) [a]	
		ρ [b], %	visually	ρ [b], %	visually
$VOCl_3$	0.005 0.01 0.05	79 62 58	Natural Light yellow Yellow	57 48 41	Yellow Light brown Brown
$TiCl_4$	0.005 0.01 0.05	82 71 62	Natural Natural Light yellow	62 55 44	Light yellow Yellow Brown
$CuCl_2$	0.005 0.01 0.05	83 71 60	Natural Natural Light yellow	63 45 40	Light yellow Light brown Brown
$NiCl_2$	0.005 0.01 0.05	83 80 78	Natural Natural Natural	76 72 63	Natural Natural Light yellow
$CoCl_2$	0.005 0.01 0.05	84 80 75	Natural Natural Natural	81 73 63	Natural Natural Light yellow

[a] 0.1 part by weight of phenol (51) per 100 parts by weight of polyethylene;
[b] ratio of luminous flux reflected by the sample to luminous flux of the compared sample.

As stabilized polymer films are ageing the intensification of their coloring occurs, it can be seen if LDPE is taken as an example (table 3.33).

In this section the characteristics of behavior of phenol stabilizers in polymers in the presence of polyvalent metals salts and their possible conversions under the action of various factors taking place during the thermal-oxidative degradation of polymers are considered [111-115]. The following possible variants should be taken into consideration.

On the one hand, it is the formation of colored complexes of phenols with ions of coordinated-unsaturated metals. As it follows from ultraviolet spectra, when $VOCl_3$ and $VOCl_4$ interact with phenol compounds, namely, 2,2'-thio-bis(4-methyl-6-α-methylbenzylphenol) (71) and 2,2'-methylene-bis(4-methyl-6-*tert*-butylphenol) (51) there is a shift of absorption maximum of initial bisphenol (λ_{max} 280 nm and 265 nm) and $VOCl_3$ (λ_{max} 270 nm) into visible region (λ_{max} 460 and 510 nm) [111].

Table 3.33. Color of LDPE containing 2,2'-methylene-bis(4-methyl-6-tert-butylphenol) (51) [a] and metal chlorides before and after ageing in air at 383 K

Metal chloride (0.1 wt%)	Color of polyethylene			
	Before ageing		After ageing	
	ρ [b], %	visually	ρ [b], %	visually
$VOCl_3$	48	Light brown	33	Dark brown
$TiCl_4$	55	Light brown	36	Dark brown
$FeCl_3$	51	Light brown	42	Brown
$CuCl_2$	45	Brown	40	Brown
$NiCl_2$	72	Natural	61	Light yellow
$CdCl_2$	73	Natural	60	Light yellow

[a] 0.1 part by weight of phenol (51) per 100 parts by weight of polyethylene;
[b] ratio of luminous flux reflected by the sample to luminous flux of the compared sample.

It was established by the method of isomolecular series that two molecules of bisphenol are a part of coordination sphere of metal during complex formation; the values of stability constants of complexes with phenols (51) and (71) which are equal to $(8.7\pm0.5)\times10^4$ and $(3.47\pm0.5)\times10^6$ l/mole, respectively are defined by the dilution method.

On the other hand, the formation of colored phenolates takes place [112-115], it is demonstrated by the interaction of 2,2'-methylene-bis(4-methyl-6-*tert*-butylphenol) with $TiCl_2(OC_4H_9)_2$ in equimolar ratio:

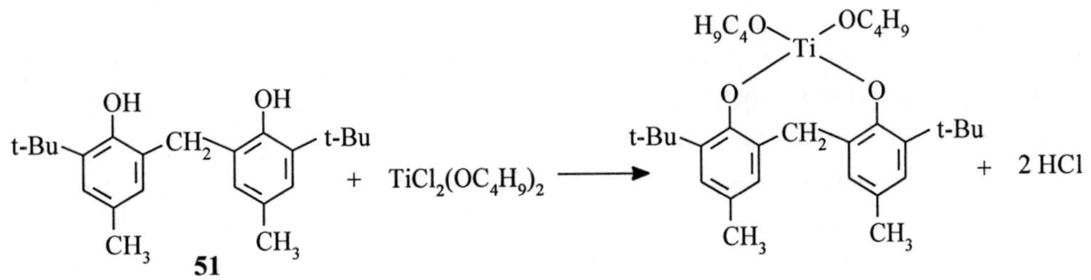

When bisphenol (51) and tetrabutoxytitanium are heated (443-448 K) the product of complete substitution of hydroxyl groups is isolated. The confirmation of this is the absence of absorption bands, which are characteristic of valence vibration of phenol OH-groups (3610 sm^{-1}–3620 sm^{-1}), in infrared spectrum, as well as the quantitative isolation of butyl alcohol and data of elemental analysis of titanium phenolate. It is important to note that the interaction of titanium derivatives and bisphenols can proceed at room temperature. Titanium phenolates obtained at room temperature and at 443-448 K are characterized by identical infrared and ultraviolet spectra [112-113].

Thus, the formation of colored products described above during the interaction of phenol antioxidants with derivatives of coordinated-unsaturated metals is one of the reasons of polymer coloring immediately after the introduction of phenol antioxidants into them.

It is important that the products of interaction of bisphenols with metal chlorides (complexes of bisphenols with $VOCl_3$ and $VOCl_4$ and titanium phenolate) are hydrolytically unstable, they convert into colorless products under ordinary conditions. Thus, 2,2'-methylene-bis(4-methyl-6-*tert*-butylphenoxy)dibutoxytitanium is hydrolyzed in solutions during standing and it decomposes practically completely at 363 K in aqueous dioxane releasing initial bisphenol (more than 80% estimating phenolate) [112]. Similar decolorization in the presence of moisture is observed for the solutions of complexes of bisphenols with vanadium chlorides. It follows from it that the formation of colorless products of hydrolysis is possible during ageing of polymers.

Since stabilization of polymers with the usage of phenol compounds takes place under rather severe conditions of thermal oxidation of polymers (the influence of oxygen and high temperatures, processing and application included), oxidative conversions of phenol antioxidants are the most practicable ones [92, 106]. Indeed, the absorption maximum (ν=480 nm) which is characteristic of quinoide structure is observed in ultraviolet spectrum of the product of oxidation (by oxygen) of hexadecane solutions of the complex $VOCl_3$ with bisphenol (51). The similar absorption maximum was identified for the oxidized product of the reaction of titanium chloride with 2,2'-thio-bis(4-methyl-*tert*-butylphenol) (72) at excess concentration of the latter in air [113].

The comparison of infrared spectra of oxidation products of bisphenol extracted from polymer and model quinone (a product of bisphenol oxidation by lead dioxide) indicated coincidence of infrared spectra in frequency region of valent vibrations (ν = 1580-1600 sm^{-1}) corresponding to quinoide structures. It is important that when bisphenol (72) is oxidized in the presence of corresponding titanium phenolate (ratio 6:1, air, 293 K, hexadecane) the formation of quinoide compounds in compliance with data of ultraviolet spectra (λ_{max} = 480 nm) takes place [113].

The kinetics of oxidation of 2,2'-methylene-bis(4-methyl-6-*tert*-butylphenol) (51) by oxygen in the presence of other polyvalent metals chlorides (Fe, Cu, Co, Ni, ratio 10:1) in hexadecane at 423, 433, 448 K is studied. Regardless of temperature conditions the catalytic action of diverse variable-valency metals in the process of bisphenol oxidation (table 3.34) [117, 118] is found, the degree of this action depends on the nature of variable-valency metals. The values of activation energy indicate that it is considerably higher in case of bisphenol oxidation individually (84 kJ/mole) in comparison with the values of activation energy for reactions in the presence of titanium salts (21 kJ/mole). The observed conversions of phenol antioxidant are followed by the change of frequencies of valent vibrations в infrared spectra in the region of 3640 sm^{-1} of an unbound hydroxyl group, simultaneously the arising of absorption bands is observed in the region of 1600 and 1580 sm^{-1} (in infrared spectrum) and 480 nm (in ultraviolet spectrum), these absorption bands are characteristic of quinoide structures [116].

Table 3.34. Rate constants of oxidation of 2,2'-methylene-bis(4-methyl-6-tert-butylphenol) (51) in hexadecane in the presence of metal chlorides

Metal chloride	$k \times 10^5$, s^{-1}	Metal chloride	$k \times 10^5$, s^{-1}
–	0.4[a] 0.7[b] 1.5[c]	$CuCl_2$	1.3[a]
$TiCl_4$	4.2[a] 4.8[b] 5.8[c]	$FeCl_3$	2.8[a]
$NiCl_2$	0.8[a]	$VOCl_3$	4.9[a]

Oxidation temperature: [a] 423 K, [b] 433 K, [c] 448 K.

From the above-stated it follows that coloring of polymers containing admixtures of derivatives of variable-valency metals inhibited by phenol antioxidants can be aroused by various reasons. At the initial stages of inhibited ageing of polymer systems the formation of colored products by direct interaction of phenol antioxidants with ions of coordinated-unsaturated metals (complexes of metal phenolate) is possible, in the course of stabilization they can undergo further transformations, namely, hydrolytic (with the formation of colorless products) and oxidative (with the formation of colored products). Coloring of polymers stabilized by phenol compounds in the course of their ageing is conditioned mainly by the formation of quinoide structures as a result of oxidation of phenol antioxidants catalyzed by variable-valency metals. The data of figure 3.10 illustrate it, the dependence of degree of LDPE coloring on the nature of introduced additives, namely, bisphenol and products of its conversions under the action of titanium salts (titanate) and model quinone is shown. It is in the presence of quinoide compounds that the maximum coloring of LDPE (ten-point color scale) when heated in the air is observed.

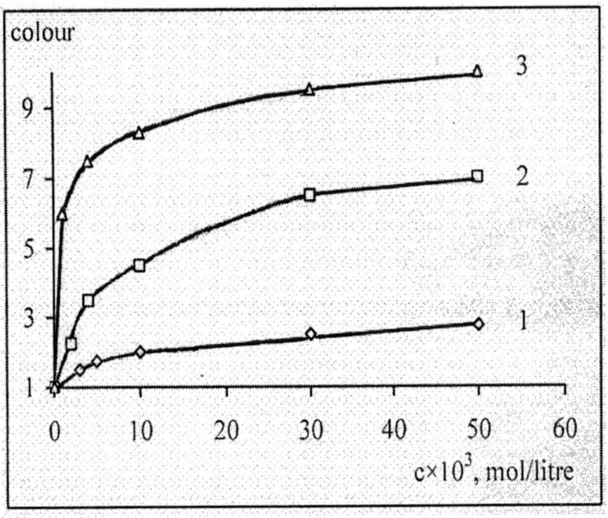

Figure 3.10. Dependence of the intensity of LDPE coloring on the concentration of an additive : 1 – 2,2'-methylene-bis(4-methyl-6-tert-butylphenol) (51);

2 –

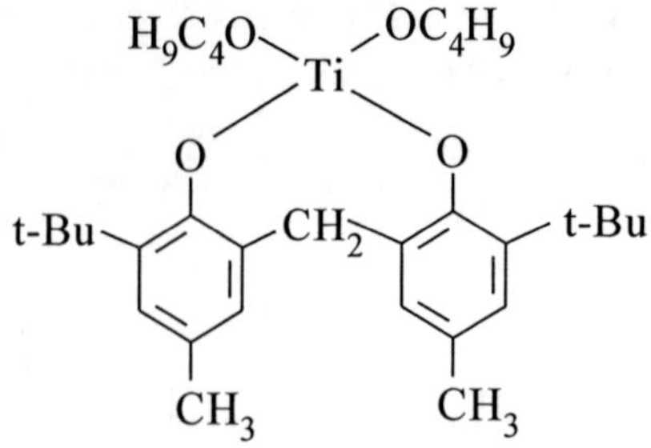

3 – model quinone (a product of bisphenol (51) oxidation).

3.2.2.1. Reactions of Phosphorous Acid Esters with Compounds of Variable-Valency Metals

As it is given in the previous section, ions of variable-valency metals initiate coloring of polymers stabilized by phenol antioxidants. Proceeding from that, their deactivation, for example, binding in the form of metal complexes with phosphites [119-121], is an efficient method of color stabilization [122-124].

Indeed, the investigation of model reactions of phosphorous acid esters (triethyl-, tributyl-, tri(2-ethylhexyl)-, tri(2-ethylhexyldiphenyl)-, tri(4-methyl-2-*tert*-butylphenyl)-, di(2-ethylhexyl)-phenyl- and tri(4-nonylphenyl)phosphites) indicated their high activity in relation to derivatives of a number of metals, particularly, titanium and vanadium – VCl_4, $TiCl_4$, $VOCl_3$, $TiCl_2(OR)_2$ [111, 121]. These reactions proceed quickly even at 275-285 K, it is illustrated by the data of table 3.35. The shift of absorption band P-O-C to 990 sm^{-1} (1050 sm^{-1} for free phosphite (i-$C_8H_{17}O$) POC_6H_5) and simultaneous decrease of its intensity indicates the presence of the *metal-phosphorus* bond and the consumption of phosphite in the course of a reaction. Electronic spectra are characterized by the shift of absorption maximum into visible region in comparison with individual phosphites ($\lambda \approx$ 260-275 nm for individual phosphites).

Table 3.35. Kinetic parameters of reactions of complexing $VOCl_3$ with $(i\text{-}C_8H_{17}O)_2POC_6H_5$

Temperature, K	k×10, m^3/mole·s	E, kJ/mole
275	1.5	
280	3.2	48.4
285	4.5	

Phosphites (1), (2) and (49) interact with metal salts, particularly, with $VOCl_3$ and VCl_4, slower and require additional heating (to 373 K).

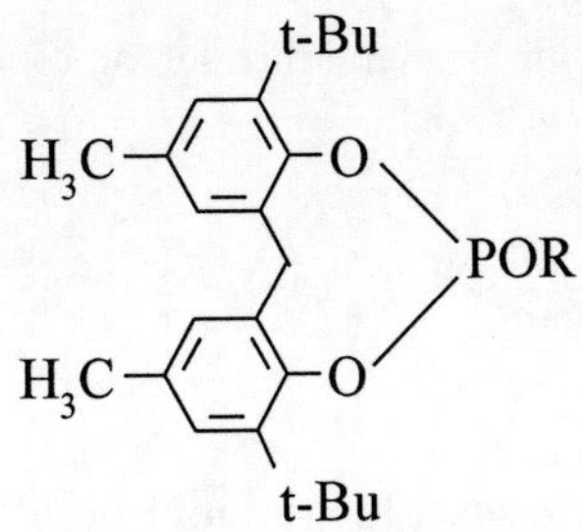

$R = C_2H_5$ (**1**), C_6H_5 (**2**); $C_{10}H_{17}$-α (**49**)

In whole, ultraviolet-, infrared and EPR-spectral characteristics confirm donor-acceptor interaction of a three-valence phosphorus atom in phosphorous acid esters with a metal atom and, thus, the possibility of the formation of phosphite complexes in polymer systems, containing derivatives of coordinated-unsaturated metals.

It should be noted that the possibility of phosphites to inhibit coloring of polymers has several essential specific features. The thing is that the phosphite complexes, obtained without the access of oxygen and moisture, have intensive color. Thus, the investigation of LDPE, stabilized and nonstabilized by phosphite, containing admixtures of compounds of coordinated-unsaturated metals, immediately after their obtaining under inert conditions demonstrated that polymer without phosphite has natural color (ρ = 72%), in the presence of phosphite polyethylene plates are intensively colored (ρ = 52%) (table 3.36).

During the atmospheric thermal exposure decolorizing of LDPE stabilized by phosphite occurs, that indicates further conversions of colored phosphite metal complexes in polymer systems under ambient conditions, moreover, the coordination with transition metals can lead to sharp strengthening of reactivity of phosphite-ligand in relation to various agents [125].

Table 3.36. Color of LDPE inhibited by phosphites after atmospheric heat treatment (10 min, 438 K)

Content of vanadium, wt%	Color of polyethylene, ρ[a] (%)					
	Without phosphite	$(C_6H_5O)_3P$		Phosphite (1)		$(i\text{-}C_8H_{17}O)_3P$
		0.3 wt%	0.8 wt %	0.3 wt %	0.8 wt %	0.1 wt %
0.03	55	71	77	67	71	
0.06	51	68	71	61	68	
0.10	42	58	67	58	65	
0.16	41					81
0.30	40	51	62	54	60	

[a] ρ is ratio of luminous flux reflected by the sample to luminous flux of the compared sample.

In this connection the factors, potencially capable of influencing the stability of formed colored compounds of phosphites with derivatives of variable-valency metals, such as the influence of oxygen, moisture, temperature, etc are considered.

When thermal action on complexes of phosphites $(C_4H_9O)_3P$, $(i\text{-}C_8H_{17}O)_3P$ and $(i\text{-}C_8H_{17}O)_2P(OC_6H_5)$ with derivatives of titanium and vanadium by the method of infrared spectroscopy was investigated it was shown, that there is intracomplex isomerization of phosphite into phosphonate with the following coordination of metal through an oxygen atom of a phosphoryl group:

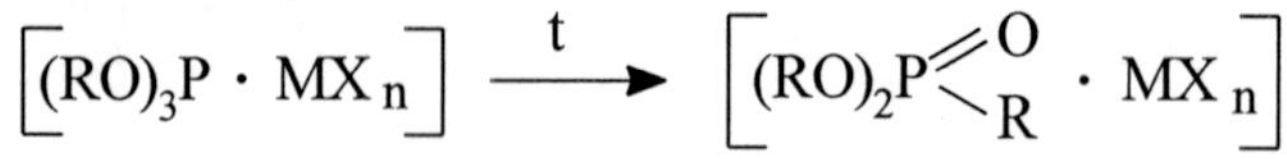

For complex of $VOCl_3$ with $(i\text{-}C_8H_{17}O)_3P$ an intensive absorption band is observed in the region of 1200 sm^{-1}. The occurrence of absorption bands in the region of 1115 sm^{-1} instead of 1250 sm^{-1} in spectra of free $(C_4H_9O)_2P(O)C_4H_9$ indicates the formation of complexes with the participation of an oxygen atom of the phosphoryl group. The investigations of conversions of complexes of tributylphosphite with chlorides of titanium and vanadium under heating showed the increase of temperature to 403-413 K leads to the reactions of condensation with the elimination, respectively, of one or two molecules of butyl chloride with the formation of compounds with Ti-O-P-bond. The occurrence of wide intensive absorption band in the region of 1050 sm^{-1} in infrared spectrum testifies to it [116]. On the basis of obtained results the interaction of phosphites with derivatives of titanium and vanadium can be demonstrated by the following scheme [124]:

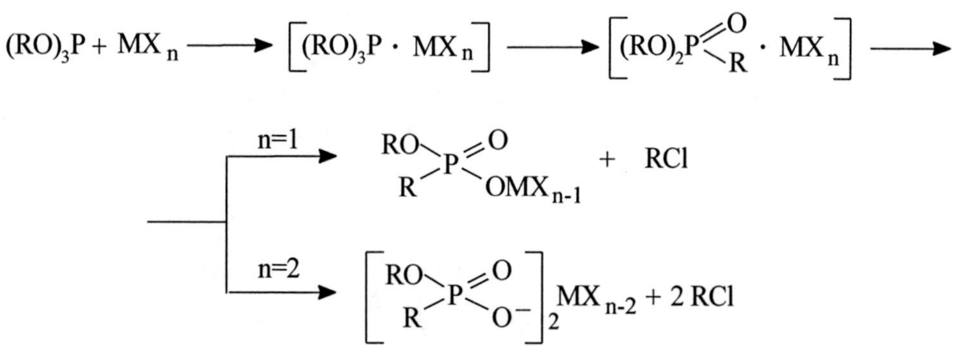

It is essential that isomerized complexes and end products of thermal transformations of complexes are intensively colored, i.e. thermal transformations of metal phosphite complexes cannot be the reason of the formation of color stabilizing effect in polymers.

Under atmospheric ageing of polymer materials the main active agents influencing conversions of phosphite complexes with ions of coordinated-unsaturated metals in the composition of polymer can be oxygen and water.

It is shown that the action of oxygen on colored solutions of complexes of phosphites with $VOCl_3$ doesn't result in the change of their color, while during their exposure to the air the decolorizing takes place. The introduction of water into colored solutions of complexes of phosphites with $VOCl_3$ leads to decolorizing, apparently, as a result of hydrolysis of the complex (figure 3.11). The considerable decrease of absorption is observed in the visible region of ultraviolet-spectrum. The occurrence of intensive wide band in the region of 2600-3600 sm^{-1} is characterized for vibration of P-OH bond [126]. Reactions of complex saponification are characterized by high rates, it is illustrated by the data of table 3.37.

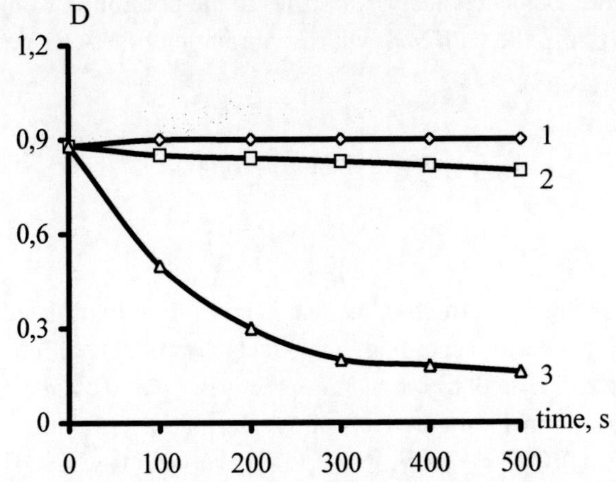

Figure 3.11. Kinetic curves of the color change of complex $VOCl_3$ with $(i\text{-}C_8H_{17}O)_3P$ in the solution of hexane under the influence of oxygen (curve 1), temperature 343 K (curve 2) and water (curve 3).

Table 3.37. Kinetic parameters of saponification of complex of $(C_4H_5O)_2P(OC_8H_{17}\text{-}i)$ with $VOCl_3$

Temperature, K	k, s^{-1}	E, kJ/mole
275	15.1	
280	26.6	48.5
285	31.7	

The saponification of coordinated phosphite can be illustrated by the scheme:

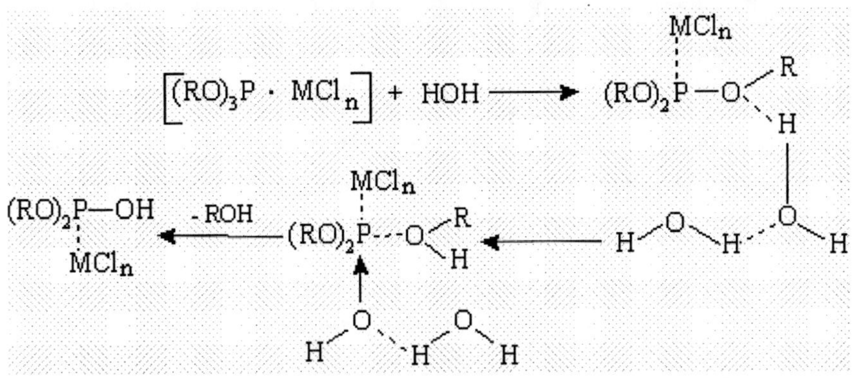

Dibutylphosphorous acid is identified as a product of hydrolysis of the complex of tributylphosphite with $VOCl_3$ by the method of gas-liquid chromatography.

On the other hand, elemental analysis testifies to the possibility of hydrolysis of products of the interaction of phosphite with MX_n with the formation of M-OH-groups, for example:

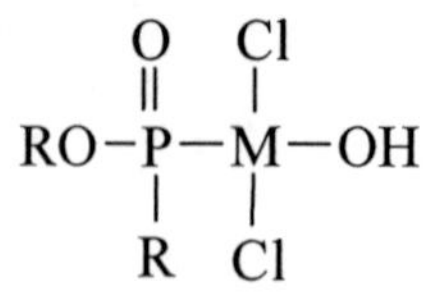

Apparently, it is with the formation of such colored little hydrolized compounds that the experimental fact is connected, according to this fact the products of interaction of phosphites with metals derivatives, particularly, titanium, practically don't color polymer. For example, color of LDPE samples containing products of transformations of phosphite metal complexes, such as $[(C_4H_9O)(C_4H_9)(O)PO]_2 \cdot TiCl_2$, $[(C_4H_9O)(C_4H_9)(O)PO]_2 \cdot TiClOH$ and $(C_4H_9O)(C_4H_9)(O)POTiCl_2OH$, is estimated at 2-3 points (according to ten-point color scale).

Thus, the inhibition (by phosphites) of color of polymers containing salts of coordinated-unsaturated metals under conditions of atmospheric ageing is conditioned by the influence of air moisture on conversions of colored metal complexes of phosphites and products of their thermal destruction, formed in polymer, into colorless compounds. It is even more important that ordinary ways of additives introduction used in practice don't provide processing of polymer under inert conditions. In whole, neutralization of ions of coordinated unsaturated metals by phosphites during thermal oxidation and destruction of polymers is one of the typical ways of their color stabilizing action [122, 124].

3.2.2.3. Competitive Reactions of Phenol Antioxidants and Phosphorous Acid Esters with Compounds of Variable-Valency Metals

As it was demonstrated above, organic phosphites can decrease or prevent coloring polymers, containing polyvalent metals, including ones in the processes of ageing inhibited by phenol antioxidants. The result of color stabilization by phosphites in such systems depends on the nature of metal, the structure of phosphites and phenol antioxidants and their concentration ratio.

Thus, trialkyl-, alkylaryl- and sterically unhindered triarylphosphites are the most efficient. For example, the color of polyethylene samples containing compounds of vanadium and titanium and 2,2'-methylenebis(4-methyl-6-*tert*-butylphenol) (**51**), when such phosphites are introduced is estimated at the range of from 1 to 3 points according to ten-point scale, while for the samples with sterically hindered aromatic phosphites results of color stabilization are much worse (from 5 to 6 points). The comparison of results for mixtures of phosphites with phenol antioxidants 2,2'-thio-bis(4-methyl-6-α-methyl-benzylphenol) (**71**) and 2,2'-methylene-bis(4-methyl-6-*tert*-butylphenol) (**51**) indicates greater efficiency in the color stabilization of the first composition.

The comparison of synergistic effects of phenols and phosphites according to thermal and color stability of polymers indicated that maximums of these factors are in different fields of phosphites content in compositions. For the optimal value of synergism according to color it is necessary to have excess of phosphite concentration of in comparison with the concentration of phenol antioxidant.

On analyzing the behavior of phosphites and phenol antioxidants in processes of inhibited ageing of polymers in the presence of admixtures of coordinated-unsaturated metals a number of regularities were determined.

The formation of primary products as a result of the direct interaction of stabilizers with ions of polyvalent metals is common for both types of stabilizers, intensive color of these primary products conditions coloring of polymer. These products are unstable under atmospheric ageing. However, the reasons leading to their conversions and the nature of products are different for phosphites and phenols.

As it was shown in the previous sections, in case of phosphite complexes and products of their thermal conversions the hydrolysis under the influence of environmental moisture alongside with the formation of colorless products is deciding in the process of their decomposition. Their inertness to oxidation should be noted. Phenol compounds are able to form complexes with variable-valency metals, but subsequent reactions of oxidation with the formation of strong chromophors, namely, quinoid structures are characteristic of them.

It follows that in case of polymer systems inhibited by binary mixtures of phosphites with phenols in the presence of variable-valency metals the common effect of color stabilization will depend on competitive ability of components in binding metal ions.

The displacement of phenols by phosphites from their complexes with $VOCl_3$ was detected by methods of infrared and electron spin resonance spectroscopy (ESR) [111, 123]. Thus, when phosphite $(i\text{-}C_8H_{17}O)_3P$ was introduced into the solution of complex $VOCl_3$ with 2,2'-methylenebis(4-methyl-6-*tert*-butylphenol) (**51**) the intensive band in the region of 3400 sm^{-1} was observed in infrared spectra of reaction mixture, that indicates the formation of free phenol [111]. It was determined by the ESR spectroscopy that the addition of phosphite into diamagnetic complex of bisphenol with $VOCl_3$ leads to the occurrence of the signal, which is typical for phosphite complex.

Spectrophotometric investigation of equilibrium of systems *$VOCl_3$: bisphenol (51) : $(i\text{-}C_8H_{17}O)_3P$* from 1:1:0 to 1:1:5 (that is at different relative excess of phosphite) indicated that there is chemical equilibrium at commensurable concentrations of bisphenol (L') and phosphite (L"):

$$VOCl_3 + L' + L'' \leftrightarrow VOCl_3 \cdot L' + L'' \leftrightarrow VOCl_3 \cdot L'' + L'$$

If there is excess of phosphite the equilibrium is shifted into the formation of phosphite complex, the trend of changing curve of optical density of systems at $\lambda = 470$ nm is similar to the trend of absorption curve of phosphite complex. Thus, in triple systems when there is excess of phosphite the phosphite-metal complexes are mainly formed or the displacement of bisphenol by phosphite in the metal complex takes place.

On the basis of the model experiment the behavior of triple compositions *metal – phosphite – phenol antioxidant* in polymer system (under process conditions) can be explained. The simultaneous introduction of phosphite together with phenol antioxidants (during the completion of polymerization process) results in the weakening of polymer color in comparison with color of polymer cleaned by alcohol and stabilized only by phenol antioxidant (table 3.38) [123]. Really, the effect of color stabilization is achieved at the phosphite concentration exceeding the bisphenol concentration 5-10 times.

When the competitive reaction of bisphenol with metal derivatives is excluded, i.e. at successive introduction of stabilizing additives (initially of phosphite, which is introduced at the stage of polymer drying, then of phenol antioxidant, which is introduced into dried polymer) the effect of color stabilizing action of phosphites increases considerably (table 3.39). In this case the maximum effect of color preserving can be achieved at much less concentrations of phosphite (0.3-0.5 wt%), than at the content of phosphite (0.5-1.0 wt%) in polymer when they are introduced simultaneously with phenols.

Table 3.38. The influence of phosphite (1) concentration on the color of HDPE inhibited by bisphenols

Phenol (0.1 wt %)	Content of phosphite, wt%	Color of polyethylene	
		ρ [a], %	Visually
(51)	–	66	Light yellow
	0.3	69	Cream
	0.5	77	Natural
	0.7	81	Natural
(71)	–	48	Lilac
	0.5	54	Light lilac
	0.7	58	Light lilac
	1.0	69	Natural

[a] ratio of luminous flux reflected by the sample to luminous flux of the compared sample.

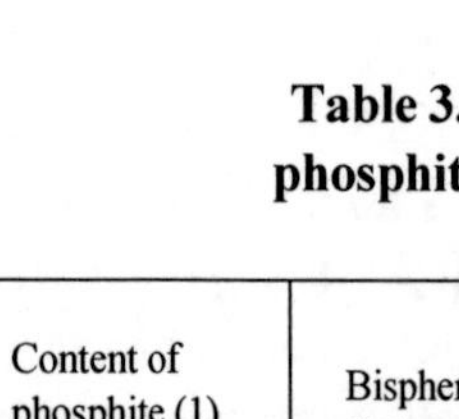
1

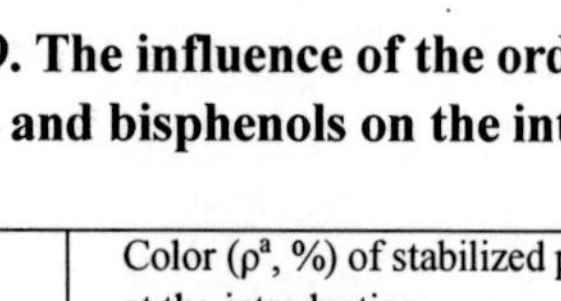
51

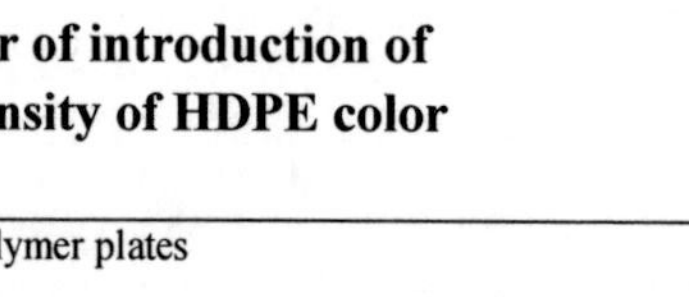
71

Table 3.39. The influence of the order of introduction of phosphites and bisphenols on the intensity of HDPE color

Content of phosphite (1) (wt %)	Bisphenol (0.1 wt %)	Color (ρ[a], %) of stabilized polymer plates at the introduction:		
		Phosphite before drying, bisphenol after drying	Phosphite and bisphenol before drying	Phosphite and bisphenol after drying
0.3	(71)	58	49	48
0.5	–	70	58	54
0.3	(51)	76	73	69
0.5	–	82	80	77

[a] ratio of luminous flux reflected by the sample to luminous flux of the compared sample.

Thus, in the triple system *phosphite – phenol antioxidant – variable-valency metal* the predominant role of phosphites is determined in the binding of polyvalent metals into colorless products.

3.2.2.4. Reactions with Products of Oxidative Conversions of Phenol Antioxidants

In the previous sections it was shown that when phenol antioxidants are oxidized, in the presence of variable-valency metals included, the main products are quinoid structures of various types. It follows that one of the variants of obtaining noncoloring stabilizing compositions on the basis of di(hydroxyphenyl)methane stabilizers can be the conversion of formed α-hydroxyphenylmethylenequinone into colorless products. It is known that sterically hindered methylenequinones have high reactivity according to nucleophilic agents [127] and form easily colorless additive compounds. Thus, for example, the chemistry of 3,5,3′,5′-tetra-*tert*-butyltoluylenequinone (73) is diverse in this regard [128].

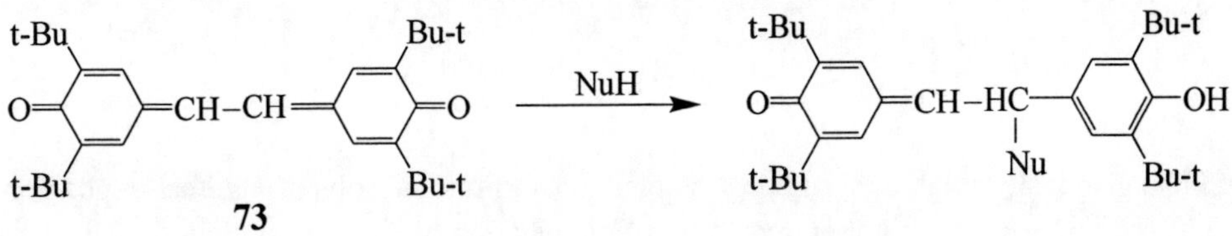

Thus, it would seem that binding of α-hydroxyphenyl substituted methylenequinones, products of oxidation of di(hydroxyphenyl)methane stabilizers (Section 3.2.1) wasn't a challenge. Commercially available alcohols, amines, mercaptane etc. could be used for this purpose.

However, when hydro-galvinoxyl (52) was used as model α-hydroxyphenyl substituted methylenequinone it was determined that long-term holding at 293 K, as well as heating its solutions with such compounds as mono- and diethanolamines, diphenylguanidine, diethyl hydroxylamine didn't produce any results. When solutions of compound (52) are dissolved in ethanol after a while the formation of a certain amount of a new product is registered by the method of thin-layer chromatography, but further standing of the solution, its heating and acidification doesn't result in exhausting of hydro-galvinoxyl (52).

Individual products of addition of water and morpholine were obtained as a result of the reaction of hydro-galvinoxyl (52) with the excess of aqua ammonia and morpholine [129].

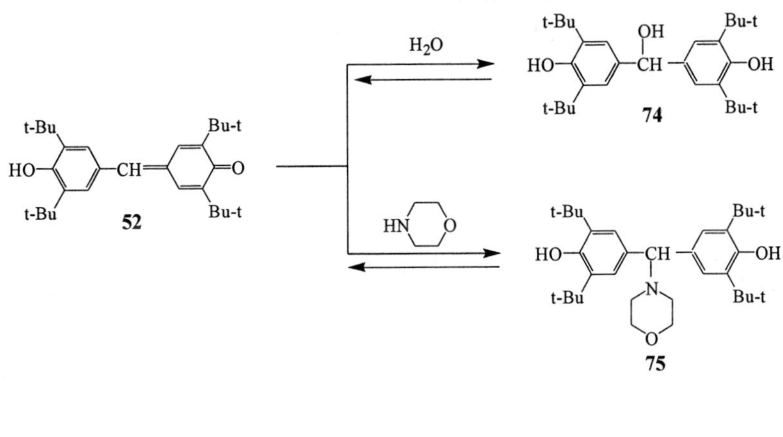

Products (74) and (75) are stable in crystalline state, but in solutions of chloroform, benzene, alcohol, acetone, DMSO they decompose more or less quickly forming hydro-galvinoxyl (52). This process proceeds more intensively in polar solvents and when acid is added.

There is a publication [130] where the process of reversible dissociation in protonic solvents of substituted di(hydroxyphenyl)methane derivatives is described, one of hydroxyl groups of these derivatives is in *ortho*–position according to a methine carbon atom. The authors consider that the dissociation takes place as a result of the concerted process of proton transfer with the participation of a solvent molecule.

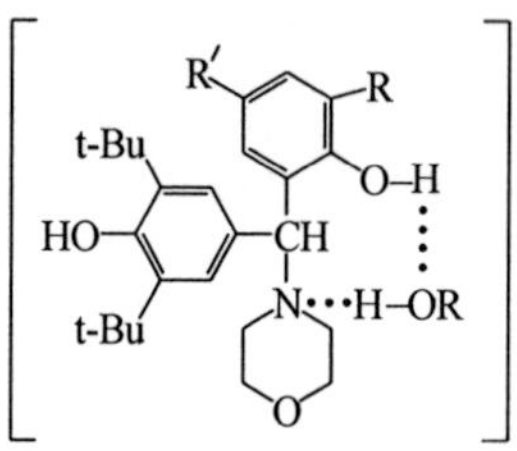

The given above data indicate that the presence of protonic solvents and *ortho*-position of one of hydroxyl groups is not prerequisites of dissociation of similar compounds. The motive force of the discussed process is likely to be high stability of formed di(hydroxyphenyl)methyl carbcation (A) as an intermediate, it conditions the proceeding of dissociation even in solvents, incapable of specific solvation.

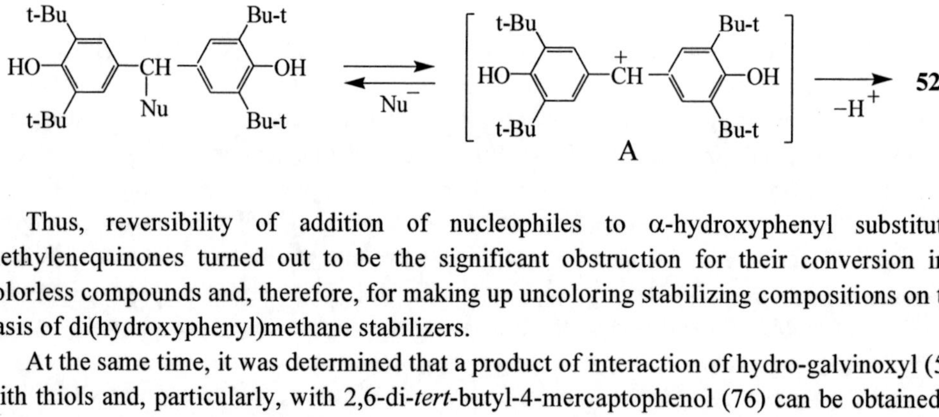

Thus, reversibility of addition of nucleophiles to α-hydroxyphenyl substituted methylenequinones turned out to be the significant obstruction for their conversion into colorless compounds and, therefore, for making up uncoloring stabilizing compositions on the basis of di(hydroxyphenyl)methane stabilizers.

At the same time, it was determined that a product of interaction of hydro-galvinoxyl (52) with thiols and, particularly, with 2,6-di-*tert*-butyl-4-mercaptophenol (76) can be obtained at equimolar ratio of reagents [129].

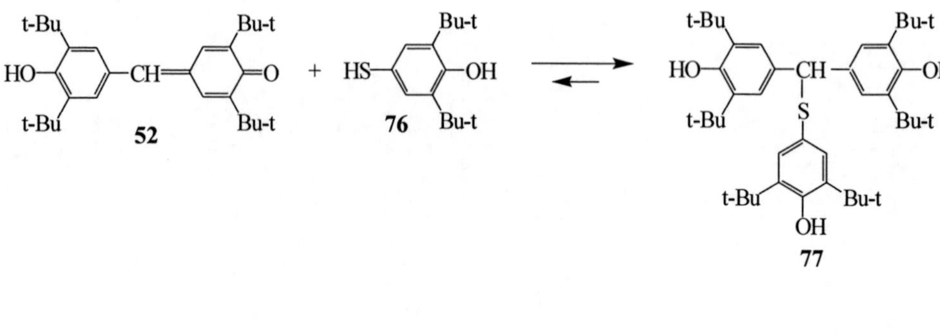

Partial dissociation in solutions up to hydro-galvinoxyl (52) and mercaptophenol (76) is characteristic of sulfide (77), but in solid phase compound (77) is thermally stable up to melting temperature (465 K). It allows to use mercaptophenol (76) as a component of noncoloring stabilizing compositions on the basis of di(hydroxyphenyl)methane stabilizers. α-Hydroxyphenylmethylenequinoid compounds formed from these stabilizers during inhibited oxidation interact with mercaptophenol (76) leading to the formation of products which are colorless and stable during the processing and service performance of polymer melt. Thus, for example, when low pressure polyethylene stabilized by 2,4,6-tris(3′,5′-di-*tert*-butyl-4′-hydroxybenzyl)resorcin (70) is processed by extrusion at 463 K, the coloring of polymer takes place. At the same time using the stabilizing composition consisting of compounds (70) and (76) allows to preserve the initial color of polymer in the extrusion process (table 3.40).

Table 3.40. Color stability of HDPE in the presence of stabilizers

Stabilizer	Content, wt%	Colority [a] after extrusion, points
(70)	0.1	3
(70)+(76)	0.05+0.05	1

[a] according to 10-point color scale.

As it is mentioned above, at present phosphorous acid esters are synergists of phenol antioxidants which are practically used for the stabilization of polymers. To identify the realistic possibility of binding quinoid compounds by phosphites in polymer the interaction of phosphites with model quinone (a product of oxidation of 2,2'-methylenebis(4-methyl-6-*tert*-butylphenol) (51) by lead dioxide) was studied.

Spectrophotometric investigation of these reactions indicated that aliphatic phosphites and phosphites, unsubstituted in an aromatic ring, and alkyl-aryl esters of phosphorous acid improved the color of LDPE better in comparison to alkylated aromatic esters of phosphorous acid (table 3.41).

Table 3.41. Influence of phosphites on the color of LDPE in the presence of 0.015 mole/kg of model quinone (the product of oxidation of 2,2'-methylenebis(4-methyl-6-tert-butylphenol (51)

Phosphite (0.18 mole/kg)	Color of polymer (473 K, air, 30 min)	
	ρ^a, %	visually
–	32	Dark brown
$(C_6H_{13}O)_3P$	80	Natural
$(C_8H_{17}O)_3P$	72	Natural
$(C_{10}H_{21}O)_3P$	72	Natural
$(C_6H_5O)_2\ POC_8H_{17\text{-}i}$	76	Natural

$(\alpha\text{-}C_{10}H_7O)_3P$	74	Natural
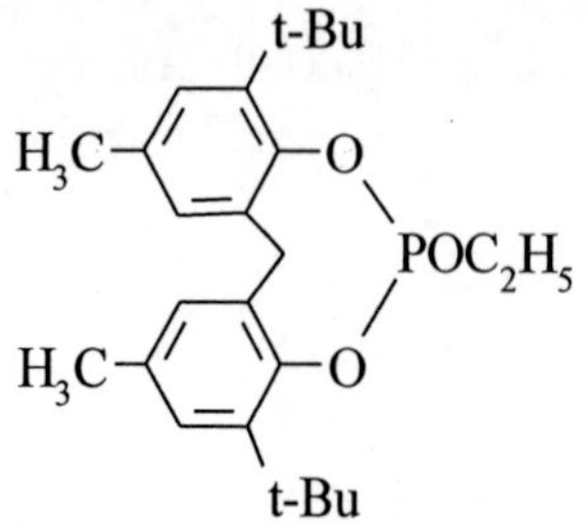	58	Yellow
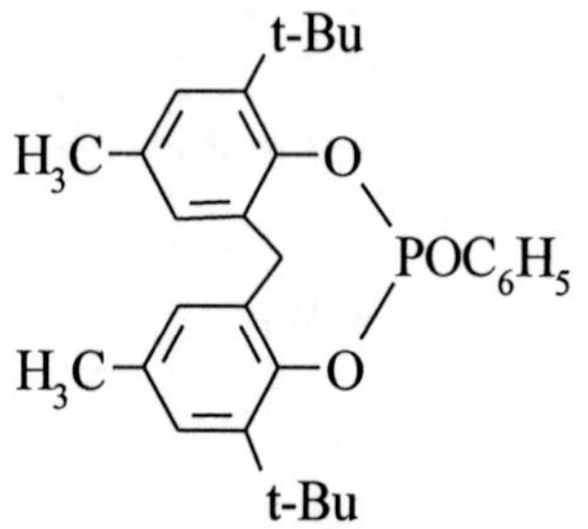	40	Brown
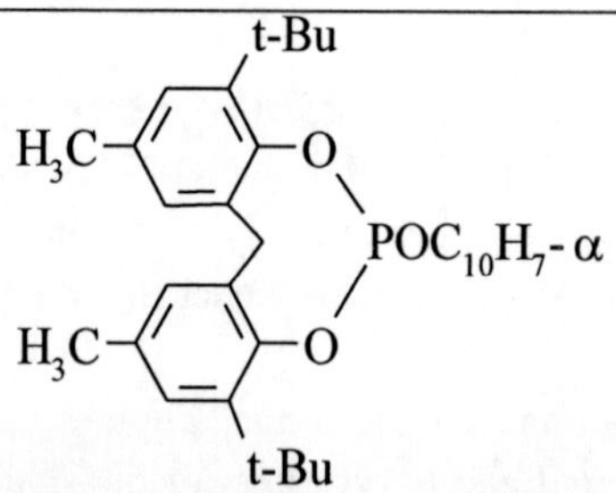	43	Brown

[a] ratio of luminous flux reflected by the sample to luminous flux of the compared sample.

Quantitative assessment of reactivity of phosphites in relation to quinoid structures was conducted by ESR method, reactions with chloranil were taken as an example.

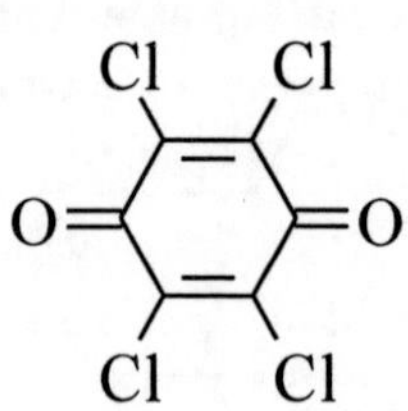

Kinetic parameters of reactions of chloranil with phosphites are given in table 3.42.

Table 3.42. Kinetic parameters of reactions of phosphites with chloranil

Phosphite	Temperature, K	k×10^3, s^{-1}	E_{act}, kJ/mole
$(C_4H_9O)_3P$	290 299 308	2.0 2.7 3.5	22.2
$(C_6H_{13}O)_3P$	290 299 313	0.9 1.2 2.1	29.1
$(i\text{-}C_9H_{17}O)_3P$	333 343 348	0.5 0.8 1.1	50.4
$(i\text{-}C_9H_{17}O)_2POC_6H_5$	313 333 353	0.3 0.8 1.7	34.6
$i\text{-}C_8H_{17}OP(OC_6H_5)_2$	313 333 353	0.1 0.4 0.9	42.3

As it follows from table 3.42, tributylphosphite and trihexylphosphite react easily with chloranil even at room temperature, tri-*i*-octylphosphite react when heated up to 333 K. In case of alkylaromatic phosphites, the reaction is initiated by photo radiation treatment along with heating. Triphenylphosphite and aromatic cyclic phosphites don't react with chloranil even under such severe conditions.

In infrared spectra of the reaction mixture of model quinone with trinaphthylphosphite there is sharp decrease of intensity in the region of absorption of carbonyl groups, which is characteristic of model quinone (1580 sm $^{-1}$), and appearance of new band in the region of absorption of phosphoryl groups (1280 sm^{-1}) [116].

On the whole, reactivity of phosphites in relation to quinoid structures (chloranil and model quinone) depending on the nature of esteric radical of phosphites and the degree of screening of a phosphorus atom in them agrees well with activity of phosphites at the inhibition of color of polymer, containing products of phenol antioxidant oxidation, and can be represented by the following scheme:

$(AlkO)_3P > (AlkO)_2POAr > AlkOP(OAr)_2 > (ArO)_3P >$

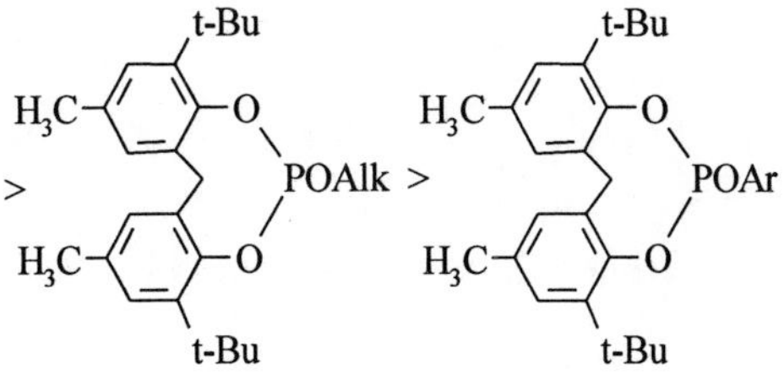

Thus, the functions of phosphorous acid esters during the color stabilization of complex polymer systems, containing phenol antioxidants and ions of coordinated-unsaturated metals were developed. The determined reactions of phosphites are regarded to be the type of chemical deactivation of admixtures and products of conversions of additives introduced into polymer. They are conditioned, on the one hand, by the ability of phosphorous acid esters to bind coordinated-unsaturated metals, changing them into inactive and uncolored products; on the other hand they are conditioned by phosphite binding of intensively colored quinoid products of oxidation of phenol antioxidants into colorless compounds.

* * *

The specific reactions and the unique properties of polyfunctional stabilizers responsible for the stabilization of polymer color under the conditions of their ageing have been illustrated by the organic phosphites.

In this connection it should be noted that ability of organic phosphites to retain the original color of polymer can be determined by the symptomatic stabilization owing to the interaction with chromophores, namely, the conjugated structures formed in the process of polymer degradation. This phenomenon is shown to be a common one for the large group of polymers characterized by the intensive coloration already at the first stages of the ageing process with the formation of chromophores of visible part of spectrum.

On the other hand, the organic phosphites can deactivate admixtures and products of transformation of additives in polymer. These admixtures and products are capable to initiate and to intensity polymer coloration.

The abovementioned reactions of organic phosphites are particularly important for such systems where the process of coloration occurs much faster than the process of polymer degradation.

High synergic effects during stabilization of polymer color by organic phosphites may be caused by their participation in the competitive binding of the catalytic admixtures of variable valency metal ions resulting to the formation of colorless compounds and in destruction of chromophores derived from the products of the oxidative transformation of the initial stabilizer (e.g. phenol antioxidant) in polymer composition.

REFERENCES

[1] Kozlov, P.V., Plate, N.A. Polietilen i drugie poliolefiny (Polyethylene and Other Polyolefines). Moscow: *Mir;* 1964.

[2] Pudov. V.S., Gromov, V.A., Shlyapnikov, Yu.A. – *Vysokomol. Soed.,* 1967, vol.9B, №2, p. 111-113.

[3] Gol'dberg, V.M., Paverman, N.G., Kashina, A.G. et al. – *Vysokomol. Soed.,* 1977, vol.19B, №11, p. 808-811.

[4] Kiryushkin, S.G., Shlyapnikov, Yu.A. – *Vysokomol. Soed.,* 1974, vol.16B, №9, p. 702-707.

[5] Kiryushkin, S.G., Shlyapnikov, Yu.A. – *Dokl. AN SSSR,* 1975, vol.220, №6, p. 1364-1367.

[6] Yasina, L.L., Silina, A.T., Pudov, V.S. – *Vysokomol. Soed.*, 1975, vol.17B, №9, p. 698-701.

[7] Bayer, R.F.A. – *J. Phys. Coll. Chem.,* 1977, vol.51, №1, p. 80-106.

[8] Oskay, G., Levai, J., Nitral, Z.S. et al. – *Eur. Polym. J.,* 1974, vol.10, №1, p. 1121-1125.

[9] Gorelov, E.P., Miller, Z.B. – *Vysokomol. Soed.,* 1978, vol.20A, №8, p. 1899-1903.

[10] Gorelov, E.P., Naumova, L.N., Miller, V.B. – *Vysokomol. Soed.*, 1978, vol.20A, №8, p. 1905-1909.

[11] Kovarskaya, B.M. Levantovskaya, I.I., Blyumenfeld, A.B. et al. – *Plast. Massy*, 1968, №5, p. 42-47.

[12] Malakhova, G.P., Lukovnikov, A.F., Kirillova, E.I. et al. – *Plast. Massy,* 1979, №4, p.24-27.

[13] Minsker, K.S., Lisitsky, V.V., Gataullin, R.F. et al. – *Vysokomol. Soed.,* 1978, vol.20B, №8, p. 571-574.

[14] Minsker, K.S., Berlin, A.A., Lisitsky, V.V. – *Vysokomol. Soed.,* 1976, vol.18B, №1, p. 54-58.

[15] Lugova, L.I., Stafeev, A.F., Selivanov, P.I. et al. – *Vysokomol. Soed.,* 1975, vol.17B, №10, p. 614-617.

[16] Grachev, V.U., Klimenko, I.B., Smirnov, L.V. et al. – *Vysokomol. Soed.,* 1974, vol.16A, №2, p. 317-322.

[17] Chien, Z.C.W. – *J. Polym. Sci.,* 1968, vol.6, Part A-1, №2, p. 375-379.

[18] Kirillova, E.I., Fratkina, G.I., Mezhevikina, E.P. et al. – *Plast. Massy*, 1976, №9, p. 61-63.

[19] Denisov, E.T. – *Vysokomol. Soed.,* 1977, vol.A19, №11, p. 1244-1249.

[20] Balaban, L., Mayer, T., Rusavy, D. et al. – *Kunststoffe,* 1970, Bd.60, №5, S. 341-343.

[21] Piotrovskii, K.B., Smirnova, V.F., Mironik, V.P. et al.– *Kauch. Resina,* 1977, №3, p. 21-26.

[22] Korobova, N.M., Lugova, L.I., Matveeva, E.N. – *Plast. Massy*, 1970, №5, p. 63-64.

[23] Kovarskaya, B.M., Sanzharovskii. A.G. – *Plast. Massy,* 1971, №8, p. 37-40.

[24] Gorbunov, B.N., Gurvich, Ya. A., Maslova, I.P. Khimiya i technologiya stabilizatsii polimernykh materialov (Chemistry and Technology of Stabilization of Polymeric Materials). Moscow: *Khimiya;* 1981.

[25] Kirpichnikov, P.A., Mukmeneva, N.A., Pobedimskii, D.G. – *Usp. Khim.,* 1983, vol.52, №11, p.1831-1851.

[26] Pobedimskii, D.G. , Mukmeneva, N.A. , Kirpichnikov, P.A. In: Developments in Polymers Stabilization – 2, Ed. by Scott, G. London: Appl. Sci. Publ.; 1980. p.125-184.

[27] Minsker, K.S., Fedoseeva, G.T. Destruktsya I stabilizatsiya polivinilkhlorida (Destruction and Stabilization of Polyvinylchloride). Moscow: *Khimiya;* 1979.

[28] Braun, D., Hepp, D. – *J. Polym. Sci.*, Part C, 1975, vol.33, №2, p. 307-312.

[29] Geddes, W.C. – *Eur. Polym. J.,* 1967, №4, p. 733-747.

[30] Thallmaier, M., Braun, D., Hepp, D. – *Angew. Makromol. Chem.,* 1968, Bd.2, S. 71-75.

[31] Kirpichnikov, P.A., Kolyubakina, N.S., Minsker, K.S. et al. – *Plast. Massy,* 1966, №12, p. 24-26.

[32] Minsker, K.S., Kirpichnikov, P.A., Kolyubakina, N.S. et al. – *Vysokomol. Soed.*, 1968, vol.A10, №11, p. 2500-2512.

[33] Kadyrova, V.Kh., Kirpichnikov, P.A., Mukmeneva, N.A. et al. – *Vysokomol. Soed.,* 1972, vol.A14, №5, p.1081-1086.

[34] Minsker, K.S., Kirpichnikov, P.A., Fedoseeva, G.T. et al. – *Plast. Massy*, 1968, №4, p. 35-38.

[35] Minsker, K.S., Mukmeneva, N.A., Kolesov, S.V. et al. – *Dokl. AN SSSR,* 1981, vol.259, №4, p. 890-894.

[36] Yamalieva, L.N., Kolesov, S.V., Mukmeneva, N.A. et al. – *Vysokomol. Soed.*, 1983, vol.25A, №10, p. 2157-2161.

[37] Mukmeneva, N.A., Yamalieva, L.N., Minsker, K.S. et al. – Abstracts of Pepers. The Conference "Processing and Long-term Stability of Hydrocarbon Polymers", Prague, Czech Republic, 1983, p. 59-61.

[38] Verizhnikov, L.V., Kirpichnikov, P.A., Kolyubakina, N.S. et al. – *Vysokomol. Soed.,* 1971, vol.13A, №3, p. 714-718.

[39] Durbetaki, A.Z. – *Anal. Chem.,* 1956, Bd.28, S. 2000-2001.

[40] Kirpichnikov, P.A., Kolyubakina, N.S., Minsker, K.S. et al – *Zh. Anal. Khim.,* 1968, vol.23, №10, p.1582-1583.

[41] Minsker, K.S., Berlin, A.A., Abdullin, M.I. – *Vysokomol. Soed.,* 1974, vol.B16, №6, p. 439-443.

[42] Minsker, K.S., Malinskaya, V.P., Abdullin, M.I. – *Plasty a Kaučuk,* 1974, №11, p. 326-331.

[43] Minsker, K.S. – *Plaste Kautsch.,* 1977, Bd.24, H.6, S. 375-379.

[44] Yamalieva, L.N., Mukmeneva, N.A., Kadyrova, V.Kh. et al. – Abstracts of Papers. The Conference "Ageing and Stabilization of Polymers", Ufa, USSR, 1983, p. 96.

[45] Troitskii, B.B., Troitskaya, L.S., Lepaev, A.D. – *Dokl. AN SSSR,* 1973, vol.210, №4, p. 877-879.

[46] Minsker, K.S., Berlin, A.A., Lisitskii, V.V. et al. – *Plast. Massy,* 1976, №10, p. 69-70.

[47] Pudov, V.S. – Plast. Massy, 1976, №2, p. 18-22.

[48] Minsker, K.S., Berlin, A.A., Lisitskii, V.V. et al. – *Vysokomol. Soed.,* 1977, vol.19A, №1, p. 32-36.

[49] Minsker, K.S., Berlin, A.A., Lisitskii, V.V. et al – *Dokl. AN SSSR,* 1977, vol.232, №1, p. 93-96.

[50] Minsker, K.S., Kazachenko, D.A., Abdullina, R.Z. et al. – *Vysokomol. Soed,* 1973, vol.15A, №4, p. 866-869.

[51] Minsker, K.S., Mukmeneva, N.A., Berlin, A.A. et al. – *Dokl. AN SSSR,* 1976, vol.226, №5, p. 1088-1091.

[52] Mukmeneva, N.a., Agadzhanyan, S.I., Kirpichnikov, P.A. et al. Abstracts of Papers. The IV All-Union conference "Ageing and Stabilization of Polymers", Tashkent, USSR, 1976, p. 36.

[53] Troitskii, B.B., Troitskaya, L.S. – *Vysokomol. Soed.,* 1978, vol.20A, №7, p. 1443-1457.

[54] Maikov, V.N., Troitskii, B.B., Razuvaev, G.G. – *Vysokomol. Soed.,* 1976, vol.18B, №4, p. 267-269.

[55] Zilberman, E.N., Perepletchikova, E.M., Getmenenko,E.N. et al. – *Plast. Massy,* 1975, №5, p. 9-11.

[56] Braun, D. – *Pure Appl. Chem.,* 1981, vol.53, №2, p. 549-566.

[57] Kirpichnikov, P.A., Verizhnikov, L.V., Mukmeneva, N.A. et al. – *Zh. Obshch. Khim.,* 1974, vol.44, №11, p. 2422-2425.

[58] Razumova, N.A., Petrov, A.A., Evtikhov, Z.L. et al. In: Khimiya I primenenie fosfororganicheskikh soedinenii (Chemistry and Application of Organophosphorus Compounds), Ed. by Grechkin, N.P. Moscow: *Nauka,* 1972, p. 100-106.

[59] Pudovik, A.N., Gur'yanova, I.V., Ishmaeva, E.A. In: Reaktsii i metody issledovaniya organicheskikh soedinenii (Reactions and Methods of Investigation of Organic Compounds), Ed. by Kazanskii, B.A. Moscow: K*himiya,* 1968, vol.19, p. 7-848.

[60] Vladimirov, I.L., Grapov, A.F., Lomakina, V.I. In: Reaktsii i metody issledovaniya organicheskikh soedinenii (Reactions and Methods of Investigation of Organic Compounds), Ed. by Kazanskii, B.A. Moscow: *Khimiya,* 1965, vol.16, p. 7-373.

[61] Ramirez, F., Patwardhan, A.V., Heller, S. R. – *J. Amer. Chem. Soc.,* 1964, vol.86, № 3, p. 514-516.

[62] Arbuzov, B.A., Zolova, O.D., Vinogradova, V.S. et al. – *Dokl. AN SSSR,* 1967, vol.173, №2, p. 235-239.

[63] Harvey, R.G. – *Tetrahedron,* 1966, vol.22, № 8, p. 2261-2573.

[64] Minsker, K.S., Mukmeneva, N.A., Kirpichnikov, P.A. et al. – *Dokl. AN SSSR,* 1977, vol.233, №3, p.375-377.

[65] Mukmeneva, N.A., Kirpichnikov, P.A., Minsker, K.S. et al. – Abstracts of Papers. The II International Symposium on Polyvinylchloride, Lion-Villerban, France, 1976, p.21.

[66] Minsker, K.S., Mukmeneva, N.A., Kolesov, S.V. et al. – *Dokl. AN SSSR,* 1979, vol.244, №5, p.1134-1137.

[67] Mukmeneva, N.A., Minsker, K.S., Kolesov, S.V. et al. – *Dokl. AN SSSR,* 1984, vol.274, №6, p.1393-1396.

[68] Mukmeneva, N.A., Cherezova, E.N., Yamalieva, L.N. et al. – Izv. AN SSSR, Ser. *Khim.,* 1985, №5, p. 1106-1108.

[69] Mukmeneva, N.A. – Abstracs of Papers. The V Conference "Ageing and Stabilization of Polymers", Vilnyus, Lithvania, 1980, p. 22.

[70] Sirota, A.G. Modifikatsiya struktury i svoistv poliolefiof (Modification of Structure and Properties of Polyolefines). Leningrad: *Khimiya;* 1974.

[71] Abramov, V.S., Semenova, N.A. – *Zh. Obshch. Khim.,* 1958, vol.28, №11, p. 3056-3059.

[72] Mukmeneva, N.A., Goldenberg, A.L., Lazareva, N.P. et al. – *Vysokomol. Soed.,* 1983, vol.25A, №8, p. 635-637.

[73] Kirpichnikov, P.A., Kolyubakina, N.S., Mukmeneva, N.A. et al. – *Vysokomol. Soed.,* 1970, vol.12B, №3, p. 189-192.

[74] Akhmadullina, A.G., Mukmeneva, N.A., Kirpichnikov, P.A. et al. – *Vysokomol. Soed.,* 1974, vol.16A, №2, p. 370-375.

[75] Schwetlick, K., Pionteck, Z., Winkler, A. et al. – *Polym. Degrad. and Stability,* 1991, №3, p. 219-228.

[76] Mukmeneva, N.A., Akhmadullina, A.G., Pobedimskii, D.G. et al. – *Vysokomol. Soed.,* 1974, vol.16B, №8, p. 635-637.

[77] Plate, N.A., Liatmanovich, A.D., Noa, O.V. Macromolekulyarnye reaktsii (Macromolecular Reactions). Moscow: *Khimia;* 1977.

[78] Kirillova, E.I. Shul'gina, E.S. Starenie i stabilizatsiya termoplastov (Ageing and Stabilization of Thermoplasics). Leningrad: *Khimiya*; 1988.

[79] Petrosyan, P.A., Ordukhanyan, K.A., Kadyrova, V.Kh. et al. – *Zh. Prikl. Khim.,* 1977, vol.10, №8, p. 1894-1895.

[80] Panov, V.A. et al. The USSR Author Certificate 378389 (1974); *Bull. Izobret.* (Rus), 1974, №19.

[81] Kirpichnikov, P.A., Kadyrova, V.Kh., Levantovskaya, I.I. et al. – *Plast. Massy,* 1971, №1, p. 44-46.

[82] Mironova, Z.I., Smirnov, L.N., Rubezhova, G.E. et al. – *Khim. Volokna,* 1975, №5, p.32-34.

[83] Kim, A.M. et al. The USSR Author Certificate 385983 (1973); *Bull. Izobret.* (Rus), 1973, №26.

[84] Militskaya, T.V. et al. The USSR Author Certificate 952923 (1982); *Bull. Izobret.* (Rus), 1982, №31.

[85] Gurvich, Ya.A., Zolotorevskaya, L.U., Kumok, S.T. Fenol'nye stabilizatory: Tematicheskii obzor (Phenol Stabilizers: Thematic review). Moscow: *CNIITENeftekhim;* 1978.

[86] Gurvich, Ya.A. et al. Fenol'nye stabilizatory, sostoyanie i perspektivy: Tematicheskii obzor (Phenol Stabilizers, Position and Prospects: Thematic Review). Moscow: *CNIITENeftekhim;* 1990.

[87] Botez, C. – *Ind. Uşoară,* 1974, B21, №12, p. 622-634.

[88] Vold'kin, A.A., Ostapez-Sveshnikova, G.D., Ershov, V.V. Sintez i issledovanie effectivnosti khimikatov dlya polimernykh materialov (Synthesis and Study of Efficiency of Chemicals for Polymer Materials). Tambov: *Tambovskaya Pravda,* 1970.

[89] Afanas'ev, S.V., Lazareva, E.P., Zhdanova, A.N. – Vestnik Moskovskogo Universiteta, *Ser. Khim.,* 1989, vol.30, №4, p. 406-408.

[90] Afanas'ev, S.V. – *Kauch. Resina,* 1994, №4, p. 20-22.

[91] Bukharov, S.V., Pod'yachev, S.N., Syakaev, V.V. – *Zh. Obshch. Khim.,* 1998, vol.3, p. 211-213.

[92] Ershov, V.V., Nikiforov, G.A., Volod'kin, A.A. Prostranstvenno-zatrudnennye fenoly (Sterically Hindered Phenols). Moscow: *Khimiya;* 1972.

[93] Iony i ionnye pary v organicheskikh reaktsiyakh (Ions and Ionic Pairs in Organic Reactions), Ed. by Beletskaya, I.P. Moscow: *Mir;* 1975.

[94] Benuell, K. Osnovy molekulyarnoi spektroskopii (The Principles of Molecular Spectroscopy). Moscow: *Mir;* 1985.

[95] Pod'yachev, S.N., Bukharov, S.V., Gubaidullin, A.T. – *Zh. Obshch. Khim.,* 1999, vol.69, №2, p.275-282.

[96] Bukharov, S.V., Pod'yachev, S.N., Zgadzai, O.E. – Abstracts of Papers. The VII Conference "Organic Reagents in Analytical Chemistry", Saratov, Russia, 1999, p.42.

[97] Garmonov, I.V. Synteticheskie kauchuki (Synthetic Rubbers). Moscow: *Khimiya;* 1976.

[98] Leont'eva, N.A., Sire, E.M., Golovacheva, O.A. – Abstracts of Papers. The IX Conference "Destruction and Stabilization of Polymers", Moscow, Russia, 2001, p. 107-108.

[99] Briescorn, S.N., Ullman, K. – *Chem.Ber.,* 1967, Bd. 100, №2, S. 618-623.

[100] Thinius, K., Reicherdt, W., Krause, H. – *Plaste Kautsch.,* 1970, Bd. 17, №10, S. 739-742.

[101] Ryshava, D., Balabam, L. – *Vysokomol. Soed.*, 1961, vol.3A, №3, p. 470-474.

[102] L'vov, Yu.A., Piotrovskii, K.B., Avdeechev, G.A.– *Kauch. Resina,* 1972, №9, p. 18-22.

[103] Stern, E.B. – *Usp. Khim.*, 1974, vol.42, №2, p.232-272.

[104] Lazareva, N.P., Matveeva, E.N., Zapletnyak, V.N. et al. – *Plast. Massy,* 1976, №2, p. 78-80.

[105] Piotrovskii, K.B., Smirnova, V.F., Mironik, V.P. et al. – *Kauch. Resina*, 1977, №7, p. 21-26.

[106] Kresta, J., Maier, J. – *Chemicky Prumisl.*, 1970, vol.20, №4, p.167-171.

[107] Lazareva, N.P., Lukovnikov, A.F. – *Plast. Massy,* 1966, №6, p. 10-11.

[108] Lazareva, N.p., Lukovnikov, A.F. – *Plast. Massy*, 1966, №8, p. 22-23.

[109] Matveeva, E.N., Lazareva, N.P., Lukovnikov, A.F. – *Vysokomol. Soed.*, 1968, vol. 6B, №6, p. 8-10.

[110] Levantovskaya, I.I., Gur'yanova, V.V., Kovarskaya, B.M. – *Vysokomol. Soed.*, 1969, vol. 11A, №5, p. 1043-1049.

[111] Mukmeneva, N.A., Sabirova, L.Kh., Polovnyak, V.K. et al. – *Zh. Prikl. Khim.*, 1980, vol. 53, №2, p. 468-470.

[112] Akhmadullina, A.G., Mukmeneva, N.A., Kirpichnikov, P.A. – *Plast. Massy,* 1971, №7, p.14-17.

[113] Mukmeneva, N.A., Akhmadullina, A.G., Kurmaeva, N.I. – *Plast. Massy,* 1975, №8, p. 64-66.

[114] Novoselova, L.V., Popova, G.S., Babel, B.I. – *Zh. Prikl. Khim.*, 1974, vol.63, №3, p. 641-644.

[115] Lerhova, Ya., Pospisil, Ya. – *Plaste Kautsch.*, 1976, Bd. 23, H.10, S. 289-294.

[116] Bellamy, L. Infrakrasnye spektry slozhnykh molecul (Infrared Spectra of Complex Molecules). Moscow: *Izdatel'stvo voennoi literatury;* 1963.

[117] Mukmeneva, N.A., Sabirova, L.Kh., Kadyrova, V.Kh. et al. – *Zh. Prikl. Khim.*, 1977, vol.10, №3, p. 604-608.

[118] Mukmeneva, N.A., Sabirova, L.Kh., Kadyrova, V.Kh. et al. – *Plast. Massy,* 1975, №1, p. 51-53.

[119] Pudovik, A.N., Muratova, A.A., Yarkova, E.T. et al. – *Zh. Obshch. Khim.*, 1971, vol.41, №7, p.1481-1488.

[120] Ginzburg, G.D., Zgadzai, E.A., Kolyubakina, N.S. et al. – *Zh. Neorg. Khim.*, 1971, vol.16, №7, p. 1023-1026.

[121] Akhmadullina, A.G., Yarkova, E.G., Mukmeneva, N.A. et al. – *Zh. Prikl. Khim.*, 1973, vol.44, №6, p. 1291-1294.

[122] Kirpichnikov, P.A., Kolyubakina, N.S., Mukmeneva, N.A. et al. – *Plast. Massy,* 1971, №7, p. 43-45.

[123] Mukmeneva, N.A., Sabirova, L.Kh., Lazareva, N.P. et al. – *Plast. Massy,* 1980, №2, p. 8-9.

[124] Mukmeneva, N.A., Akhmadullina, A.G., Kirpichnikov, P.A. – *Vysokomol. Soed.*, 1974, vol.16B, №12, p. 867-871.

[125] Kendlin, F., Taylor, K., Thompson, D. Reactsii koordinatsionnykh soedinenii (Reactions of Co-ordination Compounds). Moscow: *Mir;* 1970.

[126] Ovchinnikov, V.V., Cherkasova, O.A., Yarkova, E.G. et al. – Izv. AN SSSR, Ser. *Khim.*, 1978, №3, p. 689-691.

[127] Volod'kin, A.A., Ershov, V.V. – *Usp. Khim.*, 1988, vol.57, №.4, p. 595-624.

[128] Bradley, W., Sanders, J. – *J. Chem. Soc. Feb.*, 1962, №2, p.480-486.

[129] Bukharov, S.V., Fazlieva, L.K., Nugumanova, G.N. et al. – *Zh. Prikl. Khim.*, 2003, vol.76, №9, p. 1558-1562.

[130] Komissarov, V.N., Kharlanov, V.A., Ukhin, L.Yu. et al. – *Zh. Org. Chim.*, 1992, vol. 28, №3.

Chapter 4

THE INVESTIGATION OF POLYFUNCTIONAL STABILIZER EFFICIENCY IN POLYMERS. SYNERGISM EFFECTS

Some properties of polyfunctional compounds as multi-purpose stabilizers for various types of polymers are indicated in this chapter. Much attention is paid to the consideration of synergism for mixtures of stabilizers which are discussed in the monograph.

According to the classical definition, antioxidative synergetic effect (ASE) is exceeding of the protective action of a mixture of substances in comparison with the most effective component of the mixture taken in the concentration which is equal to the total concentration of the mixture [1, 2]. A large number of publications are devoted to the investigation of ASE [3-9]. Most of them are aimed at empirical search and selection of stabilizing compositions for polymers capable of displaying ASE.

At the present time the application of synergistic antioxidant mixtures is one of the important trends of efficient protection of polymeric materials from undesirable oxidative transformations. Alongside with that, diverse chemical nature of antioxidants hampers the development of the general theory of synergism. The establishing of main regularities binding the kinetics of antioxidative action of single antioxidants and their mixtures is very important for the creation of scientific foundations of selection of stabilizing mixtures. According to the above-mentioned the approach to the systematization of experimental data on the oxidation inhibition by stabilizer mixtures is suggested and the classification of synergetic mixtures and the mechanism of synergism is given [10].

The first two groups of synergetic mixtures include antioxidants having the similar mechanism of action, each of them react with radicals:

№1 – both inhibitors (In_1H and In_2H) interact with peroxide radicals $RO_2^{\cdot}$ (aromatic amines, phenols);

№2 – one of the inhibitors (InH) reacts with peroxide radicals RO_2, another one (Q) reacts with alkyl radicals $R^{\cdot}$ (quinones, nitroxyl radicals).

Next two groups include inhibitors having different mechanisms of action:

№3 – one of them (InH or Q) reacts with radicals, another one (S) is a hydroperoxide decomposer (sulfur- and phosphorus- containing compounds);

№4 – one of the components interacts with radicals or hydroperoxides, and another one decreases the chain initiation rate W_i (including the chain branching rate) (metal deactivators, ultraviolet absorbers);

№5 – one of the components interacts with radicals or hydroperoxides, and another one doesn't inhibit the oxidation process alone.

Table 4.1. Classification of synergetic mixtures according to the type of mechanism of inhibitors action

Group number	Synergetic mixtures	Group number	Synergetic mixtures
1	$In_1H\ (RO_2^{\cdot}) + In_2H\ (RO_2^{\cdot})$	5	InH (Q,S) + M (non inhibitor)
2	$InH\ (RO_2^{\cdot}) + Q\ (R^{\cdot})$		
3	InH, Q + S (ROOH)	a	M – catalyst
4	Antioxidant+ substance, decreasing W_i	b	M – inert substance, synergist
a	InH, Q + deactivator of metals	c	M – oxidable substrate
b	InH, Q + ultraviolet-absorber	d	M – quencher of photosensitization of an antioxidant
c	S + ultraviolet-absorber	e	M – substance increasing solubility of an antioxidant

Autosynergetic mixtures are a part of the above mentioned groups, they include an initial antioxidant and a product of its conversion, as well as antioxidants with intramolecular synergism, having various functional reaction centers in a molecule. For example, substituted *p*-phenylenediamines and pyrocatechins displaying autosynergism and reacting with $RO_2^{\cdot}$ which leads to the formation of products of quinoid structure (acceptors of alkyl radicals), belong to group №2 [3].

An express method of comparison of duration of inhibiting action of a stabilizer (induction period) (τ) [11] is widely used as a criterion of antioxidant action of compounds to test stabilizing properties of compounds, along with the stability of performance parameters of polymers and polymeric materials (strength properties, color, elasticity) to external influence (thermal oxidation, irradiation, mechanical degradation). Experiments are conducted under autooxidation conditions at increased temperature in oxygen or in the air. The prevailing usage of such tests is conditioned by the fact that they provide qualitative assessment of the efficiency of additive action when polymers are processed, this assessment correlates with the results obtained by other methods. Under autooxidation conditions an antioxidant synergetic effect (ASE) can be defined. To assess qualitatively ASE of inhibitor mixtures the value of practical synergism (S_{pr}) [8] is used, calculated according to the ratio (1), or the value of synergism effect (S) [8] is used, calculated according to the formula (2).

$$S_{pr.} = \tau_{mix.} / \tau_1^0, \quad (1)$$

where τ_1^0 is induction period of polymer oxidation for the most efficient component of the mixture at the molar concentration which is accepted for mixtures of inhibitors; τ_{mix} is induction period of polymer oxidation when the mixture of inhibitors is used.

$$S = [\tau_{mix} - (\tau_1 + \tau_2)]/(\tau_1 + \tau_2), \quad (2)$$

where τ_{mix} is induction period of polymer oxidation when the mixture of inhibitors is used; τ_1 and τ_2 are induction periods of polymer oxidation for the same component of the mixture at the concentrations corresponding to their mole fraction in the mixture.

The using of values of synergism and practical synergism allows to define sufficiently fairly the prospectivity of practical usage of this or that mixture of antioxidants for polymer.

4.1. Phosphorous Acid Esters

Mixtures of phenol antioxidants with phosphoric acid esters are widely used for the polymer stabilization [1-8, 12, 13]. At the present time various producers offer wide range of stabilizing mixtures of similar type for stabilization of various polymers. Among other producer Ciba Specialty Chemicals produces stabilizing mixtures having common name Blends of various grades depending on the nature of cocomponents and on their ratio in the composition [13]. They are either double mixtures, including phenol antioxidant (for example, Irganox 1010) and phosphorous acid ester (for example, Irgafos 168), or triple mixtures, including additionally an acceptor of alkyl radicals (for example, lacton HP 136).

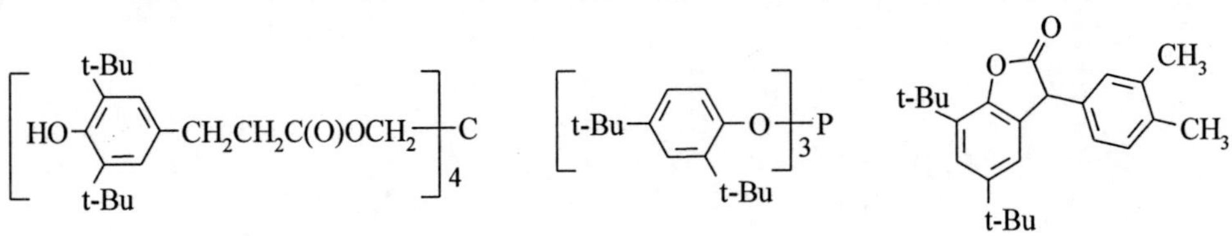

Irganox 1010 Irgafos 168 HP 136

According to the classification given in table 4.1 mixtures of phenol antioxidants with phosphoric acid esters can be included into Group №3 [10]. Synergetic effects in such systems are considerable. One component (phenol antioxidant) can terminate kinetic chains of oxidation, another one (phosphorous acid ester) can decompose hydroperoxides without forming free radicals, i.e. inhibit degenerated branching of kinetic chains [15, 16], it provides long-term protection of polymeric materials against oxidation. Below there are some examples of the dependence of the efficiency of action of stabilizing mixture of phenol antioxidant with phosphorous acid ester on the formulation of stabilizing mixture in various polymers.

The results of the study of influence of the structure of phosphite components on the efficiency of action of the stabilizing composition Irganox1010 – phosphorous acid ester are given in table 4.2. The samples of low pressure polyethylene (HDPE), polypropylene (PP), poly-4-methylpentene-1 (PMP), containing tris(2,4-di-tert-butylphenyl)phosphite(1), displaying the highest reactivity concerning hydroperoxides (Section 2.2.1.2), possess higher stability. It is indicated by the increase of induction period (τ) of oxidation, by less relative change of melt flow index (MFI) and elongation at break (ε_b) after ageing.

Table 4.2. Influence of phosphorous acid esters on aging stability of polyolefins

Parameter	Without phosphite	Phosphite[a]		
		(1)	(2)	(3)
HDPE				
[b] τ, ч (473 K)	2.0	3.0	2.2	2.1
Stability index after five-fold extrusion				
[c] MFI_5'/MFI_5	0.92	0.98	0.95	0.90
[c] $\varepsilon'_b/\varepsilon_b$	0.92	0.98	0.97	0.96
PP				
τ, ч (423 K)	0.7	1.5	0.9	0.8
Stability index after five-fold extrusion				
MFI'_5/MFI_5	0.92	0.98	0.95	0.95
$\varepsilon'_b/\varepsilon_b$	0.94	0.98	0.97	0.96
PMP				
τ, h (423 K)	2.0	4.2	2.1	2.3
Stability index after five-fold extrusion				
MFI_5'/MFI_5	0.91	0.96	0.95	0.92
$\varepsilon'_b/\varepsilon_b$	0.90	0.94	0.91	0.89
Stability index after thermal ageing (423 K, 10 hours)				
MFI_5'/MFI_5	0.90	0.93	0.91	0.91
$\varepsilon'_b/\varepsilon_b$	0.89	0.92	0.90	0.89

[a] ratio *Irganox 1010 : phosphite* is *3:2*, total concentration is 0.2 wt%;
[b] oxygen pressure is 33.3 KPa; [c] MFI' and ε'_b are values of melt flow index and elongation at break after ageing.

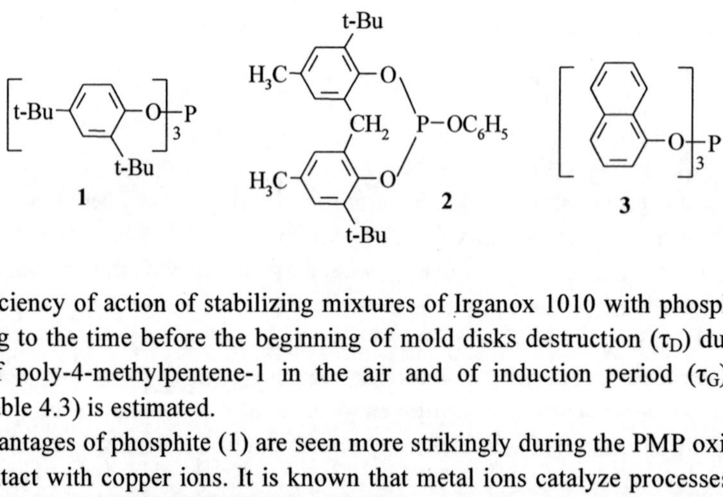

1 2 3

The efficiency of action of stabilizing mixtures of Irganox 1010 with phosphites (**1**) and (**2**) according to the time before the beginning of mold disks destruction (τ_D) during thermal oxidation of poly-4-methylpentene-1 in the air and of induction period (τ_G) of granule oxidation (table 4.3) is estimated.

The advantages of phosphite (1) are seen more strikingly during the PMP oxidation while being in contact with copper ions. It is known that metal ions catalyze processes of polymer oxidation [12]. It should be also noted that the mixture of Irganox 1010 with phosphite (1) allows to prevent coloring polymer in the course of its ageing (table 4.3).

The study of mixtures of antioxidants of various structure (Irganox 1010, Antioxidant 2246, Ethyl Antioxidant 712) with phosphite (1) indicates that the longest induction period (τ) of oxidation of HDPE and PP samples is observed for mixture of phosphite (1) with Irganox 1010 (figure 4.1).

Table 4.3. Influence of phosphorous acid esters on aging stability of poly-4-methylpentene-1

Parameter	Irganox 1010 (0.5 wt%)	Irganox 1010 (0.3 wt%)	Irganox 1010 + phosphite (1)[a]	Irganox 1010 + phosphite (2)[a]
τ_G, h (473 K)				
without copper	25–26		32	25–26
with copper (0.001%)	18	13	27	21
color	Yellow	Yellow	Natural	Light yellow
τ_D, h (438 K)	300	180	350	340
Light refraction coefficient, % (disk 1 mm)	89.0	87.0	91.0	90.5

[a] ratio *Irganox 1010 : phosphite* is 3:2, total content of stabilizers is 0.3 wt%.

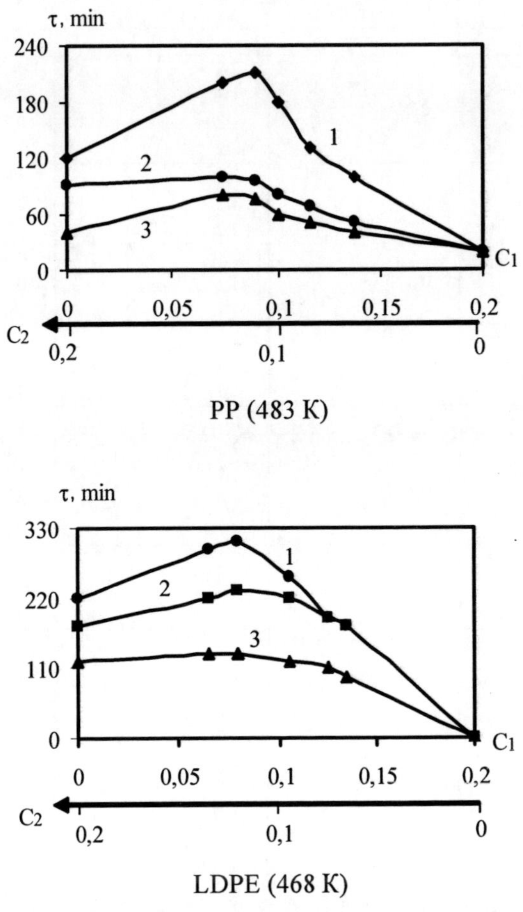

Figure 4.1. Dependence of induction period (τ, min) of polymer oxidation on the formulation of stabilizing mixture: 1 – Irganox 1010 + phosphite (1); 2 – Antioxidant 2246 + phosphite(1); 3 – Ethyl Antioxidant 712 + phosphite (1); C_1, C_2 (wt%) – content of phosphite (1) and phenol antioxidant, respectively; (oxygen pressure is 33.3 kPa).

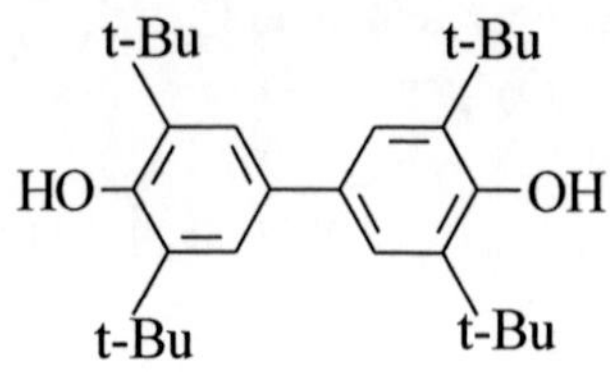

Ethyl Antioxidant 712

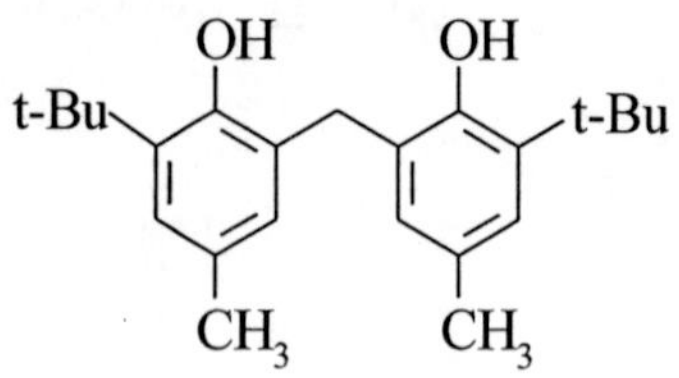

Antioxidant 2246

The values of synergism (S) and practical synergism (S_{pr}) are defined from induction periods (τ) of polyolefine oxidation (table 4.4).

Table 4.4. Influence of formulation of stabilizing mixture on the value of synergism (S) and practical synergism (S_{pr})

Stabilizing mixture	Content of a stabilizer, wt%	Ratio of stabilizers, parts by weight	S	S_{pr}
PP (438 K, oxygen pressure 33.3 kPa)				
Irganox 1010 + phosphite (1)	0.2	2:1 1:1 1:2 3:2	0.50 0.88 0.14 1.33	1.80 2.50 1.60 2.92
Antioxidant 2246 + phosphite (1)	0.2	2:1 1:1 1:2	0.27 0.33 0.23	0.92 1.16 0.68
Antioxidant 2246 + phosphite (1)	0.3	2:1 1:1 1:2	0.18 0.25 0.18	1.00 1.65 1.24
Ethyl Antioxidant 712 + phosphite (1)	0.2	2:1 1:1 1:2	0.54 0.30 0.09	2.50 2.31 1.66
HDPE (468 K, oxygen pressure 33.3 kPa)				
Irganox 1010 + phosphite (1)	0.2	2:1 1:1 1:2 3:2	0.65 1.10 0.30 0.88	1.86 2.50 2.00 2.30
Antioxidant 2246+ phosphite (1)	0.2	2:1 1:1 1:2	0.38 0.54 0.68	1.59 1.80 2.30
Ethyl Antioxidant 712 + phosphite (1)	0.2	2:1 1:1 1:2 3:2 2:3	0.66 1.00 1.17 1.08 0.35	2.19 3.00 3.26 2.08 2.19
Ethyl Antioxidant 712 + phosphite (1)	0.3	1:1 1:2 2:1	0.82 0.50 0.50	3.30 3.00 3.00

The maximum value of antioxidant synergetic effect is achieved at the ratio *phenol : phosphite* from 3:2 to 2:3 parts by weight. The highest values of S and S_{pr} are registered for the mixtures of tris(2,4-di-*tert*-butylphenyl)phosphite (1) with Irganox 1010. These effects for stabilizers Antioxidant 2246 and Ethyl Antioxidant 712 are appeared to be weaker.

The introduction of phosphite (1) together with Irganox 1010 allows not only to improve the color stability of HDPE, for the quantitative assessment of which yellowness index (γ) is used, but also to stabilize considerably tangent dielectric loss angle (tgδ) and melt flow index (MFI) (figure 4.2).

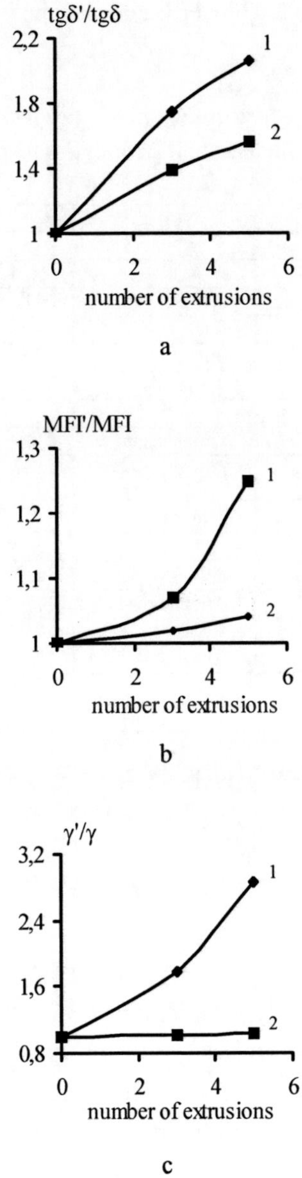

Figure 4.2. The relative change of service performances of stabilized HDPE during multiple extrusion: 1 – Irganox 1010 (0.2 wt%); 2 – Irganox 1010 (0.12 wt%) + phosphite (1) (0.08 wt%). a –change of tangent dielectric loss angle (tgδ, 500 MHz); b – the change of melt flow index (MFI, g/10min); c – change of yellowness index (γ).

It is shown in table 4.5 that substitution of the equivalent amount of tris(2,4-di-*tert*-butylphenyl)phosphite (1) (compositions №1 and №3) for a part of Irganox 1010 used in formulation №2 and №4 allows to increase the stability of physical and mechanical parameters of polyethylene composite to thermal-oxidative ageing, obviously owing to synergetic effect. The stabilizing composition for polypropylene, containing phosphite (1) along with Irganox 1010 and color stabilizer Benazol P, possesses also high efficiency (table 4.6).

Table 4.5. Influence of formulation of stabilizing mixtures on the stability of physical and mechanical properties of polyethylene composite

Formulation number	1	2	3	4
Parts by weight				
HDPE	94.4	94.4	94.4	94.4
LDPE	4.0	4.0	4.0	4.0
tris(2,4-di-*tert*-butylphenyl)phosphite	0.1	–	0.1	–
Irganox 1010	0.1	0.2	0.1	0.2
Benzon OA	0.6	0.6	0.6	0.6
Titanium oxide	–	–	0.5	0.5
Parameter	Before ageing			
Melt flow index (MFI_5), g/4 min	0.62	0.50	0.59	0.51
Yield strength (σ_{ys}), MPa	21.8	22.0	20.8	21.0
Tensile strength (σ_{ts}), MPa	33.1	30.8	34.1	33.2
Elongation at break (ε_b),%	893	820	910	864
Stability index	After thermal ageing (438 K, 10 h)			
calculated from MFI_5	1.00	0.98	0.98	0.85
calculated from σ_{ys}	0.95	0.94	0.96	0.96
calculated from σ_{ts}	0.90	0.88	0.90	0.89
calculated from ε_b	0.95	0.93	0.92	0.92

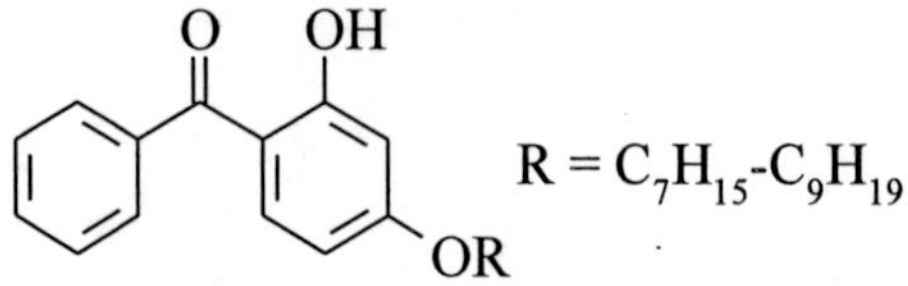

Benzon OA

Table 4.6. Influence of formulation of stabilizing mixture on stability of physical and mechanical properties of polypropylene

Formulation number	1	2	3	4	5
wt %					
Irganox 1010	0.2	0.2	0.2	0.2	0.2
Benazol P	0.4	0.4	0.4	0.6	0.6
Tris(2,4-di-*tert*-butylphenyl)phosphite	–	0.1	0.2	–	–
Parameter	Before ageing				
Melt flow index (MFI_5), g/10 min	3.23	3.38	4.49	4.09	3.72
Yield strength (σ_{ys}), MPa	39.4	36	40.1	34	35
Tensile strength (σ_{ts}), MPa	39.4	33	39.5	32	32

Formulation number	1	2	3	4	5
wt %					
Elongation at break (ε_b),%	8.5	15	11.3	20	18.3
Tangent dielectric loss angle tg δ, 10^{-4}	5.6	5.8	6.1	7.7	6.9
Stability index	Thermal ageing (423 K, 100 h)				
calculated from $MFI_{2.16}$	0.92	0.94	0.96	0.94	0.96
calculated from σ_{ys}	0.88	0.90	0.92	0.90	0.92
calculated from σ_{ts}	0.80	0.84	0.90	0.92	0.92
calculated from ε_b	0.90	0.91	0.93	0.92	0.94
calculated from tg δ	0.89	0.91	0.92	–	–
Stability index	Photodegradation (ultraviolet radiation, 100 h)				
calculated from σ_{ys}	0.92	0.94	0.96	0.94	0.96
calculated from σ_{ts}	0.92	0.94	0.97	0.97	0.97
calculated from ε_b	0.92	0.93	0.95	0.96	0.97
calculated from tg δ	0.95	0.94	0.96	0.95	0.97

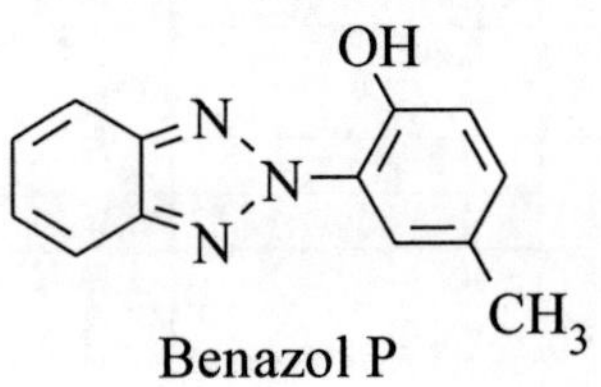

Benazol P

The ability of the mixture of tris(2,4-di-*tert*-butyl phenyl)phosphite (1) with Irganox 1010 to prevent thermal and mechanical destruction of polyolefines is illustrated by time changes of torque (M), obtained in the processing of polymer in the Brabender plasticorder (table 4.7). This parameter is especially important when polyolefines are used in the formulation of dynamic thermoplastic elastomers (DTPE), the process of vulcanization of which is combined with the process of mixing at the temperature which is not lower than the temperature of thermoplastic melting. The initial polymers are exposed to considerable thermal and mechanical destruction, worsening initial physical and mechanical properties of DTPE [18].

Table 4.7. Influence of formulation of stabilizing composition on the value of torque (M) during processing of polyolefines using Brabender plasticorder

Stabilizer	Content of a stabilizer, wt%	Brabender plasticorder torque M, (N·m) Processing time, min		
		6	12	15
HDPE (T_0=423 K)				
Without a stabilizer	–	28.50	25.50	24.00
Irganox 1010	0.10	28.50	27.80	27.00
	0.15	28.25	29.40	29.00
	0.20	28.40	26.00	25.30
	0.25	30.60	28.50	27.80
Tris(2,4-di-*tert*-butylphenyl)phosphite (1)	0.10	26.00	25.00	24.00
	0.15	28.85	29.40	26.00
	0.20	26.00	25.00	25.00
	0.25	28.70	26.30	25.75

Table 4.7. (Continued).

Stabilizer	Content of a stabilizer, wt%	Brabender plasticorder torque M, (N·m)		
		Processing time, min		
		6	12	15
HDPE (T_0=423 K)				
Irganox 1010 + phosphite (1) (3:2)	0.10	26.60	25.15	24.75
	0.15	30.30	27.80	26.90
	0.20	29.15	26.70	25.90
	0.25	29.65	28.00	27.50
PP (T_0=453 K)				
Without a stabilizer	–	14.00	10.00	9.00
Irganox 1010	0.10	16.50	15.00	14.50
	0.15	15.25	13.60	13.55
	0.20	15.25	13.75	13.60
	0.25	14.50	13.10	12.65
Tris(2,4-di-*tert*-butylphenyl)phosphite (1)	0.10	14.75	13.75	13.00
	0.15	15.25	13.25	12.90
	0.20	14.75	13.35	13.20
	0.25	14.50	13.00	12.40
Irganox 1010 + phosphite (1) (3:2)	0.10	17.25	14.00	13.75
	0.15	16.25	13.60	13.45
	0.20	14.73	12.50	12.85
	0.25	15.00	13.75	13.45

At initial time stabilizers don't influence the value of torque, when the processing time increases to 12-15 minutes the change of torque in the presence of *phenol – phosphite* mixture is insignificant (figure 4.3). The decrease of the polymer destruction degree allows to get more possibilities of obtaining thermoplastic elastomers based on polypropylene and polyethylene by dynamic vulcanization [18].

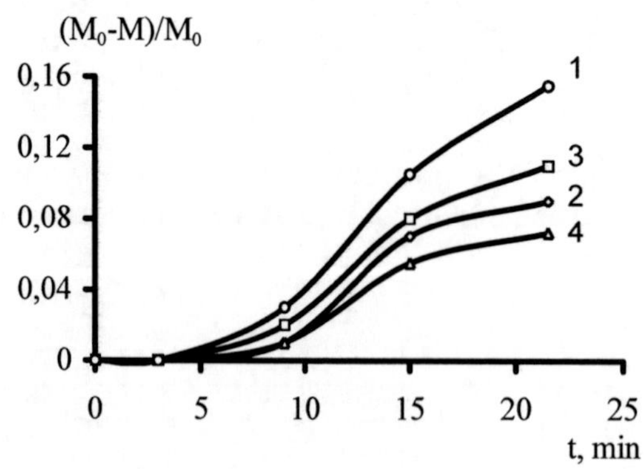

HDPE (T_0=423 K)

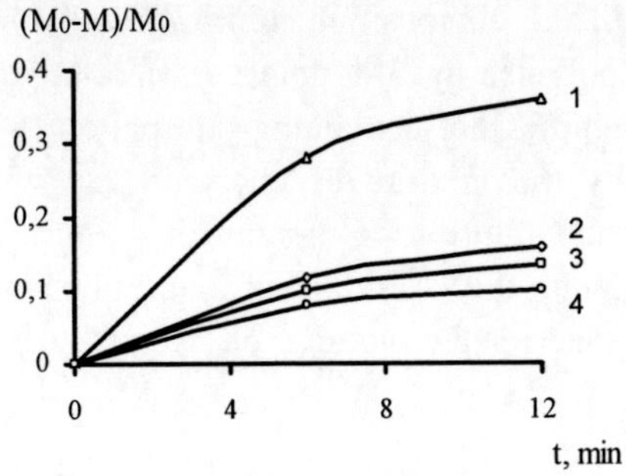

PP (T_0=453 K)

Figure 4.3. Relative time change of torque during processing of polyolefines using Brabender plasticorder: 1 – without a stabilizer; 2 – Irganox 1010; 3 – phosphite (1); 4 – mixture Irganox 1010 : phosphite (1) (3:2); (content of a stabilizer is 0.25 wt%, rotation speed is 90 rpm).

Tris(2,4-di-*tert*-butyl phenyl)phosphite (1) proved to be an efficient color stabilizer of butyl rubber in the presence of phenol antioxidants (table 4.8).

Table 4.8. Influence of stabilizers on the induction period (τ) of oxidation, color stability of butyl rubber, physical and mechanical properties of butyl rubber vulcanizates [a]

Stabilizer [b]	τ [c], min	Color [d], points		Butyl rubber vulcanizate		
		Initial	In 3 hours [e]	Module 300%, MPa	Tensile strength, MPa	Elongation at break, %
Without a stabilizer	30	1	2÷3			
Irganox 1010	275	1	3	7.6	20.1	770
Phosphite (1)	120	1	2	6.2	18.0	590
Irganox 1010 + phosphite (1) (1:1)	195	1	1	7.2	19.8	680
Irganox 1010 + phosphite (1) (2:1)	230	1	2	7.7	21.2	750
Antioxidant 2246	325	1	6	7.6	20.1	770
Antioxidant 2246 + phosphite (1) (1:1)	185	1	3	7.2	19.8	680
Antioxidant 2246 + phosphite (1) (2:1)	220	1	3	7.7	21.2	750
Antioxidant 2246 + phosphite (1) (1:2)	160	1	2	6.5	18.4	600

[a] vulcanizate formulation №1 (see Appendix to Chapter 4); [b] concentration of a stabilizer is 0.2 wt%; [c] 443 K, oxygen pressure is 33.3 kPa; [d] according to ten-point color scale; [e] 443 K, air.

The application of stabilizing composition *Antioxidant 2246 – phosphite* (1) allowed to increase color stability of polymer by 3-4 points (according to ten-point color scale) in comparison with polymer samples not containing phosphite (1) (figure 4.4 a). When butyl rubber samples stabilized by the mixture of phosphite (1) with Antioxidant 2246 were exposed in the air at room temperature for 6 months, they didn't change their color, whereas butyl rubber samples containing only Antioxidant 2246, obtained light brown color. Color stabilizing effect was also noted for the mixture of Irganox 1010 with phosphite (1) (figure 4.4 b) [19].

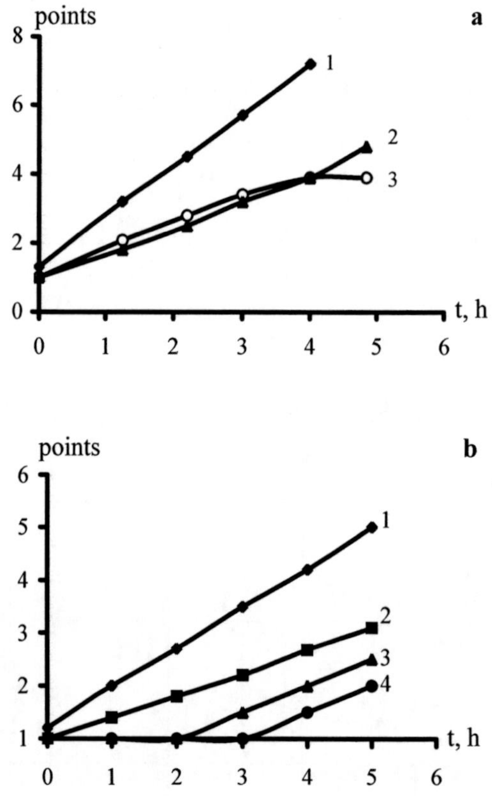

Figure 4.4. Color change (according to ten-point scale) of stabilized butyl rubber (content of stabilizers is 0.2 wt%) during thermal oxidation (air, 443 K): a: 1 – Antioxidant 2246; 2 – Antioxidant 2246 + phosphite (1) (1:1); 3 – Antioxidant 2246 + phosphite (1) (2:1); b: 1 – Irganox 1010; 2 – Irganox 1010 + phosphite (1) (1:2); 3 – Irganox 1010 + phosphite (1) (2:1); 4 – Irganox 1010 + phosphite (1) (1:1).

Based on the classification of phosphites according to their functional action under development and branching of destructive processes in polymers new types of synergetic mixtures were suggested, they allowed to conduct efficient stabilization of polymers [15, 20]. The given mixtures contain phosphites of diverse action i.e. an inhibitor of radical processes (sterically hindered aromatic phosphite) and an acceptor of molecular products of polymer conversion (aliphatic and alkylaromatic phosphites). In particular, mixtures of phosphites (4), (5), (6) with trinaphthylphosphite (7) turned out to be efficient for polyolefines.

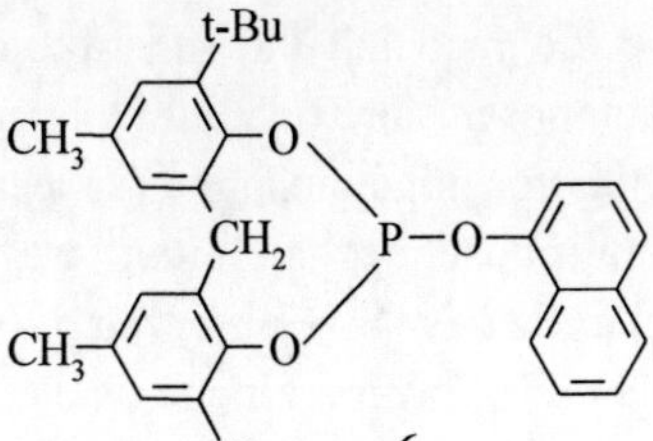

6

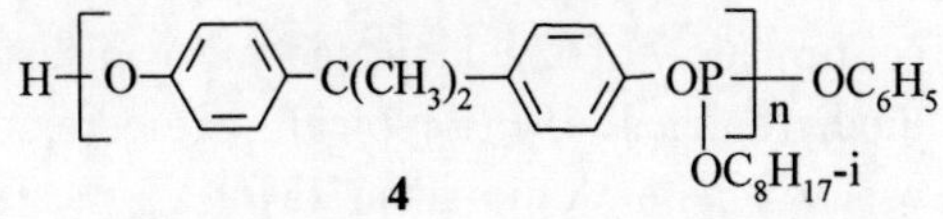

4

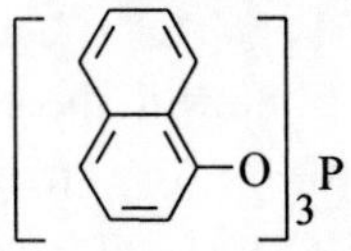

7

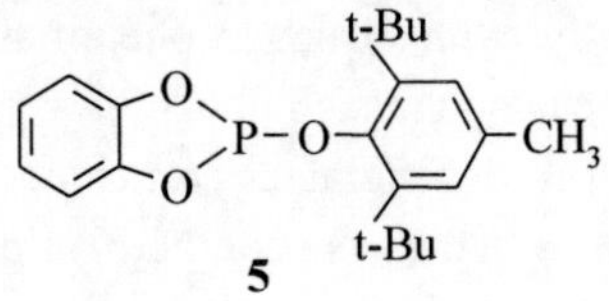

5

The mechanism of stabilizing action can be explained by the fact that one of phosphites, for example (4-6), for which the predominant way is the way of terminating of oxidation chains, accepts $RO_2^{\cdot}$ radicals, and another one, for example, trinaphthylphosphite (7), decomposing hydroperoxides, prevents degenerated branching. Kinetically it is proved by the shift of critical concentrations of corresponding phosphites (free radical inhibitors) into the region of lower values in the presence of trinaphthylphosphite (figure 4.5) under conditions of LDPE inhibited oxidation (448 K, oxygen pressure is 33.3 kPa) [20].

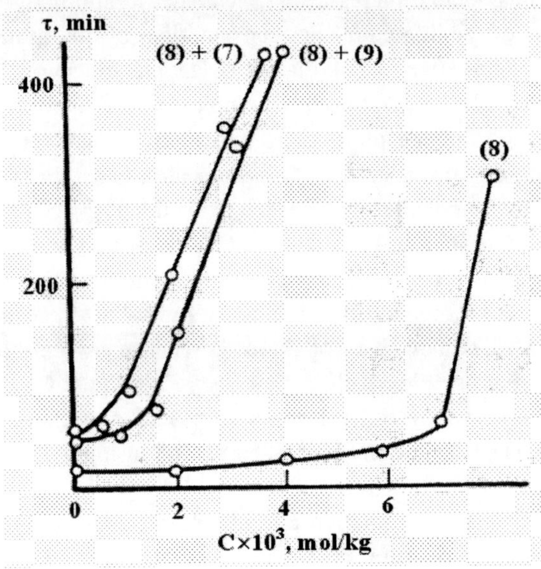

Figure 4.5. Shifting of critical concentration of phosphite (8) in the presence of compounds (7) and (9).

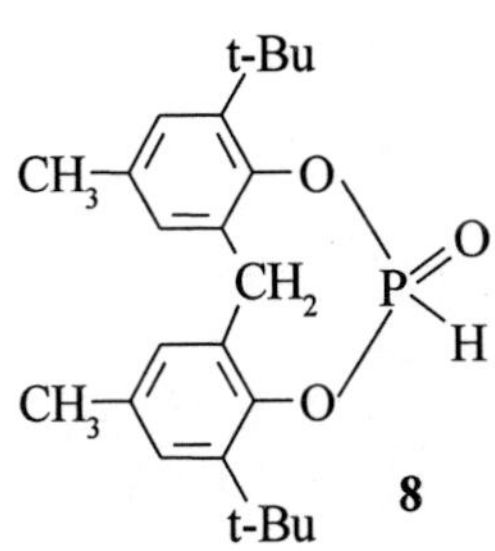

8

$(H_{24}C_{12}O(O)CCH_2CH_2)_2S$

9

The confirmation of the indicated mechanism can be the fact that the substitution of dilauryl thiodipropionate (9) (a typical hydroperoxide decomposer) for trinaphthylphosphite (7) and, respectively, of Antioxidant 2246 (a typical inhibitor) for oligophosphite (4) leads to similar results (figure 4.5). It is interesting that when some phosphites–inhibitors are used jointly with stronger phenol inhibitors the effect of synergism is also observed [20]. It indicates that such phosphites, alongside with the termination of oxidation kinetic chains, can inhibit degenerated branching. This or that trend in their stabilizing behavior will be determined by the competitive power of a compound possessing synergetic action [15, 16].

Synergism in compositions *phosphite – compound of variable-valency metals* is formed according to the principles of inhibitor generation and consists in the fact that phosphites with compounds of polyvalent metals can form corresponding complexes – new inhibitors of free radical oxidation [21].

In Section 2.2.2.1 the phenomenon of drastic strengthening of efficiency of aromatic phosphites as inhibitors of initiated oxidation of polyolefines under the action of titanium salts, acetylacetonates of transition metals (M_{acac}, where M = Co^{2+}, VO^{2+}, Cr^{3+}, Ni^{2+} etc.) is discussed [16].

Interesting regularities are developed for phosphite complexes with Cu(I), the remarkable feature of which is considerable increase of stoichiometric inhibition coefficient [22-25], it means that reactions of additional chain consumption of phosphite in a complex are practically inhibited. The study of initiated oxidation of styrene and solid isotactic polypropylene indicated [25], that phosphite complexes with an Cu(I) ion (CuP) are more efficient than free uncoordinated phosphites; quantitative characteristics of the inhibiting action of these complexes are calculated. It is indicated that the termination of oxidation kinetic chains by inhibitor molecules includes the oxidation of the central metal ion.

It should be noted that in case of initiated oxidation of polypropylene the reactions of a metal complex inhibitor or products of its conversion with polymer hydroperoxide are very important. These reactions turned out [25] to possess radical mechanism and the rate of these reactions is very high. It is they that condition the change *inhibitor* ↔ *catalyst* in kinetic behavior of the phosphite complex with Cu(I) observed in the course of oxidation. In other words, when CuP complexes are applied opportunities arise to check the mode of polymer stabilization under ambient conditions in the absence of an initiator (controlled stabilization under uncontrolled conditions)

The strenthening of stabilizing action of CuP complexes can be achieved by means of inhibition of hydroperoxide initiation.

At figure 4.6 typical time dependences of oxygen uptake by oxidizable polypropylene, stabilized by mixture of *p*-toluidine and $Cu[(PhO)_3P]Cl$ (or CuP) are given as an example, the molar ratios of the concentration of the latter being 0.5-2.0. The nature of dependences indicates the presence of synergism and, at the same time, the actual inhibition of secondary initiation of polypropylene oxidation. The value of synergetic effect depends definitely on the ratio *inhibitor – deactivator* and on the value of the concentration of metal complex additive.

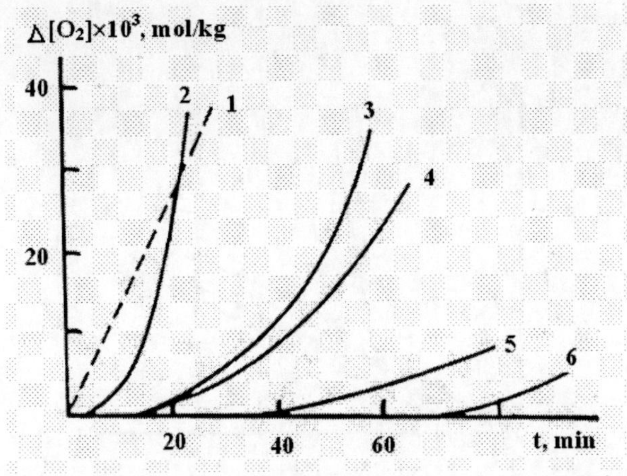

Figure 4.6. Kinetic curves of oxygen uptake by oxidizable polypropylene (358 K, $W_i = 2.5\times10^{-5}$ mole/kg s, oxygen pressure is 33.3 kPa) without an inhibitor (1) and in the presence of inhibitors: [CuP] = 5×10^{-3} mole/kg and [p-toluidine] = 5×10^{-3} mole/kg (2); [CuP] = 5×10^{-3} mole/kg and [p-toluidine] = 1×10^{-2} mole/kg (3); [CuP] = 1×10^{-2} mole/kg and [p-toluidine] = 1×10^{-2} mole/kg (4); [CuP] = 1×10^{-2} mole/kg и [p-toluidine] = 5×10^{-3} mole/kg (5).

The maximum value of stoichiometric inhibition coefficient (f) (for conditions [CuP] = $1\cdot10^{-2}$ mole/kg and [CuP]/[*p*-toluidine]=2) equals to 0.7, it exceeds more than twice the corresponding value f, obtained if CuP is used without amine adding.

The scheme of inhibition is suggested, where amine occupies vacant position in ion sphere, after phosphite converting into uncoordinating phosphate during termination of oxidation kinetic chains; formed secondary complexes decompose actively polymer hydroperoxide converting it into products, not participating in reactions of continuation of polymer oxidation chains. The scheme is confirmed kinetically and by the analysis of the reaction products [26].

4.2. Phosphorus Dithioacids

The study of stabilizing action of phosphorus dithioacids with a sterically hindered phenol fragment (10-15) under LDPE autooxidation indicated that they are comparable according to antioxidant efficiency with bis- and tetraphenol stabilizers Antioxidant 2246 and Irganox 1010 in concentrations which are applied in industry (0.001–0.01 mole/kg) (figure 4.7).

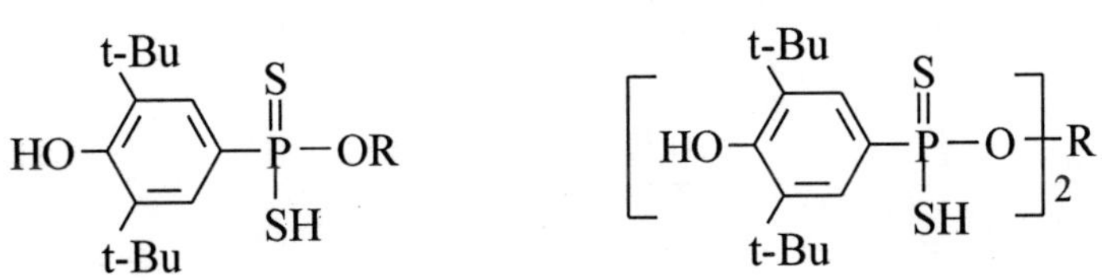

10-13 **14, 15**

R = C_2H_5 (**10**), i-C_3H_7 (**11**), i-C_8H_{17} (**12**), -CH_2-CH=CH_2 (**13**), C_3H_6 (**14**), C_4H_8 (**15**)

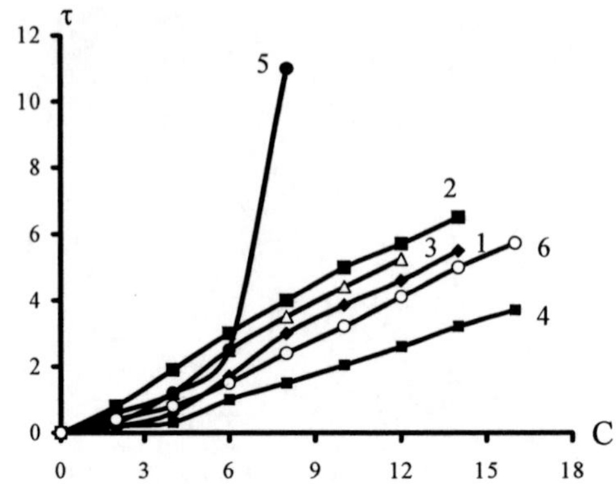

Figure 4.7. Dependence of induction periods (τ, h) of LDPE oxidation on the concentration of stabilizer ($C\times10^{-3}$ mole/kg, 473 K, oxygen pressure is 33.3 kPa): 1 – compound (10); 2 – (11); 3 – (12); 4 – (13); 5 – (14); 6 – Antioxidant 2246 (or Irganox 1010).

The observed considerable increase (under other equal conditions) of the induction period (τ) of polymer oxidation when phosphorus bis-dithioacids (14) are introduced is likely to be the result of doubling of the amount of inhibiting centers if compared with phosphorus dithioacids (10-13).

Phosphorus dialkyl(aryl)dithioacids (16, 17) don't display inhibiting properties at given concentrations. Their weak antioxidant action is registered when they are used in concentrations more than 5×10^{-2} mole/l (table 4.9). At the same time, dithioacids (16, 17), not containing a sterically hindered phenol fragment, in mixtures with the phenol stabilizer BHT display antioxidant synergetic effect, it is confirmed by the results of high-temperature LDPE oxidation (table 4.10, Test 3).

Table 4.9. Influence of phosphorus dithioacids on the induction period (τ) of LDPE oxidation (473 K, oxygen pressure is 33.3 kPa)

Stabilizer	Content of a stabilizer, mole/kg	τ, min	Stabilizer	Content of a stabilizer, mole/kg	τ, min
(10)	0.007	250	$(i\text{-}C_3H_7O)_2P(S)SH$ (16)	0.05	60
(11)	0.007	230	$(C_6H_5O)_2P(S)SH$ (17)	0.05	60
(12)	0.007	270	Antioxidant 2246	0.007	280
(14)	0.007	450			

Table 4.10. Influence of stabilizers on the induction period of LDPE oxidation (τ, 473 K, oxygen pressure is 33.3 kPa) and values of synergism (S) and practical synergism (S_{pr})

Test number	Stabilizer	τ, min	Ratio of components of stabilizing mixture (parts by weight))	S	S_{pr}
Content of a stabilizer 0.025 mole/kg					
1	$(i\text{-}C_3H_7O)_2P(S)SH$ (16)	75	–	–	–
2	BHT	90	–	–	–
3	(16) + BHT	215	1:1	1.87	2.86
		270	2:3	2.60	3.60
		300	1:4	3.00	4.00
4	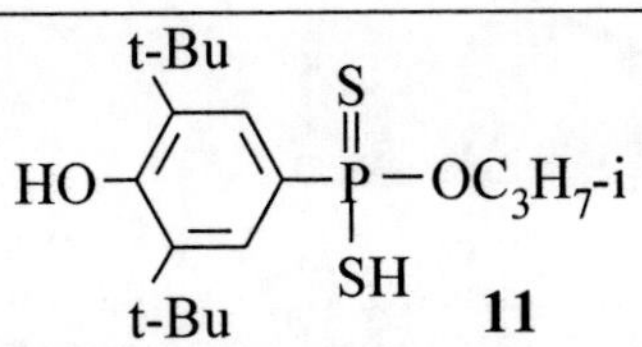 **11**	1080	–	–	–
Content of a stabilizer 0.005 mole/kg					
5	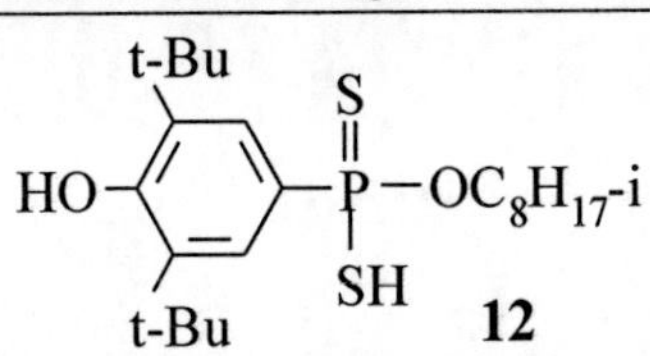 **12**	225	–	–	–
6	(12) + Antioxidant 2246	550	3:2	1.62	2.04
		630	1:1	2.15	2.33
		580	2:3	1.90	2.15
7	Antioxidant 2246	270	–	–	–
8	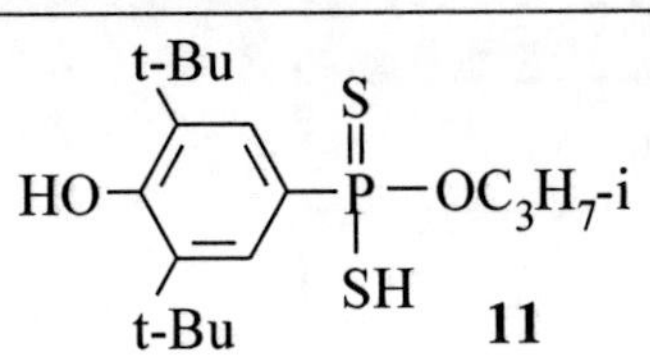 **11**	270	–	–	–
9	(11) + Antioxidant 2246	650	3:2	1.50	2.40
		700	1:1	1.60	2.59
		600	2:3	1.73	2.22

As it is seen from the data of table 4.10 phosphorus dithioacids (11) and (12), containing sterically hindered phenol fragments, increase multiply the induction period (τ) of polyethylene oxidation in comparison with τ for mixtures BHT + $(AlkO)_2P(S)SH$. The reason of the observed effect is likely to be the presence of several reaction centers in molecules of phosphorus dithioacids and, as a result, the ability of these compounds to act both as hydroperoxide decomposers and peroxide radical acceptors. Such polyfunctionality of compounds (10-15) defined the possibility of intramolecular antioxidant synergetic effect.

When mixtures of phosphorus dithioacids (16) and (17), not containing sterically hindered phenol fragment, with phenol antioxidant at nonequimolar ratio of components are used the maximum synergetic effect is established (figure 4.8a). It gives grounds to expect additional synergetic effect in mixtures of phosphorus dithioacids, containing sterically hindered phenol fragment, with phenol antioxidant, it is confirmed by the data of figure 4.8b and table 4.10 (tests 6-9).

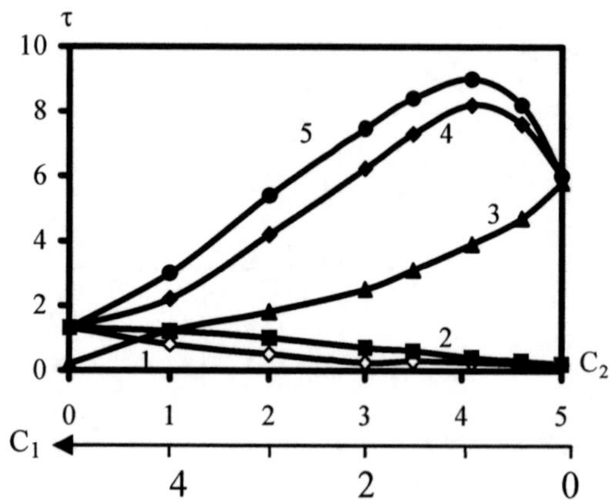

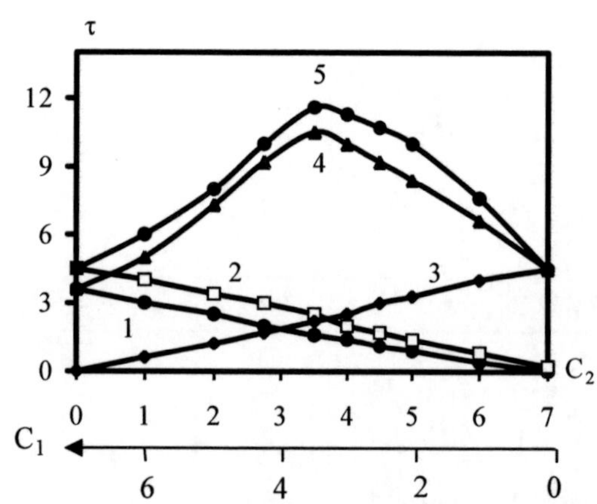

Figure 4.8. Dependence of induction period (τ, h) of LDPE oxidation on the ratio of components of stabilizing mixtures (473 K, oxygen pressure is 33.3 kPa), C_1, C_2 (wt%) is content of phosphorus dithioacids and Antioxidant 2246, respectively: a: ($C \times 10^{-2}$, mole/kg): 1 – $(i\text{-}C_3H_7O)_2P(S)SH$ (16); 2 – $(C_6H_5O)_2P(S)SH$ (17); 3 – Antioxidant 2246; 4 – $(C_6H_5O)_2P(S)SH$ (17) + Antioxidant 2246; 5 – $(i\text{-}C_3H_7O)_2P(S)SH$ (16) + Antioxidant 2246; b: ($C \times 10^{-3}$, mole/kg): 1 – compound (12); 2 – (13); 3 – Antioxidant 2246; 4 – (12) + Antioxidant 2246; 5 – (13) + Antioxidant 2246.

On the other hand, polyfunctional phosphorus dithioacids display synergism of antioxidant action not only in mixtures with a peroxide radical acceptor, but also with a hydroperoxide decomposer. The values of induction period of LDPE oxidation, stabilized by the mixture of phosphorus dithioacids (12) and (13) with dilaurylthiodipropionate (DLTDP) (a typical hydroperoxide decomposer), are given as an example (figure 4.9). Phosphorus dithioacids without sterically hindered phenol fragments in the given mixtures don't display synergetic effect.

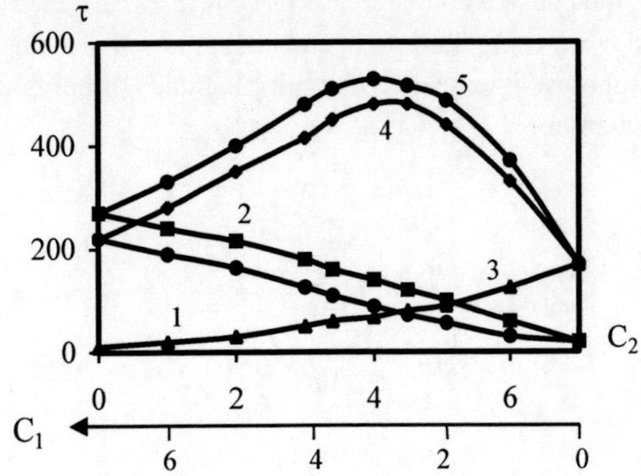

Figure 4.9. Dependence of induction period (τ, min) of LDPE oxidation on the component ratio of stabilizing mixtures ($C\times10^{-3}$, mole/kg , 473 K, oxygen pressure is 33.3 kPa): 1 – compound (12); 2 – (13); 3 – DLTDP; 4 – (12) + DLTDP; 5 – (13) + DLTDP; C_1,C_2 – content of phosphorus dithioacids and DLTDP respectively.

Compounds (11) and (14) are studied as antioxidant additives for high unsaturated isoprene rubber and vulcanizate on its basis (table 4.11). The substitution of equivalent amount of compounds (11) and (14) for amine stabilizer N-phenylnaphthylamine-1 (Neozone D) allows to increase the induction period of rubber oxidation and the rubber destruction temperature. However, the stability index of isoprene rubber vulcanizates calculated from σ_{ts} for compounds (11) and (14) is lower than when Neozone D is used. Somewhat underestimated values of stability index calculated from σ_{ts} allow to suppose that in the given case the necessary frequency of chain cross-linking is not achieved.

Table 4.11. Influence of a stabilizer on the induction period of oxidation and stability of physical and mechanical properties of isoprene rubber and vulcanizates [a] on its basis

Parameter	Stabilizer (0.6 wt%)		
	Neozone D	(11)	(14)
Induction period of oxidation (τ) [b], min	165	215	220
Destruction temperature, K	578	593	603
Tensile strength (σ_{ts}), MPa	22.09	20.36	21.26
Elongation at break (ε_b),%	520	490	520
Cohesive resistance, MPa	0.26	0.41	0.43
Thermal ageing (373 K, 72 h)			
Stability index calculated from σ_{ts}	0.76	0.62	0.62
Stability index calculated from ε_b	0.86	0.91	0.92

[a] vulcanizate formulation №1 (see Appendix to Chapter 4); [b] content of a stabilizer is 0.3 wt%, 403 K, oxygen pressure is 33.3 kPa.

At the same time, the considerable increase of cohesive resistance of the isoprene rubber vulcanizates is noticed when compounds (11) and (14) are used. This fact can be explained by the interaction of phosphorus dithioacids with multiple bonds of polymer, leading to chemical modification of isoprene rubber [27].

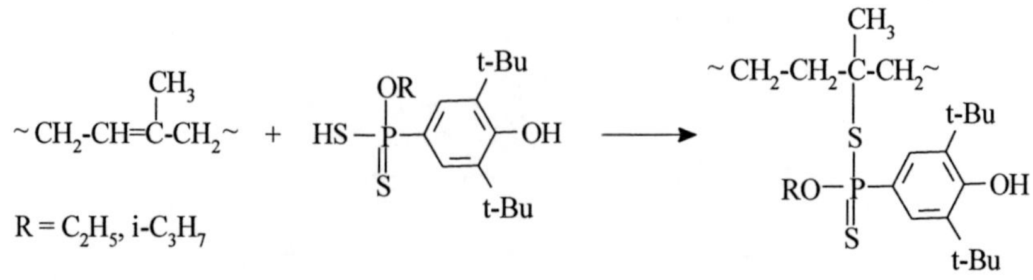

When the model polymeranalogous reaction of compounds (10), (11) and (14) with isoprene rubber was conducted according to the procedure [28] the presence of phosphorus and sulfur atoms was stated in polymer.

In the presence of phosphorus dithioacids the scorch time (t_S) of rubber vulcanizate is shortened. It can be the result of the interaction of phosphorus dithioacids with diphenylguanidine (DPG), which is used in the standard formulation as a vulcanization accelerator. The formation of corresponding complexes *DPG – phosphorus dithioacids* is possible, it results in the activation of DPG (table 4.12). However, the full replacement of DPG for compound (14) testifies to the fact that phosphorus dithioacids don't function independently as accelerators (figure 4.10).

Table 4.12. Influence of the vulcanization accelerator on rheometry characteristics of isoprene rubber vulcanizates [a]

Vulcanization accelerator	Content of an accelerator [b]	M_{min} [c]	M_{max} [d]	t_{90} [e], min	t_s [f], min
DPG	4	11.9	37.0	12.2	2.1
DPG + (14)	3+1	12.4	35.0	11.8	1.4

[a] vulcanizate formulation №1 (see Appendix to Chapter 4); [b] parts by weight per 100 parts by weight of rubber; [c] minimum torque; [d] maximum torque; [e] cure time; [f] scorch time.

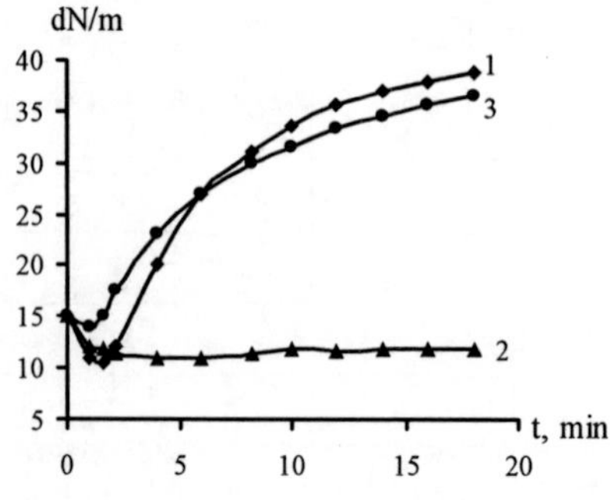

Figure 4.10. Vulcanization curves of rubber mixtures on the basis of isoprene rubber with vulcanization accelerators: 1 – DPG; 2 – compound (14); 3 – DPG + (14) in ratio 3:1 (parts by weight).

On the whole, it can be stated that phosphorus dithioacids containing a sterically hindered phenol fragment are efficient antioxidants for polymers, possessing both intramolecular synergetic antioxidant effect and intermolecular one owing to their polyfunctionality. In addition, these compounds can act as modifiers.

Betaine compounds on the basis of dithiophosphorus acids (18-20) decompose catalytically hydroperoxides (stoichiometric coefficient ν=500–3000) and efficiently accept peroxide radicals (inhibition coefficient f>1) (table 4.13).

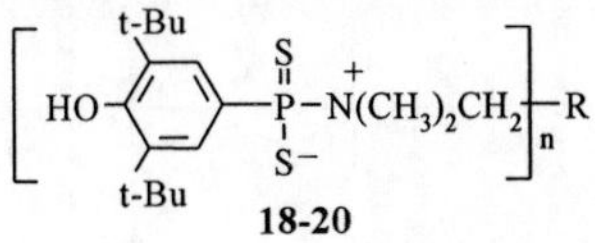

18-20

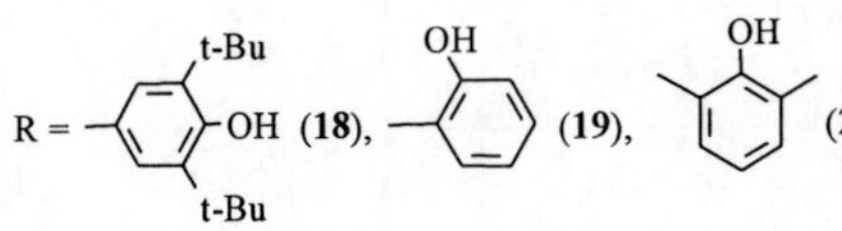

(18), (19), (20), n = 1 (18, 19), 2 (20)

Table 4.13. Kinetic and stoichiometric parameters of interaction of betaines (18-20) with cumene hydroperoxide (CHP) and peroxide radicals

Compound	Amyl alcohol (353 K)			Styrene (325 K)	
	ν_{CHP}	k×10³ l/(mole·s)	E_{act}, kJ/mole	f	k_7×10⁴ l/(mole·s)
(18)	3000	44.8	54.6	1.2	2.64
(19)	500	30.0	56.8	1.25	3.73
(20)	2000	350.0	32.5		

The characteristics of antioxidant action assume high inhibiting activity of betaines (18-20) when oxidative processes in polymers are inhibited. Thus, during high temperature oxidation of LDPE and lubricating oils in the presence of betaines (18-20) the induction periods (τ) of oxidation are considerably higher than when Antioxidant 2246 (figure 4.11, table 4.14) is used.

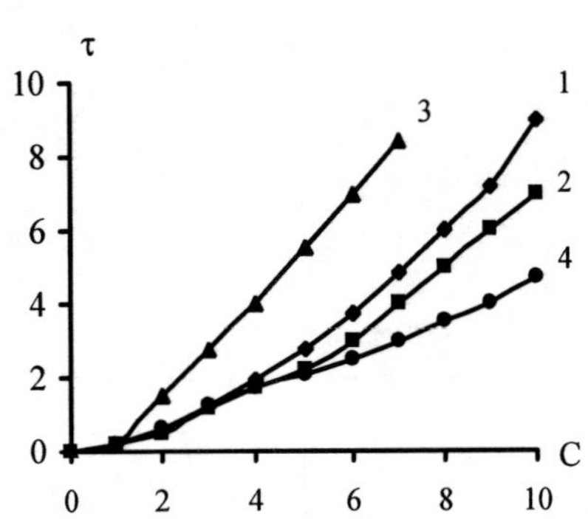

Figure 4.11. Dependence of induction period (τ, h) of LDPE oxidation on the concentration of compounds (18-20): 1 – compound (18); 2 – (19); 3 –(20); 4 – Antioxidant 2246; (C×10^{-3} mole/kg , 473 K, oxygen pressure is 33.3 kPa).

Table 4.14. Influence of the concentration of compounds (18-20) on induction period (τ) of lubricating oils oxidation (473 K, oxygen pressure is 33.3 kPa)

Compound	Paraffin oil		Di-*i*-octyladipate	
	Content of a stabilizer, wt%	τ, min	Content of a stabilizer, wt%	τ, min
–	–	15	–	20
(18)	0.5 1.0	350 730	0.25 0.5	300 980
(19)	0.5 1.0	310 640	0.25 0.5	290 820
(20)	0.5 1.0	540 900	0.5	1100
Antioxidant 2246	0.5	270	0.5	390

Being antioxidants of combined action betaines (18-20), as well as phosphorus dithioacids, in the combination with Antioxidant 2246 and dilauryl thiodipropionate (DLTDP) form mixtures possessing synergetic antioxidant effect (figure 4.12).

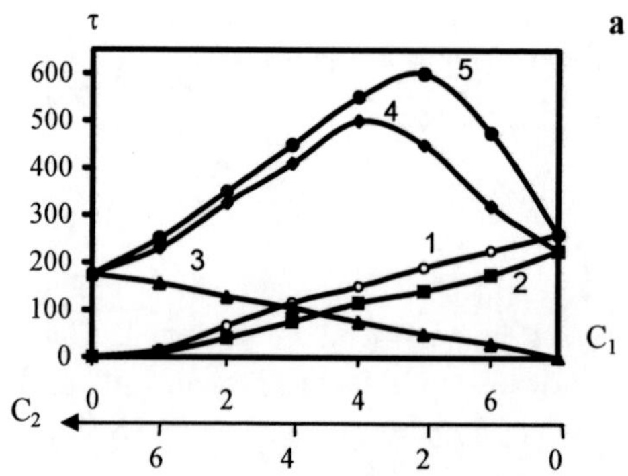

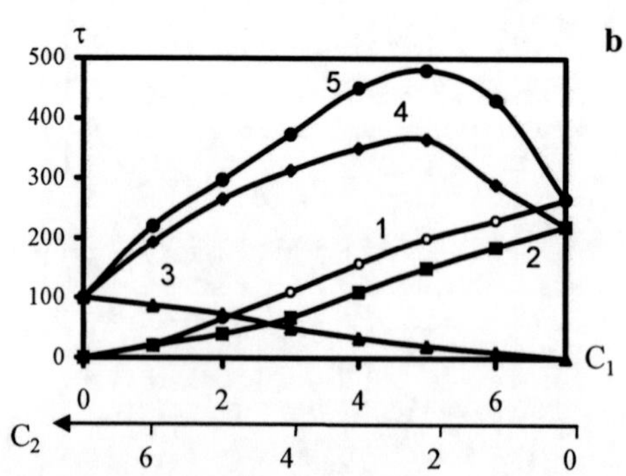

Figure 4.12. Dependence of induction period (τ, min) of LDPE oxidation on the component ratio of stabilizing mixtures ($C\times10^{-3}$, mole/kg, 473 K, oxygen pressure is 33.3 kPa): C_1, C_2 is content of betaine and DLTDP (a) or Antioxidant 2246 (b), respectively: a: 1 – compound (18); 2 – (19); 3 – DLTDP; 4 – (18) + DLTDP; 5 – (19) + DLTDP; b: 1 – compound (18); 2 – (19); 3 – Antioxidant 2246; 4 – (18) + Antioxidant 2246; 5 – (19) + Antioxidant 2246.

4.3. Dithiophosphonates of 3d-Metals

Dithiophosphonates of 3d-metals (ML_2, 21-24) are also polyfunctional stabilizers (Section 2.2.1.4), that predetermines developing "intramolecular" antioxidant synergism during the polymer stabilization.

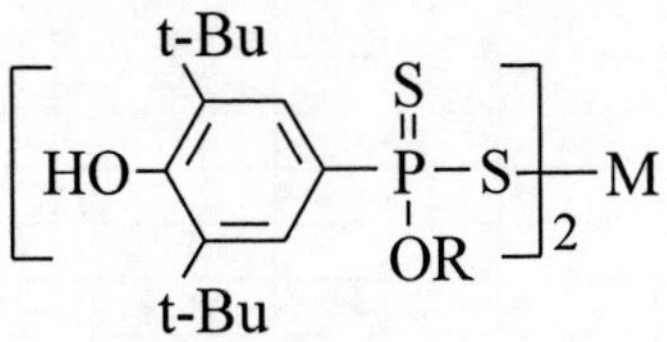

21-24

R= C_2H_5 (**21, 23**), i-C_3H_7 (**22, 24**),
M = Zn (**21, 22**), Ni (**23, 24**)

Values of oxidation induction period of LDPE, stabilized by dithiophosphonates (22) and (24), increase considerably in comparison with the same values for the mixture of Antioxidant BHT with zinc (or nickel) diisopropyldithiophosphate (table 4.15).

Table 4.15. Influence of 3d-metal dithiophosphonates on the induction period (τ) of LDPE oxidation

Stabilizer (0.01 mole/kg)	τ^a, min	Stabilizer (0.01 mole/kg)	τ^a, min
BHT	70	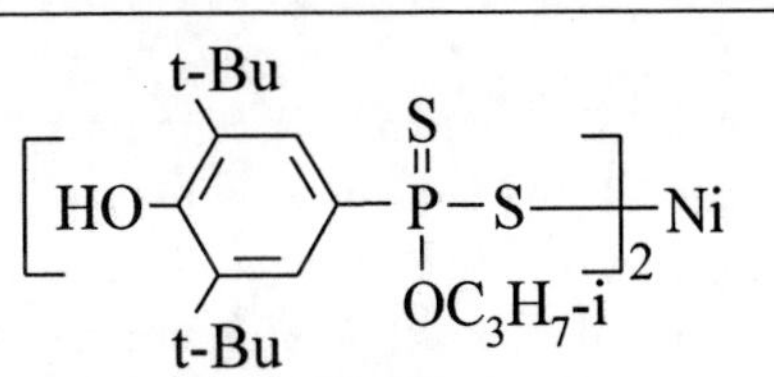 **(24)**	1244
$[(i\text{-}PrO)_2P(S)S]_2Zn$	480		
$[(i\text{-}PrO)_2P(S)S]_2Ni$	470		
$[(i\text{-}C_3H_7O)_2P(S)S]_2Ni$ + BHT (1:1) [b]	880	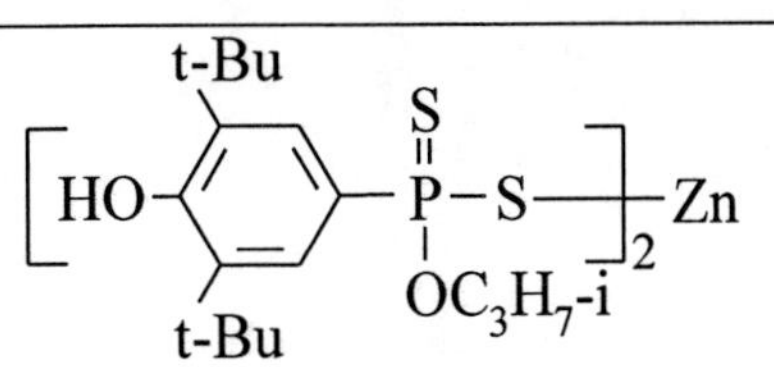 **(22)**	1264
Irganox 1010	360		

[a] 473 K, oxygen pressure is 33.3 kPa; [b] S = 2.2, S_{pr} = 1.87.

Dithiophosphonates (22) and (24) containing sterically hindered phenol fragment are efficient antioxidant additives for lubricating synthetic and mineral oils, they exceed commercial additive zinc isobutylisooctyldithiophosphate 1.5-2 times according to their antioxidant activity (table 4.16).

Table 4.16. Influence of 3d-metal dithiophosphonates on stability of lubricating oil properties

Compound	Content of a stabilizer, wt%	τ [a], min	Q_{cr} [b], kg	Q_w [c], kg	Burr index	Wear factor
Paraffin oil						
–	–	15	15	20	20	30
(22)	0.5	120				
	1.0	560	100	248	25.2	25.4
(24)	0.5	300				
	1.0	540	100	230	26.5	28.1
Zinc *i*-butyl-*i*-octyl dithiophosphate	0.5	90				
	1.0	240	100	200	26.7	22.1
Di-*i*-octyladipate						
–	–	20				
(22)	0.25	320				
	0.5	1000	100	220		26.1
(24)	0.25	390				
	0.5	1000	100	220		27.3
Zinc *i*-butyl-*i*-octyl dithiophosphate	0.25	70				
	0.5	170	100	200		23.0

[a] τ – oxidation induction period, 473 K, 33.3 kPa; [b] Q_{cr} is critical load; [c] Q_w – welding load.

When antioxidant action of phosphorus dithioacids and metal dithiophosphonates is compared the increase of induction period of polymer oxidation in the presence of metal dithiophosphonates is observed (figure 4.13). It should be connected with the participation of metal ions in the antioxidant action of inhibitors owing to the activation of ligand in reactions responsible for the polymer stabilization.

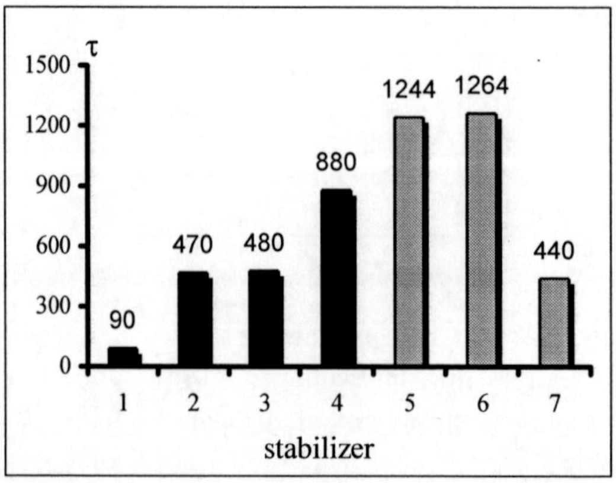

Figure 4.13. Influence of the stabilizer structure on induction period (τ, min) of LDPE oxidation (content of a stabilizer is 0.01 mole/kg ; 473 K, oxygen pressure is 33.3 kPa): 1 – BHT; 2 – $[(i\text{-}C_3H_7O)_2P(S)S]_2Ni$; 3 – $[(i\text{-}C_3H_7O)_2P(S)S]_2Zn$; 4 – $[(i\text{-}C_3H_7O)_2P(S)S]_2Ni$ + BHT (1:1); 5 – compound (24); 6 – compound (22); 7 – compound (11).

It is determined that metal complexes (22) and (24) provide long induction periods of butyl rubber oxidation (table 4.17). Rheometry characteristics of the butyl rubber vulcanizate indicate that compounds (22) and (24) participate also in the process of vulcanization, reducing the time of getting the optimum of vulcanization t_{90} and increasing scorch time.

Table 4.17. Influence of a stabilizer on the induction period (τ) of butyl rubber oxidation and the stability of physical and mechanical properties of butyl rubber vulcanizates [a]

Parameter	Stabilizer				
	0.3 wt%			0.5 wt%	
	Neozone D	NiL_2 (24)	ZnL_2 (22)	NiL_2 (24)	ZnL_2 (22)
τ [b], min	90	300	780		
Minimum torque (M_{min})	10.5			11.0	10.5
Maximum torque (M_{max})	31.5			28.0	27.4
Cure time (t_{90}), min	45			40	42
Module 300%, MPa	4.4	4.3	4.2	4.1	4.0
Tensile strength (σ_{ts}), MPa	19	23	21	23	23
Elongation at break (ε_b),%	680	720	680	700	710
Residual elongation, %	36	32	30	36	34
Shore hardness	62	60	60	61	60
Abradability, $sm^3/kW{\cdot}h$	307	350	300	340	320
Thermal ageing (373 K, 72 h)					
Stability index calculated from σ_{ts}	0.70	0.70	0.73	0.72	0.73
Stability index calculated from ε_b	0.70	0.72	0.74	0.74	0.75

[a] vulcanizate formulation №2 (see Appendix to Chapter 4); [b] 443 K, oxygen pressure is 33.3 kPa.

It is known that the joint use of dialkyldithiophosphates with phenol or amine antioxidants can provide the efficient stabilization of isoprene rubber vulcanizates [8]. Combined introduction of metal complex (24) and N-nitrosodiphenylamine permits to increase stability index calculated from main physical and mechanical parameters and also to increase adhesion to metal cord both before and after ageing of vulcanizate (table 4.18).

Table 4.18. Influence of modifier structure on physical and mechanical properties of isoprene rubber vulcanizates [a]

Parameter	Modifier		
	N-nitrozodi-phenylamine	$NiCl_2 + ZnO$ (1:1) [b]	(24) + ZnO (1:1) [c]
Adhesion, N	386	478	440
Module 300%, MPa	13.8	10.8	12.3
Plasticity	0.25	0.35	0.30
Tensile strength (σ_{ts}), MPa	24.8	23.4	21.7
Elongation at break (ε_b),%	460	525	460
Tear resistance, kN/m	12.0	10.6	10.0
Thermal ageing (373 K, 72 h)			
Stability index calculated from σ_{ts}	0.80	0.82	0.90
Stability index calculated from ε_b	0.81	0.83	0.89
Stability index calculated from adhesion	0.81	0.83	0.90

[a] vulcanizate formulation №3 (see Appendix to Chapter 4); [b] ($NiCl_2$+ZnO) (1:1) – 3.5 parts by weight, sulfur – 3.4 parts by weight; [c] 3.5 parts by weight.

In whole, polyfunctional dithiophosphonates of 3d-metals, containing a sterically hindered phenol fragment, display properties of antioxidants, vulcanization accelerators, modifiers and adhesives. They were stated as additives of wide range of action for a number of polymers, exceeding commercial stabilizers according to their properties in most cases. High efficiency of their action is connected with intramolecular synergetic antioxidant effect.

4.4. Phosphonates Containing a Sterically Hindered Phenol Fragment

Results of testing phosphonates (25-30) under high temperature oxidative ageing of LDPE characterize them as efficient antioxidants (table 4.19). The length of alkyl substitute influences little on the value of induction period (τ) of LDPE oxidation, however phosphonate solubility in polymer increases with the increase of carbon atoms number in the substitute. The high solubility (δ_S) of *i*-octyl-3,5-di-*tert*-butyl-4-hydroxybenzylphosphonate (30) in polyethylene should be noted, it exceeds the given parameter for phosphonates with alkyl substitutes C_1–C_4 (27-29) for an order.

Table 4.19. Induction periods (τ) of LDPE oxidation in the presence of phosphonates (25-30) and their solubility in polymer

Stabilizer (0.2 wt%)	τ [a], min	Solubility (δ_s, wt%)	
		At 293 K	At 333 K
(25)	180	0.10	0.15
(26)	150	0.17	0.25
(27)	150	0.18	0.27
(28)	150	0.20	0.30
(29)	280	0.45	0.50
(30)	180		3.35
Antioxidant 2246	175		
BHT	50		
tris(*p*-nonylphenyl)phosphite	40		

[a] oxygen pressure is 33.3 kPa, 473 K.

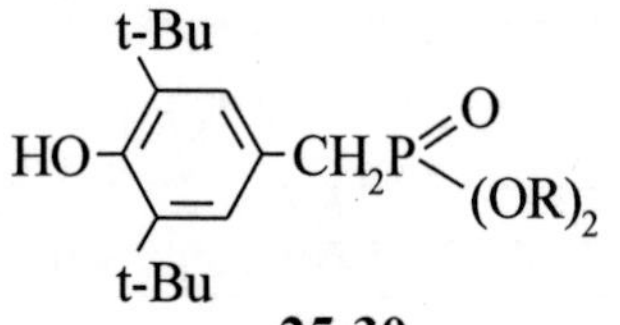

25-30

R=CH_3 **(25)**, C_2H_5 **(26)**, i-C_3H_7 **(27)**, C_4H_9 **(28)**, C_6H_5 **(29)**, C_8H_{17} **(30)**

However, the phosphonates (25-30) are of interest not only as additives able to accept peroxide radicals (it is this mechanism of action that is typical of them (see Chapter 2)), but also as efficient color stabilizers. The possibility of their action in the process of color stabilization is provided by a phosphonate fragment, conditioning impossibility of the formation of quinoid structure.

It should be noted that stabilizing mixtures of phosphonates (25-30) with phenol antioxidant Irganox 1010 provide not only high thermal stability, but transparency of poly-4-

methylpentene-1 (PMP) as well (table 4.20). Binary charts "*content – property* "(figure 4.14) indicate the appearance of antioxidative synergetic effect when stabilizer mixtures are used.

Table 4.20. Influence of a stabilizer on the physical and mechanical properties of poly-4-methylpentene-1

Parameter	Stabilizer		
	Irganox 1010 (0.5 wt%)	Irganox 1010 (0.2 wt%)	
		Phosphonate (25) (0.3 wt%)	Phosphonate (29) (0.3 wt%)
[a] τ_D, h (438 K)	300	310	360
Melt flow index ($MFI_{2.16}$), g/4 min (553 K, load is 2.16 kg)	3.3	3.25	3.33
Yield strength (σ_{ys}), MPa	27.5	28.5	28.5
Light refraction coefficient, % disk of 1 mm disk of 2 mm	 88.0 85.0	 90.5 87.0	 91.0 87.0
Thermal ageing (423 K, 10 hours)			
Stability index calculated from MFI	0.68	0.75	0.80
Stability index calculated from σ_{ys}	0.87	0.91	0.92

[a] time before the destruction of molded disks.

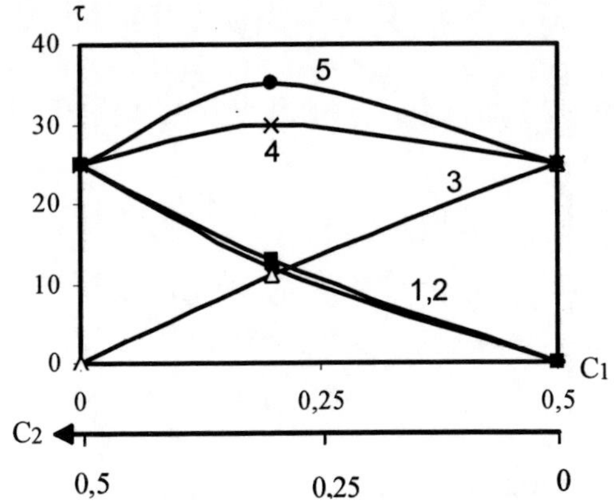

Figure 4.14. Dependence of induction period of PMP oxidation (τ, h) on the ratio of components of stabilizing mixtures (C_1 C_2 (wt%), content of Irganox 1010 and phosphonate respectively, 473 K, air): 1 – compound (25), 2 – (29), 3 – Irganox 1010; 4 – Irganox 1010 + compound (25), 5 – Irganox 1010 + compound (29).

Stabilizing action of phosphonates (25-30) in plasticized polyvinylchloride (PVC) composites is evaluated under high-temperature oxidation of plasticizer, as PVC stability to thermal-oxidative destruction is defined considerably by plasticizer stability as well [29-32].

Oxidation of a nonstabilized plasticizer starts practically straight away, while the introduction of phosphonates (25-30) leads to the occurrence of oxidation induction period (figure 4.15). An alkoxyl substitute at a phosphorus atom influences insignificantly the oxidation induction period. After the completion of the induction period the reaction rate of substrate oxidation is somewhat lower than the reaction rate of noninhibited oxidation, it indicates the formation of a new antioxidant during oxidation. The induction period (τ) is proportional to the content of phosphonates (25-30) in polymer (figure 4.16).

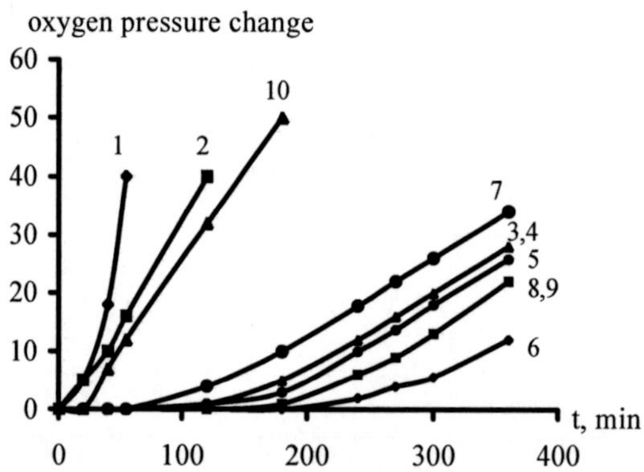

Figure 4.15. Dependence of the rate of oxygen uptake during oxidation of plasticizers dioctylphthalate (DOP) (curves 1,3,5) and dioctylsebacate (DOS) (curves 2,4,6,7,8,9,10) on the structure of stabilizing additives (0.25 wt%, oxygen pressure is 33.3 kPa, 473 K): 1 – DOP without a stabilizer, 2 – DOS without a stabilizer, 3, 4 – compound (25), 5, 6 – (26), 7 – (27), 8 – (28), 9 – Santonox R, 10 – phosphite $(C_6H_5O)_2P(OC_8H_{17}\text{-i})$.

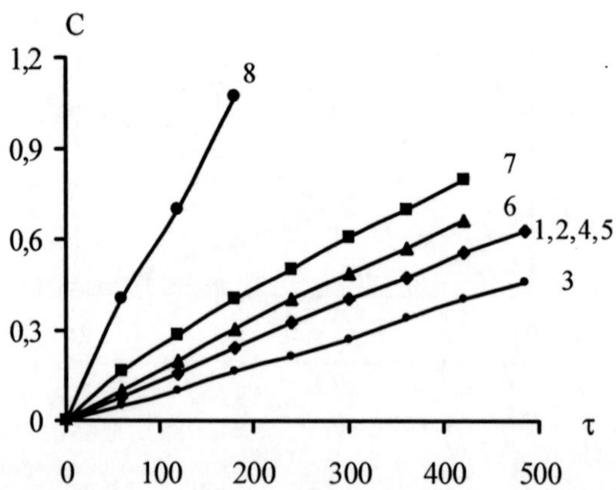

Figure 4.16. Dependence of oxidation induction period (τ, min) of plasticizers dioctylsebacate (DOS) (curves 1,2,5,7,8) and dioctylphthalate (DOP) (curves 3,4,6) on the concentration (C, wt%) of phosphonates (25-27) (oxygen pressure is 33.3 kPa, 473 K). DOS: 1 – compound (26), 2– (25), 5 – Santonox R, 7 – (27), 8 – phosphite $(C_6H_5O)_2P(OC_8H_{17}\text{-i})$. DOP : 3 – compound (26), 4 – (25), 6 – (27).

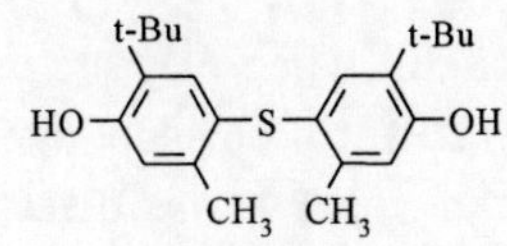

Santonox R

Antioxidant efficiency of phosphonates (25, 26) in plasticized PVC-composite exceeds the efficiency for mixtures of widely used additives Santonox R and diphenyl-*i*-octylphosphite (table 4.21). In addition, phosphonates (25, 26) increase film whiteness and preserve better elasticity of polymeric material in comparison with test stabilizing mixture.

Table 4.21. PVC-composite properties

Formulation number	1	2	3	4	5	6	7	8
Parts by weight								
PVC	100	100	100	100	100	100	100	100
Dioctylphthalate	50	50	50	50	50	50	50	50
DAFF [a]	20	20	20	20	20	20	20	20
Cadmium stearate	–	–	–	–	–	2	2	2
SKS K-17 [b]	2	2	2	2	2	–	–	–
ED-16 [c]	3	3	3	3	3	2.7	2.7	2.7
Santonox R	1	1	–	–	–	1	1	–
$(C_6H_5O)_2P(OC_8H_{17}\text{-i})$	0.5	–	–	–	–	0.5	–	–
Phosphonate (25)	–	0.5	0.5	1.0	2.0	–	–	–
Phosphonate (26)	–	–	–	–	–	–	0.5	1.5
Parameters								
Thermal stability at 448 K, min	100	120	100	115	125	99	120	120
Thermal ageing (433 K, 4 h)								
Stability index calculated from								
– Tensile strength	0.68	0.69	0.65	0.69	0.80	0.75	0.63	0.69
– Elongation at break	0.58	0.71	0.71	0.68	0.68	0.41	0.79	0.89
– Whiteness	0.22	0.46	0.44	0.43	0.42	0.31	0.47	0.31
– Rigidity	2.68	2.58	3.32	2.30	2.43	–	–	–
Light and thermal ageing (473 K, 100 h)								
Stability index calculated from								
– Tensile strength	1.19	1.04	1.22	1.13	1.14	0.74	0.79	0.77
– Elongation at break	0.54	0.49	0.52	0.52	0.61	0.58	0.83	0.78
– Whiteness	0.31	0.46	0.43	0.45	0.47	0.91	0.94	0.97
– Rigidity	2.56	2.97	4.53	4.10	3.48	1.97	1.17	1.00

[a] dialkyl ester of alkylphosphorus acid;

[b] stabilizer on the basis of Ba-, Cd-, Zn-salts of synthetic fatty acid C_{10}-C_{13};

[c] epoxide on the basis of epichlorhydrin and diphenylolpropane.

The ability of phosphonates (25-30) when they are used jointly with Antioxidant 2246 and dilaurylthiodipropionate (DLTDP) to develop synergism of antioxidant action is important from the practical point of view, it was detected when the stabilizing action of phosphonates (25-30) in polyethylene was studied (figure 4.17).

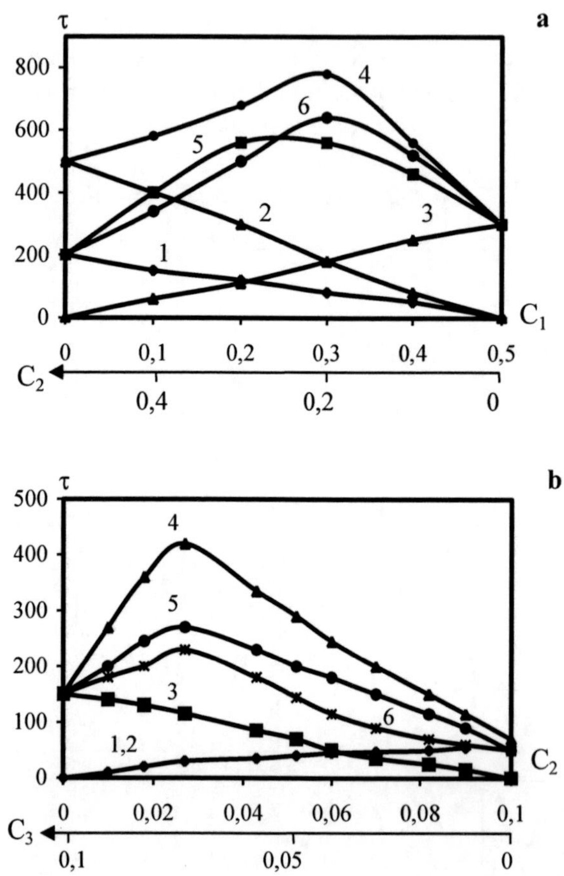

Figure 4.17. Dependence of induction period (τ, min) of LDPE oxidation on the ratio of components of stabilizing mixtures C_1, C_2, C_3 (wt%) is the concentration of DLTDP, phosphonate and Antioxidant 2246, respectively (473 K, oxygen pressure is 33.3 kPa); a: 1 – phosphonate (25), (27), 2 – phosphonate (29), 3 – DLTDP; 4 – DLTDP + (29); 5 – DLTDP + (27); 6 – DLTDP + (25); b: 1 – phosphonates (27), (28), 2 – phosphonate (29), 3 – Antioxidant 2246; 4 – Antioxidant 2246 + (29); 5 – Antioxidant 2246 + (27); 6 – Antioxidant 2246 + (28).

Apparently, in the first case the mechanism of homosynergism ($RO_2^{\cdot} + RO_2^{\cdot}$) is realized, the considerable difference of rate constants of interaction of the given stabilizers during the initiated styrene oxidation (323 K) counts in its favour, the difference is as follows: $k_7 = 1.38\times10^{-4}$ l/(mole·s) for Antioxidant 2246; $k_7 = 4.4\times10^{-5}$ l/(mole·s) for phosphonate (26). It is under this condition that synergism of inhibiting action is stated most frequently. In the second case the mechanism of heterosynergism ($RO_2^{\cdot}$ + ROOH) is realized.

Thus, the investigated phosphonates (25-30) display sufficiently high stabilizing ability, performing simultaneously functions of thermal and color stabilizers, they have sufficient solubility in polymers and can be successfully used for polyolefines and PVC stabilization.

4.4. THIOCARBAMIDE DERIVATIVES

According to the conclusions made on the basis of the study of model reactions (Section 2.2.1.5), arylthiocarbamides (31-44) are secondary antioxidants in stabilization processes, namely, hydroperoxide decomposers (ν=40÷500).

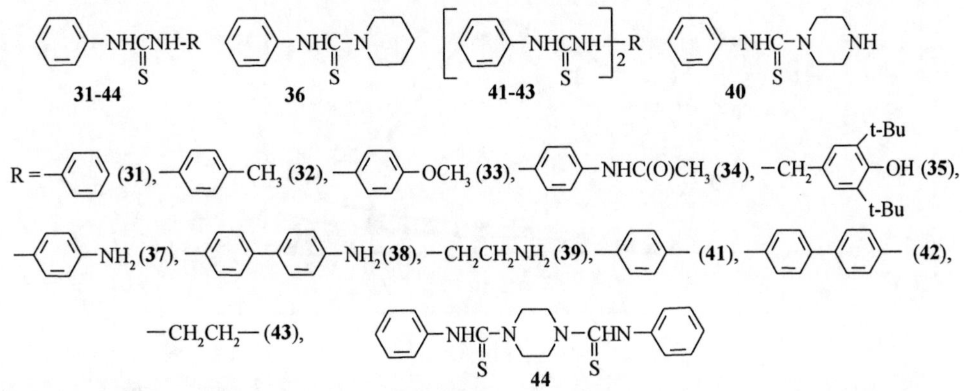

According to the antioxidant activity in the order of its increasing arylthiocarbamides can be arranged in the sequence (36) ≈ (33) ≈ (32) ≈ (31) < (44) ≈ (34) < (43) < (27) < (45), coinciding with the sequence indicated during model tests of arylthiocarbamides (Section 2.2.1.5). However, it is inexpediently to use these compounds individually, as their stabilizing efficiency is lower than that of commercial stabilizers (figure 4.18) [33]. The efficiency of arylthiocarbamides during inhibition of polymer thermal-oxidative destruction increases substantially when a sterically hindered phenol fragment (an additional inhibiting center) is introduced into arylthiocarbamide molecule (compound 35, 45) [34].

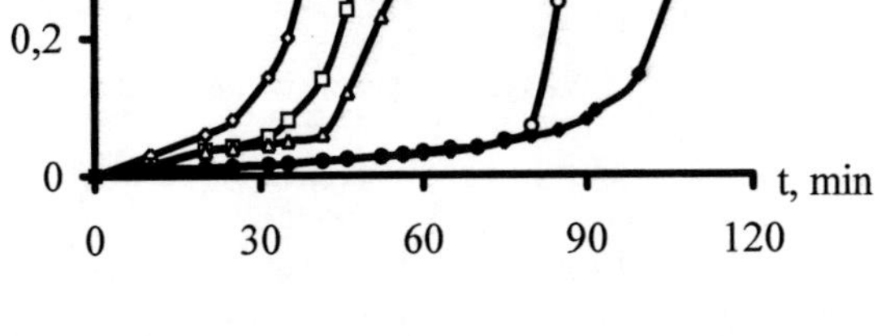

Figure 4.18. Kinetic curves of oxygen uptake ($\Delta O_2 \times 10^2$, mole/l) by oxidable LDPE (473 K, oxygen pressure is 33.3 kPa) in the presence of arylthiocarbamides (0.2 wt%): 1 – compound (31-33), (36); 2 – (34), (44); 3 – (43); 4 – (45), Antioxidant 2246; 5 – (35).

On the other hand, the addition of a peroxide radical acceptor (for example Antioxidant 2246) to arylthiocarbamides as a cocomponent allowed also to increase induction period of oxidation of HDPE and synthetic rubber ethylene-propylene (SREP, a copolymer of ethene with propene-1 and tricycle[5.2.1.0^{5,9}]decadiene-2,6) as a result of antioxidant synergetic effect (figure 4.19, 4.20).

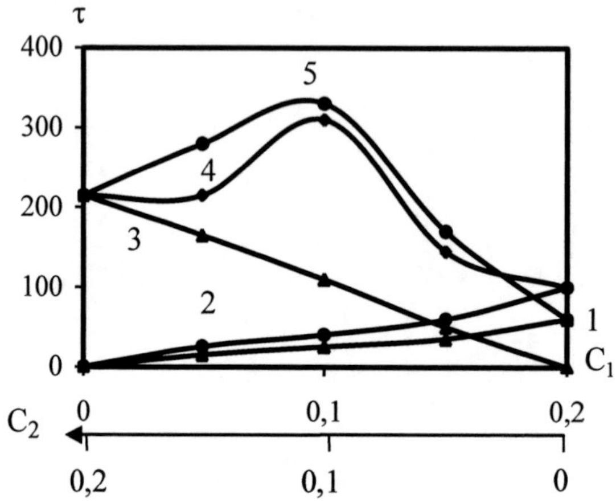

Figure 4.19. Dependence of induction period (τ, min) of HDPE oxidation on the content of components of stabilizing mixtures (473 K, oxygen pressure is 33.3 kPa). C_1,C_2 (wt%) are contents of arylthiocarbamide and Antioxidant 2246, respectively: 1 – compound (35); 2 – (46); 3 – Antioxidant 2246; 4 – mixture of Antioxidant 2246 and (35); 5 – mixture of Antioxidant 2246 and (46).

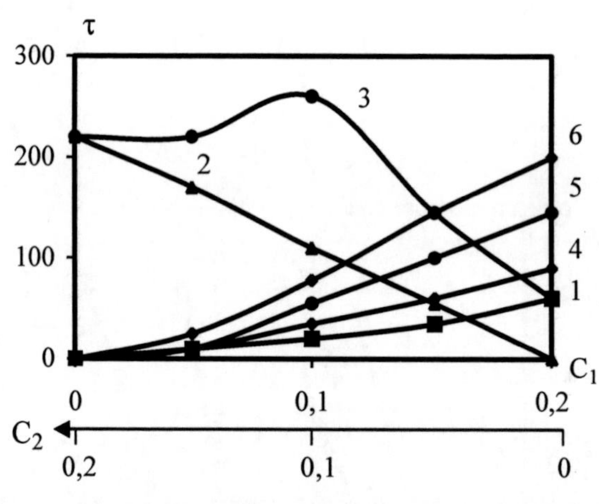

Figure 4.20. Dependence of induction period of SREP oxidation (τ, min) on the content of components of stabilizing mixtures (473 K, oxygen pressure is 33.3 kPa). C_1,C_2 (wt%) are contents of arylthiocarbamide and Antioxidant 2246, respectively: 1 – compound (31), (33), (36); 2 – Antioxidant 2246; 3 – mixture of Antioxidant 2246 and (31); 4 – compound (33), (44); 5 – (45); 6 – (35).

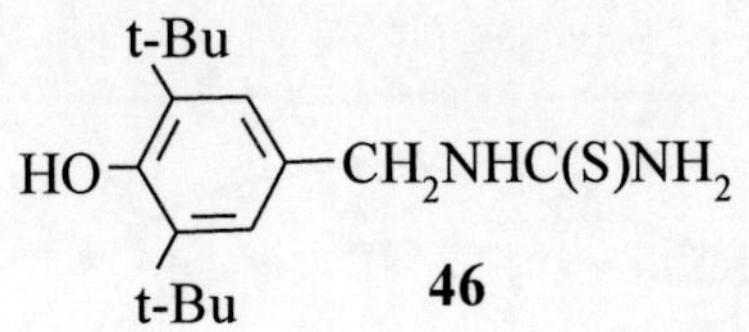

46

Benzoylthiocarbamide derivatives (47-55), unlike arylthiocarbamides, are efficient as hydroperoxide decomposers and peroxide radical acceptors (ν=1000÷13000; *f*=0.5÷2.7). Therefore, arylthiocarbamides should be considered as antioxidants of combined action, they possess higher stabilizing efficiency during thermal oxidation of polymers, it is confirmed by the increase of induction period of polyethylene oxidation (figure 4.21) [34].

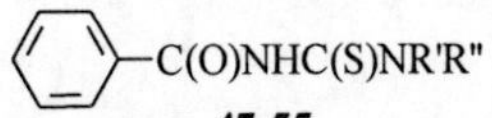

47-55

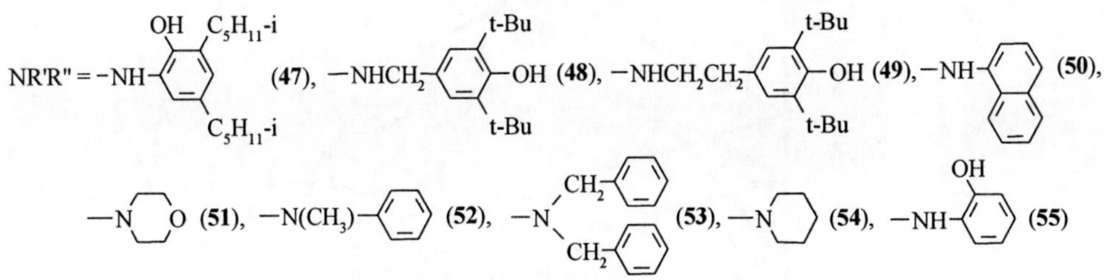

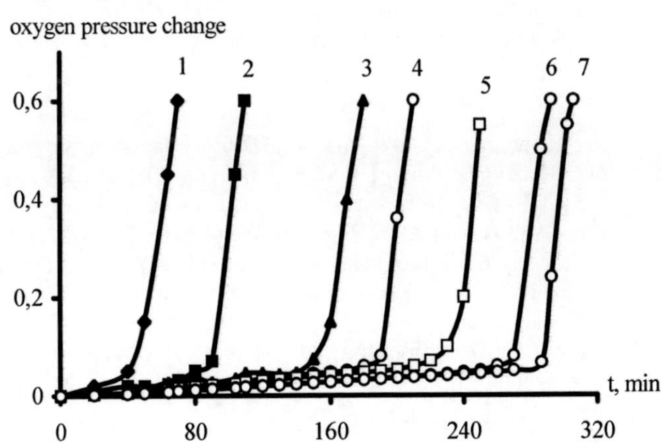

Figure 4.21. Kinetic curves of oxygen uptake ($\Delta O_2 \times 10^2$, mole/l) by oxidable LDPE (473 K, oxygen pressure is 33.3 kPa) in the presence of benzoylthiocarbamides (0.2 wt%): 1 – compounds (50), (51); 2 – (47), (55); 3 – (53), (54); 4 – (52); 5 – Antioxidant 2246; 6 – (48); 7 – (49).

Benzoylthiocarbamides as polyfunctional stabilizers during HDPE autooxidation develop antioxidant synergetic effect when they are used jointly both with peroxide radical acceptors (Antioxidant 2246) and hydroperoxide decomposer (dilaurylthiodipropionate (DLTDP) (figure 4.22).

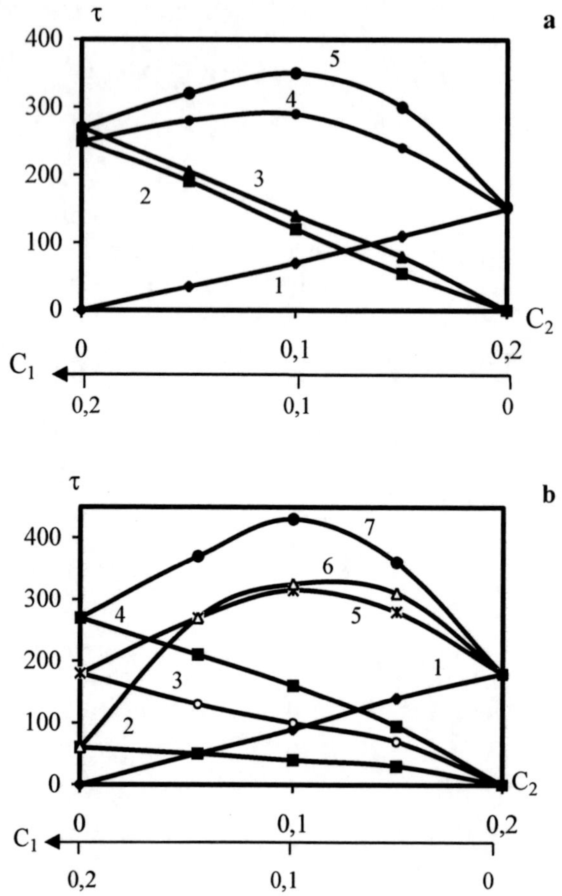

Figure 4.22. Dependence of induction period (τ, min) of HDPE oxidation on the content of components of stabilizing mixtures (473 K, oxygen pressure is 33.3 kPa). a: 1 – DLTDP; 2 – compound (48); 3 – (49); 4 – DLTDP + (48); 5 – DLTDP + (49); C_1,C_2 (wt%) are contents of benzoylthiocarbamide and DLTDP, respectively; b: 1 – Antioxidant 2246; 2– compound (47), (55); 3 – (52), (53); 4 – (48), (49); 5 – Antioxidant 2246 + (47) (or 55); 6 – Antioxidant 2246 + (52) (or 53); 7 – Antioxidant 2246 + (48) (or 49); C_1,C_2 (wt%) are contents of benzoylthiocarbamide and Antioxidant 2246, respectively.

Along with antioxidant action, thiocarbamide derivatives develop color stabilizing effect which is especially evident when they are used jointly with phenol antioxidants. As it is seen from table 4.22, the usage of arylthiocarbamides together with Antioxidant 2246 allows to increase polymer color stability by 3-4 points.

It is known that thiocarbamide derivatives are used for the protection of colorless polymethylmethacrylate (PMMA) from radiation exposure [35, 36], they are also used as photostabilizers of polymers applied in laser equipment. This problem is rather urgent, as the application of conventional stabilizers is considerably limited in this case, since they often worsen optical properties of transparent materials [37, 38].

The influence of thiocarbamide derivatives (31) and (54) on photostability of copolymer of methylmethacrylate (MMA) with methacrylic acid (MAA) is studied; for comparison N,N'-diphenylurea (56) and 1,6-hexamethylene-bis(N,N-dioxyethylurea) (57) are used as compounds having no sulfur atom.

Table 4.22. Influence of arylthiocarbamides on color stability of HDPE and SREP

Stabilizer	Content of a stabilizers, wt%	Color, points [a]			
		Oxidation time, h			
		0	1	2	3
HDPE (air, 423 K)					
Antioxidant 2246	0.2	1	4	5	7
Thiocarbamide (46)	0.2	1	3	3	4
(46) + Antioxidant 2246	0.1 + 0.1	1	3	3	4
Thiocarbamide (36)	0.2	1	3	4	5
(36) + Antioxidant 2246	0.1 + 0.1	1	3	4	5
Thiocarbamide (33)	0.2	1	2	2	3
(33) + Antioxidant 2246	0.1 + 0.1	1	2	2	3
SREP (air, 443 K)					
Antioxidant 2246	0.3	2	5	7	8
Thiocarbamide (46)	0.3	2	2	2	3
(46) + Antioxidant 2246	0.15 + 0.15	2	3	4	4
Thiocarbamide (33)	0.3	2	3	3	4
(33) + НГ 2246	0.15 + 0.15	2	3	4	4
Thiocarbamide (43)	0.3	2	3	3	4
(43) + НГ 2246	0.15 + 0.15	2	3	4	4

[a] according to 10-point color scale.

The kinetics of photoageing of test and doped copolymers of MMA with MAA is illustrated by the dependence of relative change of optical transmission coefficient (K/K_0) of copolymer on the time of photoexposure (t) given at figure 4.23. K_0 and K are optical transmission coefficients (%) of copolymer at λ=300 nm, where the largest change is observed before and after photoexposure. As it follows from the obtained kinetic curves, noticeable decrease of optical transmission takes place during photoexposure, it is conditioned by the formation of new chromophoric groups, absorbing ultraviolet light, in the course of photo-oxidative destruction of copolymer [39]. The introduction of additives of carbamide derivatives into copolymer leads to the decrease of degree and the rate of optical transmission coefficient lowering. Among all studied thiocarbamide derivatives more noticeable photostabilizing effect is achieved by means of sulfur containing additives (31) and (54).

The influence of carbamide derivatives on the kinetics of accumulation of free radicals ($[R^{\cdot}]$) in copolymer of MMA with MAA, and relative change of solution viscosity (η/η_0) during photoexposure is given in figure 4.24, the additive (31) is taken as an example. It can be seen from the comparison of curves that the nature of kinetic dependence η/η_0 – (t) is practically antibate to the nature of dependence $[R^{\cdot}]$ – (t). At the same time the number of breaks of macrochains (it can be judged by the value η/η_0) and the number of free radicals, formed in doped copolymer, is considerably less than in test one. It is the confirmation of photostabilizing influence of compound (31) on copolymer.

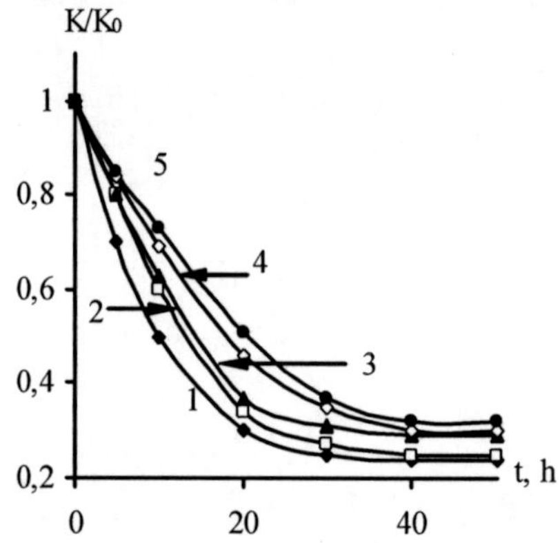

Figure 4.23. Dependence of relative change of optical transmission coefficient (K/K_0) of copolymer MMA with MAA on the photoexposure time (t): 1 – without additives, 2 – compound (57), 3 – (56), 4 – (55), 5 – (31). Content of additives is 0.004 mol%.

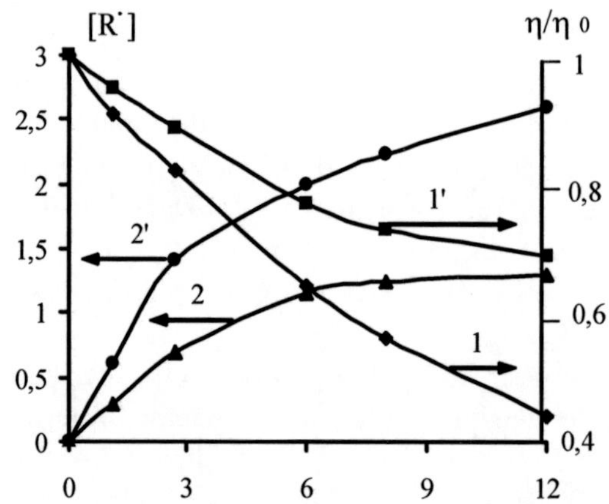

Figure 4.24. Kinetics of free radical accumulation in copolymer of MMA with MAA (1,1') and the dependence of relative change of copolymer solution viscosity in DMF (2,2') η/η_0 (sm^3/g) during copolymer photoexposure: 1,2 – without an additive; 1',2' – with compound (31). Content of an additive is 0.008 mol%. (η and η_0 – intrinsic viscosity of copolymer solution before and after photoexposure of samples, respectively).

At the same time carbamide derivatives can display properties of ultraviolet absorbers [33]. The ability of investigated carbamide derivatives to absorb light of ultraviolet and visible regions is estimated according to the difference in values of optical transmission coefficient (ΔK_{ab}) of test and doped copolymer, found respectively at λ=300 nm and λ=500 nm (table 4.23). All carbamide derivatives decrease optical transmission of copolymer in ultraviolet spectral region in a varying degree, displaying properties of weak ultraviolet absorbers (the change of optical transmission coefficient (K) in the visible region is far less noticeable).

Table 4.23. Influence of carbamide derivatives (0.004 mol%) on optical transmission coefficient of copolymer MMA with MAA

Additive	ΔK_{ab}, %		$(K/K_0)/(K'/K'_0)$ [a]
	λ=300 nm	λ=500 nm	
(31)	24	3	1.80
(54)	10	3	1.70
(56)	17	2	1.28
(57)	19	3	1.14

[a] K/K_0 and K'/K'_0 are obtained for doped and control copolymers at λ=300 nm and 293 K, respectively.

However, the correlation between the ability to absorb ultraviolet radiation and photostabilizing properties of thiocarbamide derivatives is not traced. Thus, additive (54), having the least value ΔK_{ab}, makes a comparatively large photostabilizing impact on copolymer, than compounds (31) and (57). The tests also indicated that the value ΔK_{ab}, found both for UV and visible spectral regions, is practically directly proportional to the concentration of additives when the concentration of thiocarbamide derivatives is equal to 0.02 mol%. It is illustrated by figure 4.25, where dependence $\Delta K_{ab.}$ (at λ=300 nm) on the concentration of an additive in copolymer obtained for the compound (31) is given. The dependence *(K/K₀) – the additive concentration* has practically linear character: the increase of compound (31) content in copolymer leads to the increase of copolymer photostability.

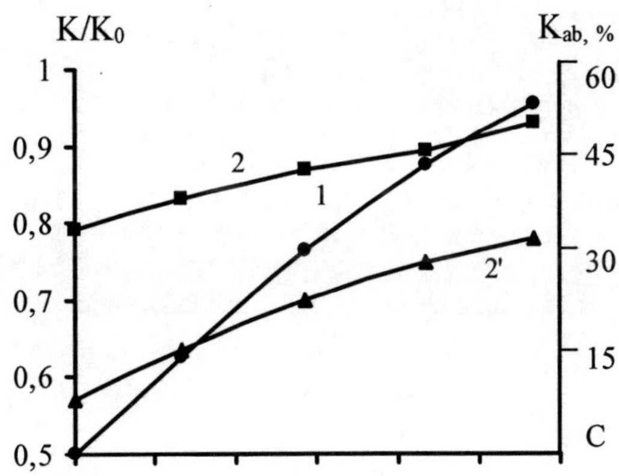

Figure 4.25. Dependence ΔK_{ab} (1) and K/K_0 (2, 2'), obtained at λ=300 nm for copolymer of MMA with MAA, on the additive concentration ($C\times10^{-3}$ mol%) (31). 2 – photoexposure time 10 h, 2' – photoexposure time is 50 h.

The observed relatively high protective efficiency of compounds (31) and (54) agrees with the data obtained in [40]. It can be connected with the ability of thiocarbamide to tautomeric transformations:

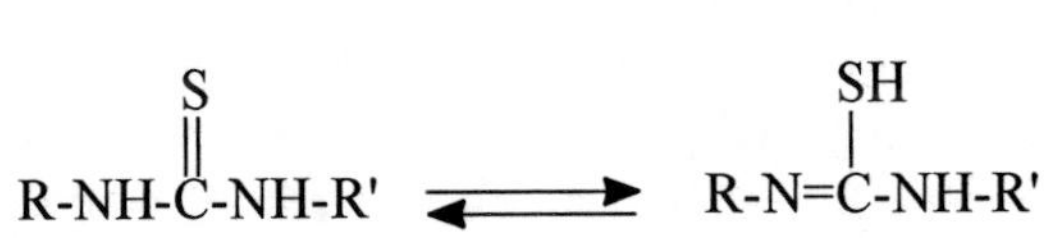

owing to it the quenching of photoexitation of chromophoric macromolecule groups is likely to be possible.

It is also revealed that additives (31) and (54) are able to inhibit the process of photodestruction of dyestuff of polymeric matrix. In figure 4.26 the dependence $D/D_0 - (t)$, reflecting kinetics of photodecolorization of Rhodamine 6G in copolymer of MMA with MAA without additives and with additives of carbamide derivatives is given, where D_0 and D are initial and current optical density in the maximum of absorption of colored samples.

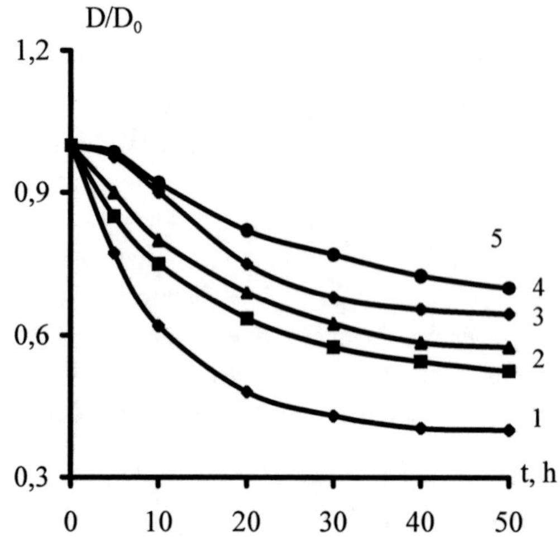

Figure 4.26. Kinetic curves of photodecolorization of Rhodamine 6G in copolymer of MMA with MAA: 1 – without additives, 2 – compound (57), 3 – (56), 4 – (31), 5 – (54). Additives concentration is 0.004 mol%; Rhodamine 6G concentration is 2×10^{-4} mole/l.

In the course of photoexposure the decrease of D/D_0 with gradual rate lowering was registered, as it takes place in the course of photoageing of colorless copolymer (figure 4.23). Photochemical stability of Rhodamine 6G in doped copolymer is appreciably higher than in control one. At the same time, as it was assumed, the degree and rate of photodecolorization have the least values in the presence of additives (31) and (54), making larger photostabilizing impact concerning also colorless copolymer.

Analyzing kinetic dependences, given in figure 4.26 and figure 4.23, it can be noted that photostabilizing action of carbamide derivatives towards dyestuff is considerably higher than the one obtained towards copolymer itself. It is seen from the comparatively large difference in values D/D_0 than in values K/K_0 for control and doped copolymer. Thus, over 10 hours of exposure the decrease of dyestuff optical density in copolymer containing 0.004 %mol of compound (31) makes 10%, that is 3.8 times less than in control copolymer. At the same time the decrease of optical transmission coefficient of colorless copolymer containing this additive makes 28%, it is only 1.8 times less than the decrease of optical transmission coefficient of control copolymer. Therefore, the increased photostability of Rhodamine 6G in doped copolymer is the result not only of the increasing of photostability of copolymer itself (as matrix), but also of straight inhibiting influence of carbamide derivatives on dyestuff photodestruction. The occurrence of induction period, which is characteristic of inhibitor action, at curves of dyestuff photodecolorization in copolymer containing additives (31) and (54) indicates it definitely. Seemingly, the movable mutually regenerable system *dyestuff (Kp) – additive (RSH)* appears in the presence of additives (31) and (54) in colored copolymer. Radicals of Rhodamine 6G semirestored form (Kp·H) are formed when activated dyestuff molecules interact with polymer medium [37, 41]: $Kp^* + PH \rightarrow Kp^{\cdot}H + P^{\cdot}$.

Photostabilizing influence of additives (31) and (54) on copolymer and dyestuff is confirmed by the results of laser investigations.

Laser investigations indicated that by means of additives (27) and (50) the generation photostability of dyestuff in copolymeric matrix (which was defined by the number of impulses up to generation energy decrease by 50%), can be noticeably increased (upon the average ~ 13 times), raising operation life of colored copolymer as active laser materials.

The investigation results of the influence of additives (31) and (54) on laser stability of copolymer of MMA with MAA are given in figure 4.27 as concentration dependence of N_d/N, where N_d and N are a number of laser impulses, withstood by samples of doped and control copolymer, respectively, at constant laser fluence. The most considerable increase of laser stability was also registered when additives (31), (54) were introduced into copolymer. As it is seen, in the given case the stabilizing influence of thiocarbamide derivatives is much more considerable than the one obtained by them when mercury lamp light impacts on samples (maximum value of N_d/N ~ 60). Judging the value of stabilizing effect, the inhibition of free radical transformations is likely to be dominant in copolymer stabilization by the indicated additives under laser induced destruction. It is also indicated by the considerable increase of the ratio N_d/N as the content of compounds (31) and (54) in copolymer increases, and, at identical concentrations of additives the values of this parameter are comparatively bigger for compound (31). It can be connected with the greater ability of (31) to oxidation owing to the presence of aromatic substitutes [42]. Less increase of laser stability (~ 10 times) is observed when an additive (57) is introduced into copolymer. In comparison with additives (31) and (54) the stabilizing influence of (56) on the stated parameter is negligible.

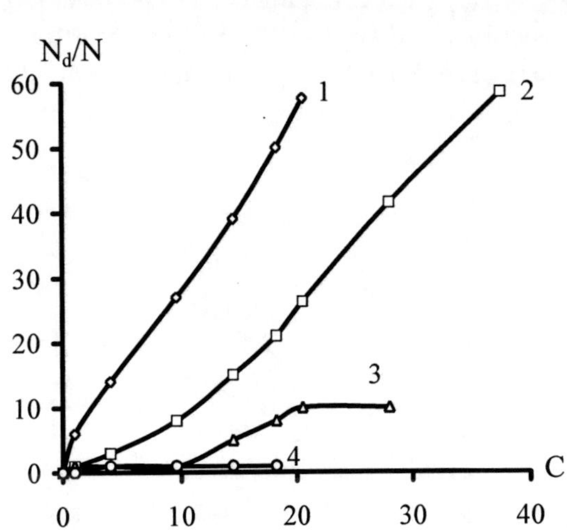

Figure 4.27. Dependence of abundance quantity of laser impulses (N_d/N), withstood by copolymer of MMA with MAA, on the additive concentration ($C\times10^{-3}$ mol%): 1 – compound (31), 2 – (54), 3 – (57), 4 – (56). (λ=252 nm, laser fluence is 0.5 J/sm^2).

On the whole, it can be concluded that thiocarbamide derivatives, being a part of stabilizing mixtures, can be used for polymer stabilization displaying color stabilizing effect alongside with high antioxidant properties. The possibility of using thiocarbamide derivatives for the increase of photochemical (including laser) stability of copolymer of MMA with MAA and generating dyestuff Rhodamine 6G, introduced into it, was stated. It is conditioned (alongside with other mechanisms) by the ability of thiol form of thiocarbamide derivatives to inhibit free radical transformations both of copolymer and dyestuff.

4.6. Arylamines Containing Sterically Hindered Phenol Fragments

Arylamines having a sterically hindered phenol fragment (58-73) contain two types of functional centers which are potentially able to participate in reactions of peroxide radicals accepting. According to the results of modeling investigations (Chapter 2), these compounds possess high inhibiting activity.

t-Bu, HO, CH_2NH, R, t-Bu — **58-63**

t-Bu, HO, $(CH_2)_2N$, R, t-Bu — **64-68**

t-Bu, HO, $CH_2NH\text{-}R\text{-}NH_2$, t-Bu — **69, 70**

t-Bu, HO, $CH_2NH\text{-}R\text{-}NH$, t-Bu, OH, t-Bu — **71-73**

R = OH (**58, 64**), CH_3(**59, 65**), OCH_3(**60, 66**), NHC(O)CH_3(**61, 67**), NH-C_6H_5-CH-C_6H_5(**62**), H (**63, 68**) (**69, 71**), (**70, 72**), -CH_2-CH_2-(**73**)

Dependence of induction period of butyl rubber oxidation on the concentration of compounds (58-63, 69, 71, 73) (figure 4.28) indicates that regarding antioxidant efficiency they exceed phenol stabilizers Antioxidant 2246, Irganox 1010 and amine antioxidant Neozone D, used for the stabilization of the given polymer industrially.

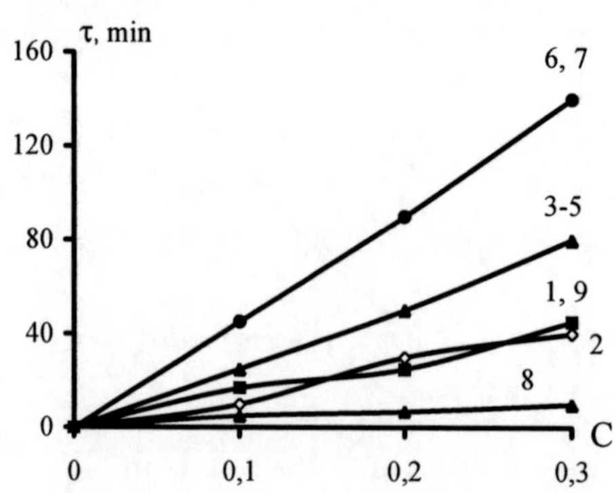

Figure 4.28. Dependence of induction period (τ, min) of butyl rubber oxidation on the concentration of additives (C, wt%) (443 K, oxygen pressure is 33.3 kPa): 1, 9 – Antioxidant 2246, Irganox 1010; 2 – compound (73); 3 – (58); 4 – (59); 5 – (60); 6 – (69); 7 – (71); 8 – Neozone D.

The high antioxidant efficiency of arylamines containing sterically hindered phenol fragments is connected with the effect of "intramolecular" synergism realized by the regeneration of more efficient inhibiting center, namely, amine owing to an exchange reaction with OH-group of a phenol fragment, similar to stabilizing mixtures *phenol – amine*. The formation of antioxidant synergetic effect is illustrated by the comparison of butyl rubber oxidation in the presence of amines (58) or (69) with the mixture of 4-methyl-2,6-di-*tert*-butylphenol (BHT) and arylamine (*p*-aminophenol or *p*-phenylenediamine, respectively). The induction period (τ) of polymer oxidation in the presence of stabilizer (58) increases practically two times in comparison with τ for the mixture BHT with arylamine (table 4.24).

Table4.24. Influence of a stabilizer on the induction period (τ) of butyl rubber oxidation (443 K, oxygen pressure is 33.3 kPa)

Stabilizer	τ, min	Content of a stabilizer, wt%	S	S_{pr}
p-Aminophenol	200	0.2	–	–
BHT	50	0.2	–	–
p-Aminophenol + BHT	280	0.1 + 0.1	1.24	1.40
Compound (58)	500	0.2		
p-Phenylenediamine	250	0.2	–	–
p-Phenylenediamine + BHT	350	0.1 + 0.1	1.33	1.40
Compound (69)	900	0.2		

It should be noted that the efficiency of antioxidative action of arylamines containing sterically hindered phenol fragments increases considerably with the occurrence of the second aminogroup in an aromatic ring of the molecule (69), as well as when two sterically hindered phenol groups are joined by means of aromatic diamine (71). The joining of two groups in the molecule by means of diethylamine bridge (73) doesn't lead to the increase of induction period of polymer oxidation in comparison with Antioxidant 2246. It is likely to be connected with the fact that radical $\dot{N}$-Alk, forming at the inhibition of oxidation, is sufficiently active and enters into various side reactions, leading to unreasonable consumption of antioxidant. In case of radical $\dot{N}$-Ar formation the stabilization of a radical is implemented at the expense of electron delocalization along an aromatic fragment, as a result of it, its activity in the oxidation chain transfer seemingly decreases.

The investigated arylamines also provide higher protection of performance properties of butyl rubber vulcanizate in comparison with stabilizers Neozone D and Antioxidant 2246 (table 4.25).

Table 4.25. Influence of a stabilizer on properties and stability of butyl rubber vulcanizates [a]

Parameter	Neozone D (0.3 wt%)	Stabilizer (0.2 wt%)					
		AO 2246	(58)	(59)	(60)	(69)	(73)
Tensile strength (σ_{ts}), MPa	16.6	21.6	18.8	18.5	20.3	21.6	18.8
Elongation at break (ε_b),%	600	690	560	590	608	671	625
Residual elongation, %	36	30	33	34	34	34	35
Module 300%, MPa	4.1	5.8	4.8	5.2	5.3	4.6	4.0
Shore hardness	62	61	63	63	61	62	63
Thermal ageing (373 K, 72 h)							
Stability index calculated from σ_{ts}	0.72	0.75	0.80	0.80	0.80	0.82	0.70
Stability index calculated from ε_b	0.68	0.74	0.80	0.80	0.80	0.82	0.68

[a] vulcanizate formulation №4 (see Appendix to Chapter 4).

When amines (63, 71, 73) were investigated as antioxidants for isoprene rubber the longest oxidation induction period (as during butyl rubber oxidation) is registered for the stabilizer (71), having two sterically hindered phenol groups and two secondary arylamine groups in the molecule (figure 4.29).

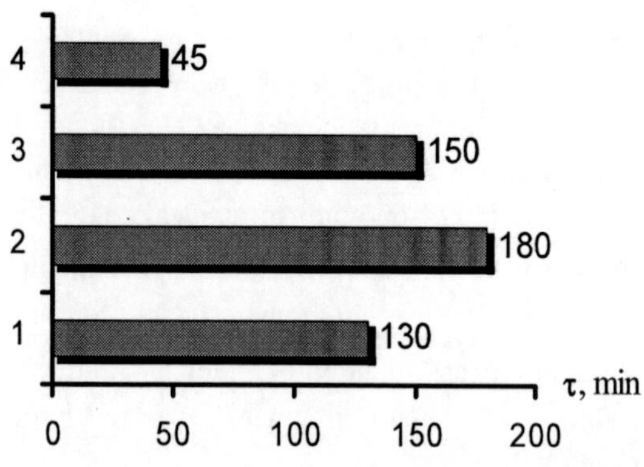

Figure 4.29. Induction period (τ, min) of isoprene rubber oxidation in the presence of stabilizers (0.3 wt%, 438 K, oxygen pressure is 33.3 kPa): 1 –Vulkanox 4010 NA; 2 – compound (71); 3 – (63); 4 – (73).

Arylamines (63) and (71) are capable of displaying antioxidant synergetic effect when they are jointly used with compounds of quinoid structure, which are efficient alkylradical acceptors [43, 44]. Isoprene rubber vulcanizates, containing 3,3',5,5'-tetra-*tert*-butyl-4,4'-diphenoquinone (74) in the formulation of a stabilizing mixture, are characterized by high coefficients of resistance to thermal ageing (table 4.26) [45, 46].

Table 4.26. Influence of a stabilizer on properties and stability of isoprene rubber vulcanizates [a]

Parameter	Stabilizer (0.6 wt%)					
	Vulkanox 4010 NA	(63)	(71)	(74)	(63)+(74)	(71)+(74)
Tensile strength (σ_{ts}), MPa	25.1	26.6	23.5	26.9	25.4	24.9
Module 300%, MPa	6.6	6.2	6.3	7.1	6.4	7.3
Elongation at break (ε_b),%	598	652	598	611	628	593
Residual elongation, %	22.7	26.7	24.7	25.3	22.7	24.0
Tear resistance, N/m	70.7	62.9	71.6	73.8	73.5	68.1
Adhesion, N	71.5	51.0	85.3	51.0	56.8	69.6
Stability index (343 K, 72 h)						
Calculated from σ_{ts}	0.89	0.85	0.90	0.80	0.96	0.96
Calculated from ε_b	0.90	0.80	0.90	0.75	0.97	0.91
Stability index (373 K, 72 h)						
Calculated from σ_{ts}	0.70	0.62	0.65	0.58	0.83	0.83
Calculated from ε_b	0.64	0.62	0.68	0.56	0.78	0.75
Stability index (298 K, 72 h, 10 %-solution of NaCl)						
Calculated from σ_{ts}	0.73	0.77	0.84	0.70	0.84	0.75
Calculated from ε_b	0.84	0.86	0.88	0.80	0.90	0.88

[a] vulcanizate formulation №5 (see Appendix to Chapter 4).

The action efficiency of a number of stabilizers in polypropylene (PP) and isoprene rubber/polypropylene thermoplastic elastomer (TPE) were studied. TPE was obtained by vulcanization, combining the stage of mixing with the stage of vulcanization, which is inevitably accompanied by mechanical destruction of polymer [40], as it was mentioned above.

Arylamines (71) and (73) prevent efficiently mechanical destruction of isotactic polypropylene: the decrease of the torque (M) value after 10 minutes polymer processing slows down with the increase of stabilizer concentration (figure 4.30).

Positive influence of amines containing sterically hindered phenol fragments on strength properties of polypropylene, undergone mechanical destruction, is noticeably stated already at small concentrations of stabilizers (figure 4.31), it is a precondition for their using as additives slowing down thermal-mechanical destruction of polymer.

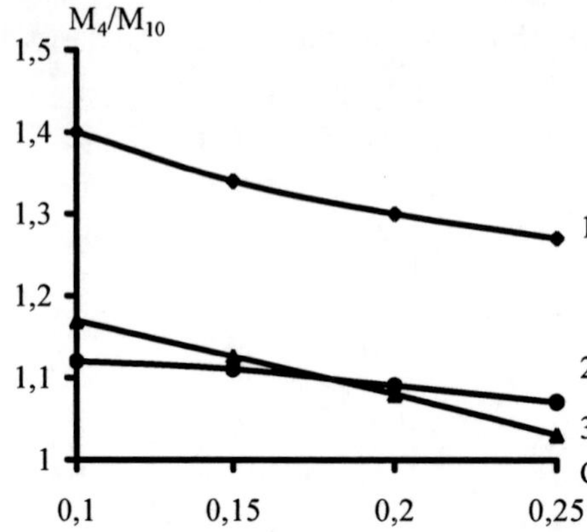

Figure 4.30. Dependence of torque relative change (M_4/M_{10}) on the concentration of a stabilizer (C, wt%) after 10 min polypropylene processing in Brabender plasticorder: 1 – compound (73); 2 – (71); 3 – (63) (T_0=453 K, n=90 rpm); M_4 and M_{10} are torque values after 4 min and 10 min processing, respectively.

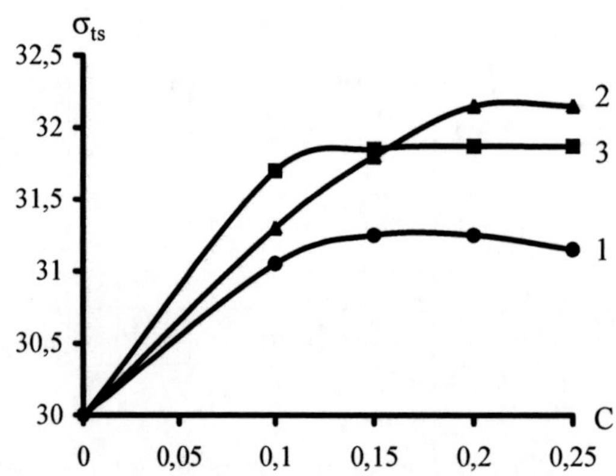

Figure 4.31. Dependence of polypropylene tensile strength (σ_{ts}, MPa) on the stabilizer concentration (C, wt%) after polymer processing in Brabender plasticorder (T_0=453 K, n=90 rpm, processing time is 10 min): 1 – compound (63); 2 – (73); 3 – (71).

The estimation of stabilizing efficiency of arylamines (71) and (63) indicates that they exceed Vulkanox 4010 NA according to their ability to preserve dynamic thermoplastic elastomer (DTPE) properties when polypropylene content is less 30 parts by weight (figure 4.32).

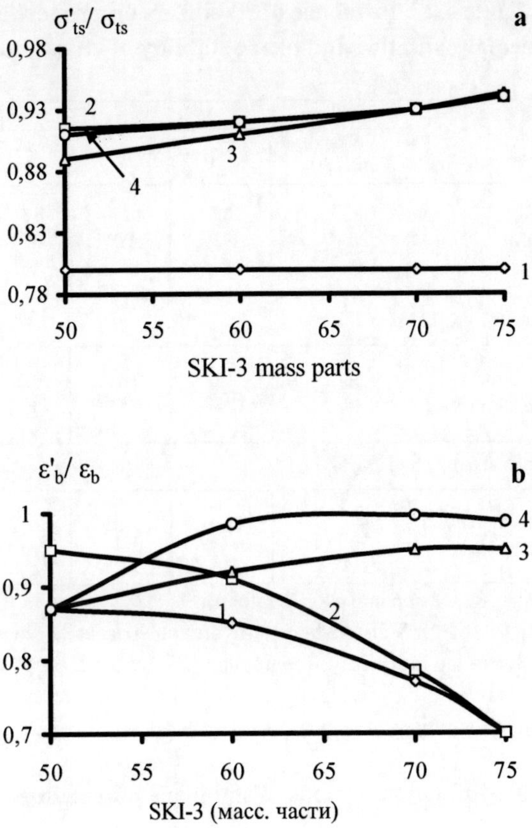

Figure 4.32. Dependence of DTPE stability index (373 K, 72 hours) on a stabilizer type (1 wt%) at various ratio isoprene rubber – polypropylene: 1 – without a stabilizer; 2 – Vulkanox 4010 NA; 3 – compound (71); 4 – (63); a) according to tensile strength index (σ_{ts}); b) according to elongation at break index (ε_b).

In table 4.27 the values of thermal-oxidative and ozone stability of vulcanizates on the basis of isoprene, butadiene and butadiene-methylstyrene rubbers, stabilized by aromatic amines, containing sterically hindered phenol fragments, are given. Compounds (75-79) are synthesized by the interaction of 3,5-di-*tert*-butyl-4-hydroxybenziacetate with diphenylamine, phenyl-β-naphthylamine and N-phenyl-N'-phenylphenylenediamine-1,4 (Section 1.3.3.3).

As it is seen from the given data, benzylated aromatic amines (75-79) are capable of protecting vulcanizates from thermal-oxidative destruction not worse than Vulkanox 4010 NA. At the same time the given compounds don't provide the necessary degree of vulcanizate protection from ozone ageing.

In table 4.28 the values of oxidation induction periods (τ) of polypropylene and low pressure polyethylene, stabilized by benzylated phenylhydrazines (80-82) are given.

Table 4.27. Influence of stabilizers on properties, thermal-oxidative and ozone stability of vulcanizates [a]

Parameter	Stabilizer [b]					
	Vulkanox 4010 NA	(75)	(76)	(77)	(78)	(79)
Tensile strength (σ_{ts}), MPa	18.2	18.6	17.6	18.7	18.4	16.5
Elongation at break (ε_b),%	520	490	500	530	510	600
Residual elongation, %	16	18	18	16	16	18
Thermal-oxidative ageing (373 K, 72 h)						
Stability index calculated from σ_{ts}	0.66	0.66	0.64	0.62	0.68	0.66
Stability index calculated from ε_b	0.48	0.59	0.43	0.34	0.53	0.56
Ozone ageing (323 K, 72 h at tensile by 10%)						
Stability index calculated from σ_{ts}	0.49	0.14			0.16	0.15
Stability index calculated from ε_b	0.76	0.57			0.59	0.62

[a] vulcanizate formulation №6 (see appendix to Chapter 4);

[b] condensation products of 3,5-di-*tert*-butyl-4-hydroxybenziacetate with N-phenyl-N'-phenyl-phenylenediamine-1,4 in the ratio (1:1) – compound (75), in the ratio (3:1) – compound (76); with phenyl-β-naphthylamine in the ratio (1:1) – compound (77), in the ratio (2:1) – compound (78); with diphenylamine in the ratio (2:1) – compound (79) (Section 1.3.3.3).

Table 4.28. Induction periods of stabilized polyolefines oxidation

Stabilizer (0.2 wt%)	τ [a], min	
	PP (438 K)	HDPE (468 K)
–	20	20
(80)	150	180
(81)	150	270
(82)	100	310

[a] oxygen pressure is 33.3 kPa.

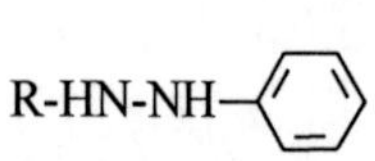

80

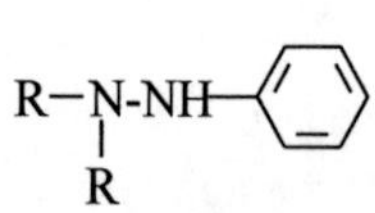

81

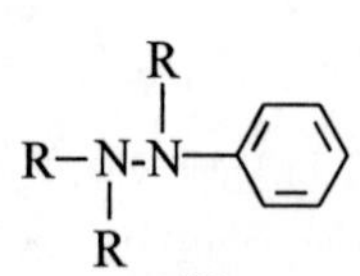

82

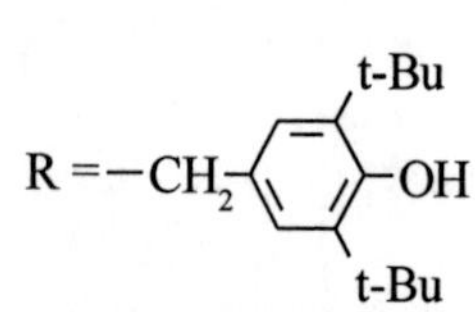

As it is seen from the given data, in polypropylene less stable to thermal-oxidative destruction (the oxidation of polypropylene was conducted at lower temperature), stabilizers able to display the effect of intramolecular synergism of amine and phenol groups possess maximum activity. The similar trend can be seen when oxidation induction periods of polypropylene stabilized by products of benzylation of hydrazine (83), benzidine (84) and aniline (85) are compared, monobenzylated aniline (85) is the most active antioxidant among them (figure 4.33).

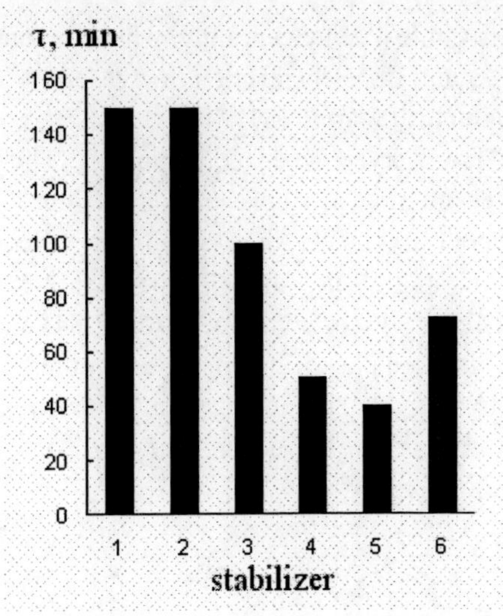

Figure 4.33. Induction periods of polypropylene oxidation in the presence of stabilizers (0.2 wt%): 1 – (80), 2 – (81), 3 – (82), 4 – (84), 5 – (83), 6 – (85); (438 K, oxygen pressure is 33.3 kPa).

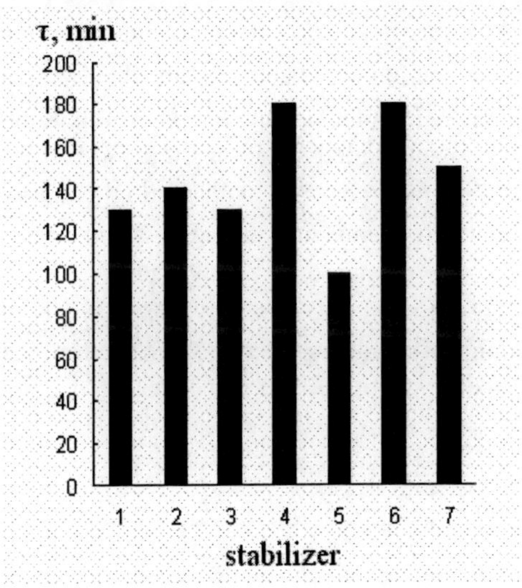

Figure 4.34. Induction periods of isoprene rubber oxidation in the presence of stabilizers (0.3 wt%): 1 – (Vulkanox 4010 NA), 2 – (80), 3 – (82), 4 – (81), 5 – (84), 6 – (83), 7 – (85); (438 K, oxygen pressure is 33.3 kPa).

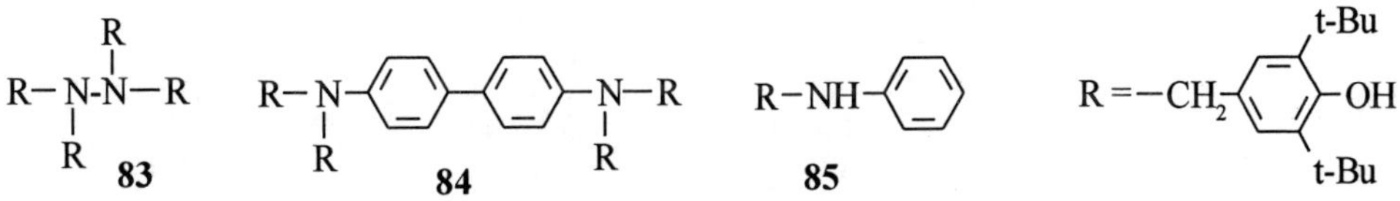

At the same time in HDPE which is more stable to thermal-oxidative destruction the antioxidant activity of stabilizers is directly proportional to the number of sterically hindered fragments in their molecules (table 4.28). Seemingly, when the temperature grows the activity of amine radicals in the chain transfer reaction increases, it results in the decrease of antioxidant activity of compounds (80) and (81).

Intermediate situation is observed when compounds (83-85) and (80-82) are used to prevent thermal-oxidative destruction of isoprene rubber (figure 4.34). In this case both the number of sterically hindered fragments in a stabilizer molecule and the possibility of intramolecular synergism of amine and phenol groups influence the value of oxidation induction period. The relative activity of compounds (82), (84) and (83) agrees with the ratio of the number of sterically hindered phenol fragments and molecular mass of stabilizer.

N-(3,5-di-*tert*-butyl-4-hydroxybenzyl)mercaptobenzothiazole-2 (86) are considered to be antioxidants, which are able both to decompose hydroperoxides and interact with peroxideradicals (Chapter 2).

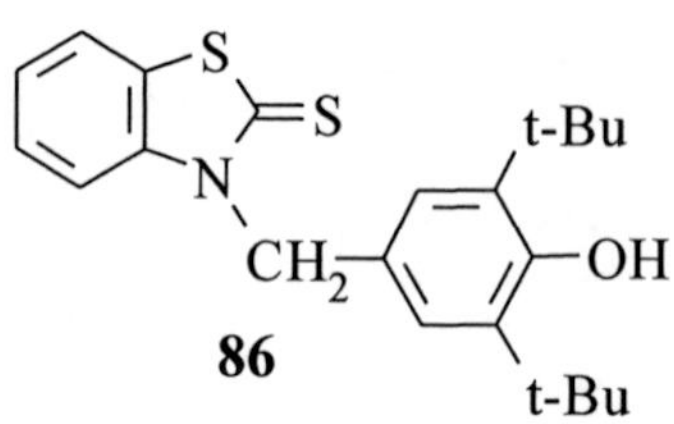

86

During the LDPE stabilization it was detected that amine (86) has lower critical concentration equal to $0{,}3 \cdot 10^{-4}$ mole/kg and it is comparable to bisphenol stabilizer Antioxidant 2246 according to antioxidative action (figure 4.35).

When the mixture of compound (86) with antioxidants of various mechanism of stabilizing action (phenol antioxidants are peroxide radicals acceptors, sulfur-containing antioxidants are hydroperoxide decomposers) was used antioxidant synergetic effect was observed. Thus, when LDPE was stabilized by mixtures containing amine (86) with Antioxidant 2246, or amine (86) with Irganox 1035 induction period of polymer oxidation increased 3 times (figure 4.35).

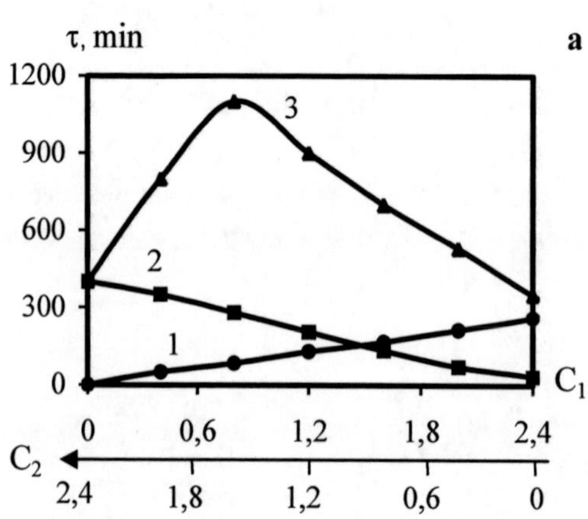

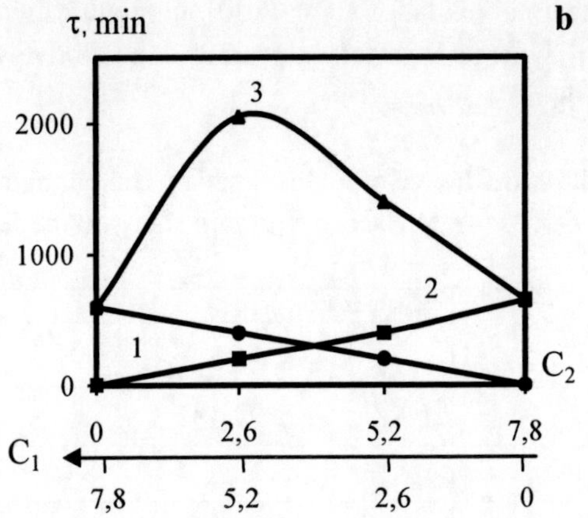

Figure 4.35. Dependence of induction period (τ, min) of LDPE oxidation on the ratio of components of stabilizing mixtures (468 K, oxygen pressure is 33.3 kPa). C_1, C_2 (10^{-4} mole/kg) are contents of compound (86) and phenol antioxidant, respectively: a: 1 – compound (86); 2 – Irganox 1035; 3 – (86) + Irganox 1035; b: 1 – compound (86); 2 – Antioxidant 2246; 3 – (86) + Antioxidant 2246.

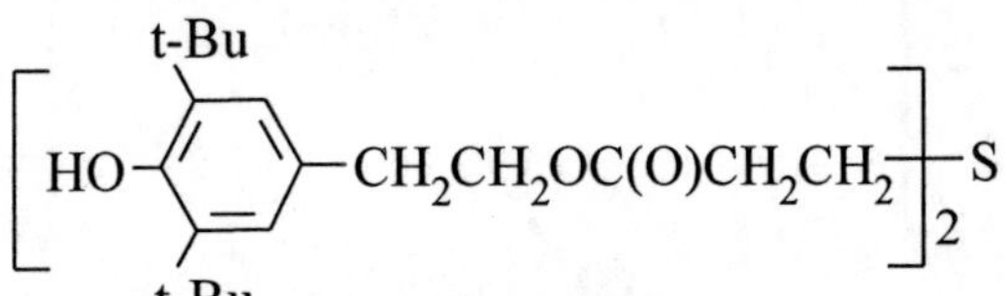

Irganox 1035

Technical carbon possessing antioxidant properties is of essential importance as a component extending stability of polyethylene products [47]. When a stabilizing mixture containing amine (86) was tested, technical carbon (in the amount of 2 wt%) acted as an efficient synergist (table 4.29). The similar stabilizing effect was found by the authors [47] during thermal oxidation of LDPE stabilized by mixtures containing amine and technical carbon.

Table 4.29. Influence of a stabilizer on the time of brittleness beginning (t) and induction periods (τ) of HDPE oxidation

Stabilizer	Formulation (wt%)				
	1	2	4	5	6
Irganox 1010	0.2	0.2			
Amine (86)				0.2	0.2
Carbon black (ASTM №220)		2.5	2.5		2.5
Parameter					
τ^a, min	520		80	420	
t^b, h	15.0	28.0	5.0	14.0	30.0

[a] 443 K, oxygen pressure is 33.3 kPa; [b] milling, 433 K.

Benzothiazole (86), as a rubber stabilizer, exceeds Antioxidant 2246 and Neozone D according to antioxidant efficiency, it is indicated by the least decrease of butyl rubber solution viscosity after thermal oxidation (table 4.30). In mixtures with Neozone D compound (86) displays antioxidant synergetic effect, increasing non-additively the induction period of butyl rubber oxidation (figure 4.36).

Table 4.30. Influence of a stabilizer on the change of butyl rubber Mooney viscosity during thermal oxidation

Parameter	Stabilizer		
	Neozone D (0.15 wt%)	Antioxidant 2246 (0.1 wt%)	Compound (86) (0.1 wt%)
Mooney viscosity	47	47	47
Thermal ageing (373 K, 14 h)			
Mooney viscosity	44	42.3	45.5
Stability index	0.94	0.90	0.97

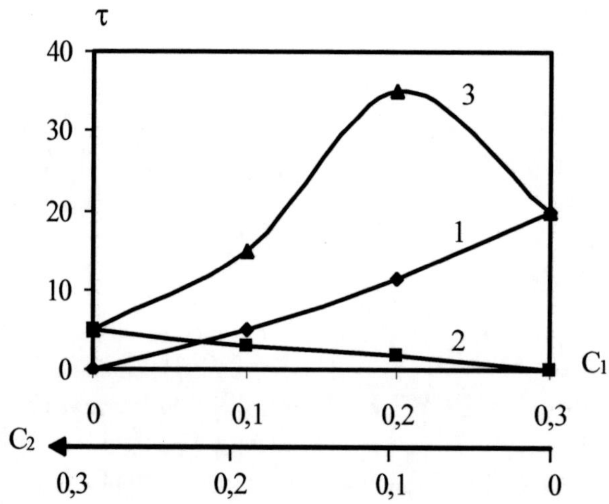

Figure 4.36. Dependence of induction period (τ, h) of butyl rubber oxidation on the ratio of components of stabilizing mixture (423 K, oxygen pressure is 33.3 kPa): 1 – compound (86); 2 – Neozone D; 3 – (86) + Neozone D; C_1,C_2 (10^{-3} mole/kg) are contents of compound (86) and Neozone D, respectively.

Compound (86) can also be used for the stabilization of carbon black-extended vulcanizates providing high stability of physical and mechanical properties of vulcanizates on the basis of butyl rubber and isoprene rubber during thermal ageing (table 4.31). In comparison with control standard carbon black-extended vulcanizate containing Neozone D, stability factors increase from 0.7 to 0.9.

Table 4.31. Influence of antioxidants on stability of physical and mechanical properties of vulcanizates on the basis of butyl rubber and isoprene rubber

Parameter	Stabilizer		
	Neozone D	Compound (86)	Neozone D + compound (86)[a]
Butyl rubber vulcanizate [b]			
Module 300%, MPa	4.1	4.44	4.5
Tensile strength (σ_{ts}), MPa	16.6	20.5	21.1
Elongation at break (ε_b),%	650	610	621
Residual elongation, %	36	40	
Thermal ageing (373 K, 72 h)			
Stability index calculated from σ_{ts}	0.7	0.94	0.96
Stability index calculated from ε_b	0.65	0.93	0.98
Isoprene rubber vulcanizate [c]			
Module 300%, MPa			
Tensile strength (σ_{ts}), MPa	22.1	19.8	
Elongation at break (ε_b),%	520	520	
Thermal ageing (373 K, 72 h)			
Stability index calculated from σ_{ts}	0.90	0.95	
Stability index calculated from ε_b	0.86	0.85	

[a] content of Neozone D is 0.1 wt%, compound (86) is 0.2 wt%; [b] vulcanizate formulation №2, stabilizer content is 0.3 wt%; [c] vulcanizate formulation №1, stabilizer content is 0.6 wt% (see Appendix to Chapter 4).

According to the notion about the formation of cross-links in sulfur vulcanizates the interaction of the vulcanization accelerator with a macromolecule and with free sulfur is a stage preceding cross-linking. At the same time side branches which consist of polysulfide bonds and accelerator fragments *(rubber-S_x–accelerator)* are formed. It is known that when thiazol vulcanization accelerators are used side branches of the following type are formed [48]:

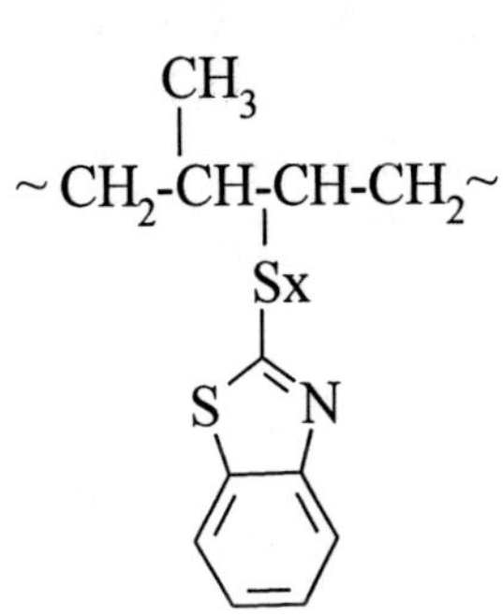

Stabilizer (86), containing thiazol fragment, is likely to display additional function as vulcanization accelerator according to the same principle.

Tests using Monsanto rheometer testify that compound (86) may perform as sulfur vulcanization accelerator. When there is partial or complete substitution of compound (86) for 2-mercaptobenzothiazole in the standard formulation of a vulcanization accelerator (table 4.32) the vulcanization process is characterized by the high rate of cross-linking. At the same time vulcanizates are characterized by high stability to thermal ageing.

Table 4.32. Influence of stabilizer on stability of physical and mechanical properties of butyl rubber vulcanizates [a]

Parameter	Vulcanization accelerфtor		
	2-mercapto-benzothiazole [b]	Compound (86) [c]	2-mercaptobenzo-thiazole + (86) [d]
Minimum torque (M_{min})	10.5	10.0	
Maximum torque (M_{max})	31.5	28.0	
Cure time (t_{90})	45	40	
Module 300%, MPa	4.1	3.9	4.3
Tensile strength (σ_{ts}), MPa	16.6	17.0	18.1
Elongation at break (ε_b),%	650	680	651
Residual elongation, %	36	40	
Rebound elasticity, %	9	9	9
Thermal ageing (373 K, 72 h)			
Stability index calculated from σ_{ts}	0.70	0.72	0.75
Stability index calculated from ε_b	0.75	0.74	0.76

[a] vulcanizate formulation №2 (see Appendix to Chapter 4); [b] accelerator content is 0.65 wt%; [c] compound (86) content is 0.6 wt%; [d] accelerator content is 0.25 wt%, compound (86) content is 0.4 wt%.

As it follows from the investigation results of a number of carbochain polymers, stabilizer (86) can be used as an antioxidant, a metal passivator, a vulcanization accelerator, i.e. it is an additive of polyfunctional action for a large number of polymers. Amine (86) forms synergetic mixtures with antioxidants acting according to various mechanisms, owing to it the antioxidant efficiency of compound (86) can be increased.

On the whole, perspective of using polyfunctional arylamines containing sterically hindered phenol fragments for the increase of stability of various types of polymeric materials to thermal-oxidative ageing was determined.

4.7. Polyphenol Stabilizers

In recent years much attention is paid to the development of efficient methods of obtaining and study of protective action of polyphenol stabilizers, containing several 3,5-di-tert-butyl-4-hydroxybenzyl fragments. In table 4.33 the values of oxidation induction periods (τ) of polypropylene stabilized by polyfunctional stabilizers (87-91) are given.

Table 4.33. Induction periods (τ) of polypropylene oxidation in the presence of stabilizers (0.3 wt%, 456 K, oxygen pressure is 33.3 kPa)

Stabilizer	n[a]	τ, min
(87)	3,88	155
(88)	3.93	190
(89)	3.54	60
(90)	2.82	40
(91)	4.47	155
Compound (89) (30%) + compound (90) (70%)	–	50
Irganox 1010	3.40	150

[a] number of sterically hindered phenol fragments in a molecule, multiplied by 1000 and divided by molecular mass of a stabilizer.

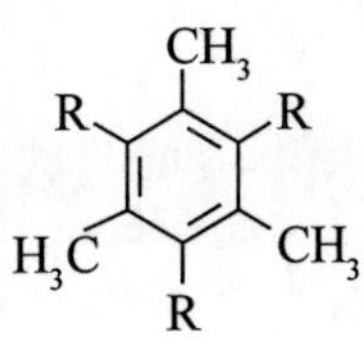

87

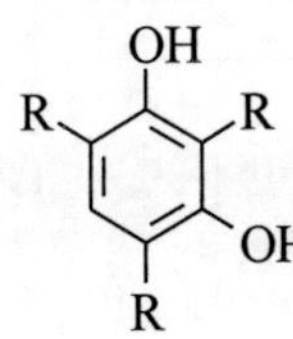

88

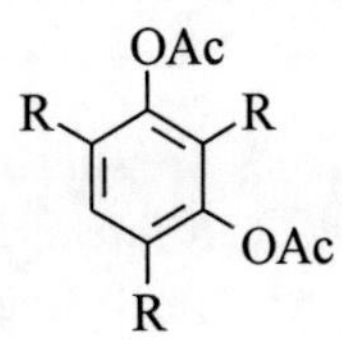

89

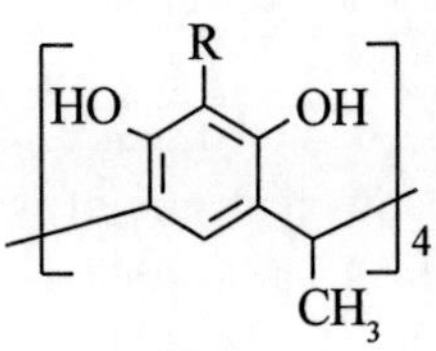

90

NR_3

91

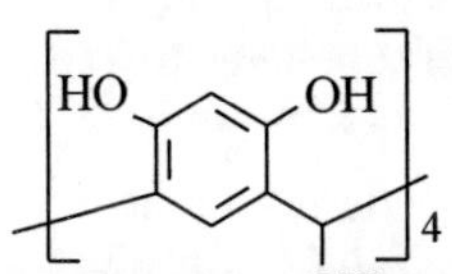

92

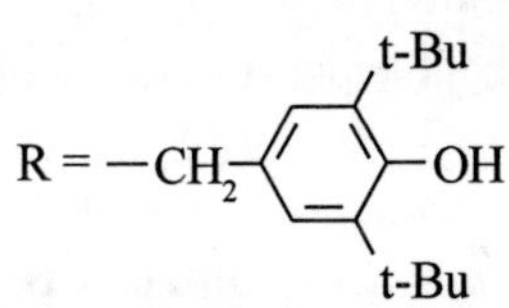

It is indicated that the presence of two types of hydroxyl groups, differing in their reactivity, in the molecule of benzylated resorcin (88) promotes the increase of antioxidant activity of this compound in comparison with other polyphenol stabilizers, having similar ratio of the number of sterically hindered phenol fragments and molecular mass. The similar trend is traced in polyethylene composites, oxidation induction period of which is substantially higher in the presence of benzylated resorcin (88), than in the presence of benzylated mesitylene (87). It may be assumed that compound (88) is able to display the intramolecular synergism according to the mechanism of regeneration of more active peroxide radical acceptor (resorcinol hydroxyl) with the formation of more stable sterically hindered phenoxyl radical:

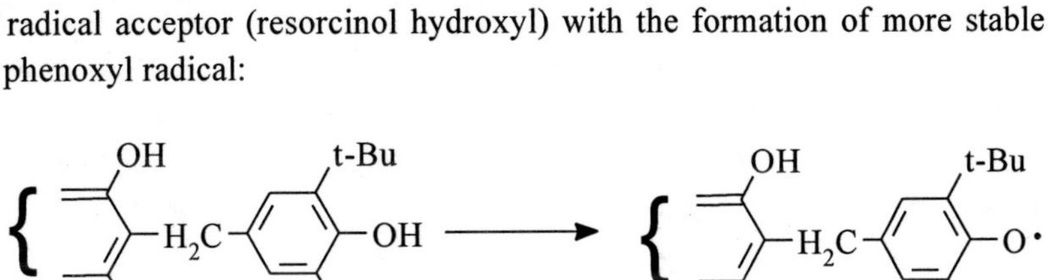

Relatively low antioxidant activity of benzylated calix[4]resorcin (90) in polypropylene (table 4.33) can be explained by less ratio of the number of sterically hindered phenol fragments and the molecular mass of this compound. At the same time during processing and accelerated ageing of HDPE calixarene (90) exceeds compounds (89) and (88) in stabilizing activity and it is comparable with stabilizer Irganox 1010 (table 4.34).

Table4.34. Influence of stabilizers on properties and thermal-oxidative stability of HDPE

Parameter	Stabilizer (0.2 wt%)			
	Irganox 1010	(88)	(89)	(90)
Tensile strength (σ_{ts}), MPa	24.8	26.0	23.9	24.1
Elongation at break (ε_b),%	758	779	798	767
Thermal ageing (448 K,·8 h)				
Stability index calculated from σ_{ts}	0.95	0.75	0.85	1.05
Stability index calculated from ε_b	0.96	0.85	0.60	1.00

It should be noted that resorcinarenes are suitable little for the polyolefine stabilization considering low compatibility with them. It refers to compound (90) in less degree, as the presence of 3,5-di-*tert*-butyl-4-hydroxybenzyl fragments in its molecule improves its solubility and compatibility with polyolefines. Resorcinarenes are much more compatible with rubbers. As it seen from table 4.35, benzylated calixarene (90) is a more efficient inhibitor of thermal-oxidative destruction of butadiene-nitrile rubber vulcanizate in comparison both with unbenzylated calixarene (92) and Vulkanox 4010 NA [49].

Table 4.35. Influence of stabilizers on properties and thermal-oxidative stability of butadiene-nitrile rubber vulcanizates [a]

Parameter	Stabilizer				
	Vulkanox 4010 NA	(92)	(90)	(88)	(92)+(88)
Tensile strength (σ_{ts}), MPa	21.2	23.5	20.1	20.5	21.6
Elongation at break (ε_b),%	314	320	308	310	280
Residual elongation, %	8	9	8	10	8
Thermal ageing (373 K, 72 h)					
Stability index calculated from σ_{ts}	0.87	0.78	0.95	0.95	0.88
Stability index calculated from ε_b	0.69	0.60	0.68	0.77	0.69

[a] vulcanizate formulation №7 (see Appendix to Chapter 4).

As it is seen from the data of table 4.36, 3,5-di-(3′,5′-di-*tert*-butyl-4′-hydroxybenzyl)-2,4-dihydroxybenzophenone (93) is a more efficient stabilizer of butyl rubber and vulcanizate on its basis in comparison with phenols (94-98) and commercial stabilizer Antioxidant 2246.

Table 4.36. Influence of stabilizers on properties and thermal-oxidative stability of butyl rubber and vulcanizates [a] on its basis

Parameter	Stabilizer (0.2 wt%)						
	(94)	(95)	(96)	(97)	(98)	(93)	Antioxidant 2246
Induction period of thermal oxidation (τ) [a], h	24	24	23	24	24	31	22
Vulcanizates							
Tensile strength (σ_{ts}), MPa	21.9	21.7	21.0	22.3	22.3	22.2	21.6
Elongation at break (ε_b),%	730	780	710	660	700	730	690
Thermal-oxidative ageing (373 K, 72 h)							
Stability index calculated from σ_{ts}	0.89	0.87	0.86	0.89	0.89	0.96	0.88
Stability index calculated from ε_b	0.85	0.81	0.85	0.86	0.93	0.91	0.86

[a] defined by the UV spectroscopy according to the absorption band of valence vibrations of C=O group (λ=1720 sm^{-1}).

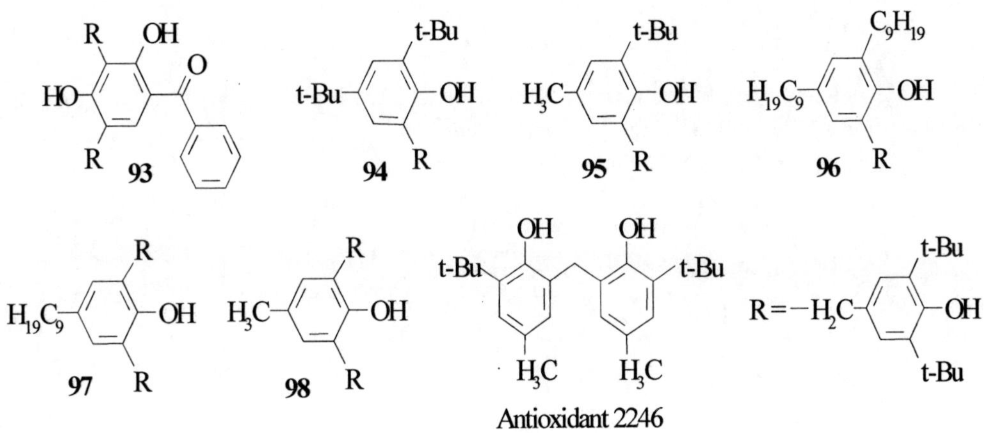

According to antioxidant activity in LDPE compound (93) exceeds compound (98) and is comparable with stabilizer Irganox 1010, though preserves polyethylene color slightly worse. The similar trend is observed for stabilizing mixtures of phenols (93), (98) and Irganox 1010 with commercial organophosphorus stabilizers Irgafos 168, Irgafos 126, Irgafos TNPP (table 4.37).

Table 4.37. Properties of LDPE containing 0.1 wt% phenol compound, 0.2 wt% phosphite and 0.05 wt% calcium stearate

Stabilizer		Melt flow index			Yellowness index			τ[a], min
Phenol	Phosphite	Extrusion number						
		1	3	5	1	3	5	
(93)	–	2.61	2.59	2.57	25	26	28	36
	Irgafos 126	2.59	2.59	2.59	20	21	21	137
	Irgafos 168	2.58	2.58	2.57	25	26	26	70
	Irgafos TNPP	2.62	2.61	2.60	23	24	24	76
Irganox 1010	–	2.79	2.71	2.62	12	17	19	30
	Irgafos 126	2.60	2.62	2.64	12	14	16	138
	Irgafos 168	2.59	2.59	2.58	11	13	15	68
	Irgafos TNPP	2.58	2.58	2.58	11	13	15	73
(98)	–	2.62	2.52	2.61	18	22	25	38
	Irgafos 126	2.58	2.58	2.60	12	14	16	114
	Irgafos 168	2.63	2.61	2.59	17	26	28	37
	Irgafos TNPP	2.61	2.62	2.64	14	21	23	70

[a] oxidation induction period, 473 K, oxygen pressure is 33.3 kPa.

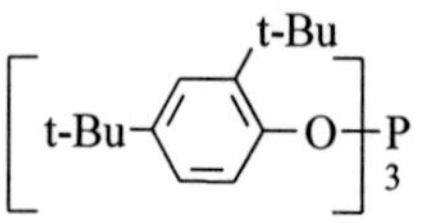

Irgafos 168

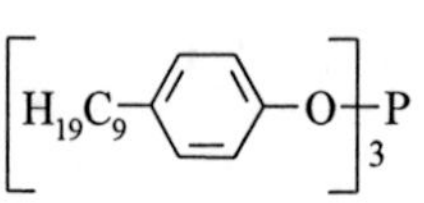

Irgafos TNPP

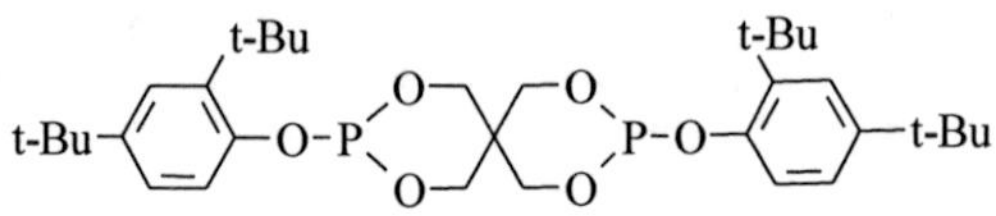

Irgafos 126

The efficiency of light-protective action of benzylated hydroxybenzophenones (93) and (99) is studied in polyethylene, polypropylene and styrene-butadiene-styrene triblock thermoplastic elastomer (SBSTPE) [50] in comparison with commercial light stabilizers 2,4-dihydroxybenzophenone (100) and 2-hydroxy-4-octoxybenzophenone (101).

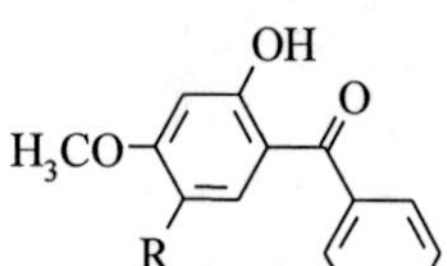

99

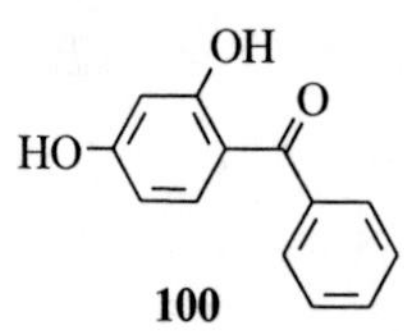

100

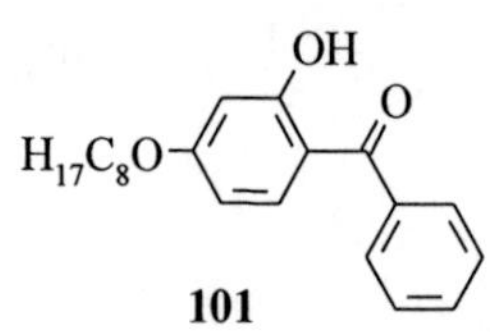

101

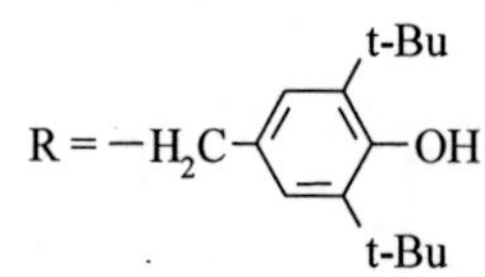

Stabilizer (93), and especially (99) increase considerably light stability of polypropylene, however they are less efficient than commercial stabilizers (figure 4.37). Similar results (from the qualitative point of view) were obtained for stabilizers (93) and (99) when polyethylene was tested (figure 4.38).

The comparison of the data of figure 4.37 and 4.38 indicates that stabilizers (93) and (99) are more efficient during polypropylene photooxidation than during polyethylene photooxidation, it corresponds to higher light stability of unstabilized polyethylene in comparison with polypropylene. Therefore it can be suggested that relatively low efficiency of light-protective action of stabilizers (93) and (99) is conditioned by their own low light stability. The study of kinetic of stabilizer consumptions during light stability tests confirmed this assumption. The analysis of the change of UV-absorption spectra of stabilized films during irradiation indicates that stabilizers (93) and (99) are consumed noticeably even at early stages (figure 4.39), while commercial stabilizers (100) and (101) are substantially more stable.

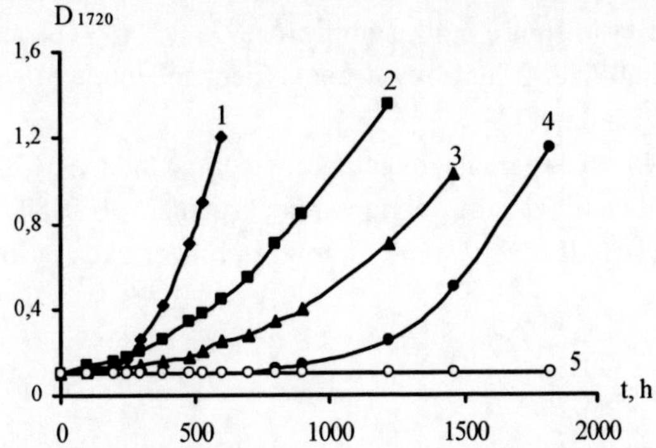

Figure 4.37. Kinetic curves of carbonyl group accumulation during photooxidation of polypropylene films (thickness is 0.3 mm), containing 0.5 wt% stabilizer: 2 – (93), 3 – (99), 4 – (100), 5 – (101), 1 – unstabilized polypropylene.

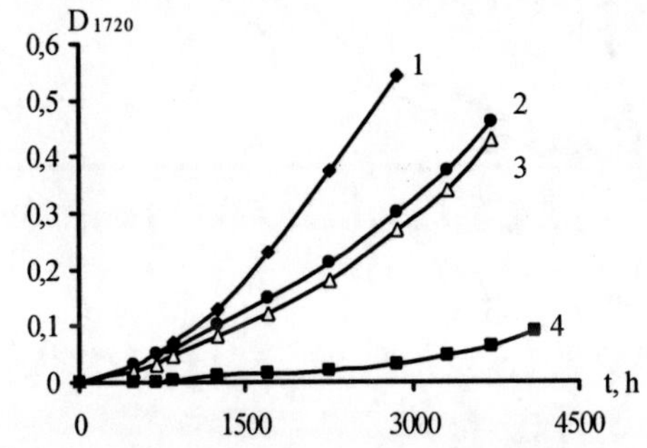

Figure 4.38. Kinetic curves of carbonyl group accumulation during photooxidation of HDPE films (thickness is ≈ 80 μm), containing 0.5 wt% stabilizer: 2 – (93), 3 – (99), 4 – (101), 1 – unstabilized polyethylene.

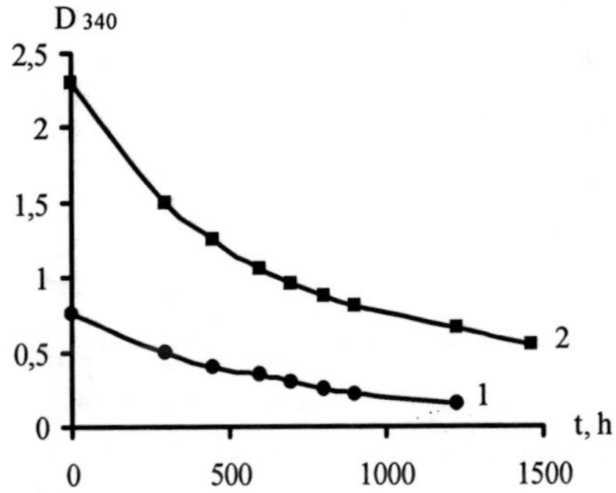

Figure 4.39. Kinetic curves of stabilizer consumption during thermal oxidation of polypropylene films (thickness is 0.3 mm), initially containing 0.5 wt% stabilizer: 1 – (93), 2 – (99).

The mentioned above results, testifying to insufficient light stability of compounds (93) and (99) for the efficient stabilization of polyolefines, lead to the assumption that the given compounds will be more efficient stabilizers in easier oxidable polymers.

Indeed, as it is given in figure 4.40, stabilizers (93) and (99) exceed noticeably the most efficient commercial stabilizer 2-hydroxy-4-octoxybenzophenone (101) in styrene-butadiene-styrene thermoplastic elastomer (SBSTPE).

Polyfunctionality is an apparent advantage of compounds (93) and (99). They develop high activity as thermal stabilizers of polymers, in particular, of SBSTPE, while 2-hydroxy-4-octoxybenzophenone (101) is a very weak thermal stabilizer even at low temperatures (figure 4.41).

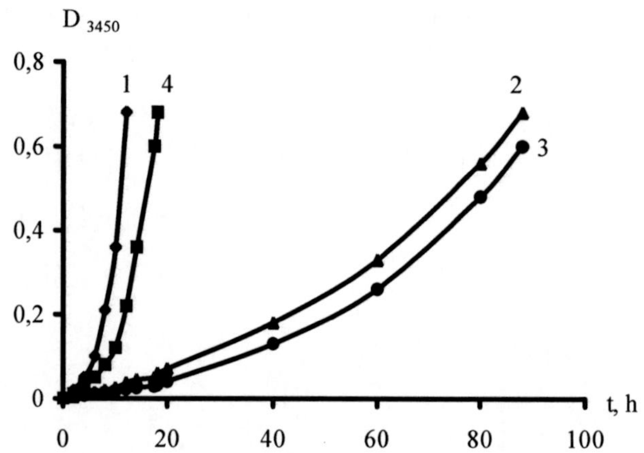

Figure 4.40. Kinetic curves of hydroxyl group accumulation during photooxidation of SBSTPE films (thickness is ≈ 150 μm), containing 0.5 wt% stabilizer: 2 – (93), 3 – (99), 4 – (101), 1 – unstabilized SBSTPE.

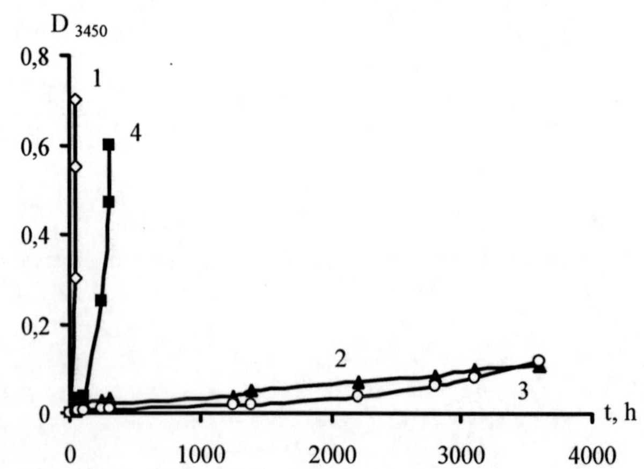

Figure 4.41. Kinetic curves of hydroxyl group accumulation during thermal oxidation of SBSTPE films at 323 K (thickness is ≈ 150 μm), containing 0.5 wt% stabilizer: 2 – (93), 3 – (99), 4 – (101), 1 – unstabilized SBSTPE.

* * *

The specific ways of polymer stabilization are determined on the basis of disclosed mechanisms of polyfunctional stabilizer action (Chapters 2 and 3). It is shown that polyfunctional stabilizers are efficient inhibitors of thermal-oxidative destruction and color stabilizers of polymers. The perspective means of improving key functions of polyfunctional stabilizers and of increasing their overall efficiency (synergism effects) during the stabilization of various polymer types are developed.

APPENDIX TO CHAPTER 4

The vulcanizate formulations used during tests of polyfunctional stabilizers are given in table 4.38.

Table 4.38. Vulcanizate formulation

Formulation number	Ingredient	Content (wt%)
1	Synthetic *cis*-isoprene rubber SKI-3 (Russia)	100.0
	Stabilizer	0.6
	Stearic acid	1.0
	1,3-Diphenylguanidine	3.0
	Zinc oxide	5.0
	2,2'-Benzothiazyl disulfide	1.0
	Carbon black (ASTM №326)	50.0
	Sulfur	1.0
2	Synthetic butyl rubber BK-1675N (Russia)	100.0
	Stearic acid	3.0
	2-Mercaptobenzothiazole	0.65
	Thiuram TMTD	1.3
	Zinc oxide	5.0
	Carbon black (ASTM №220)	50.0
	Sulfur	3.0
3	Synthetic *cis*-isoprene rubber SKI-3 (Russia)	100.0
	Sulfur	3.0
	Zinc oxide	5.0
	Stearic acid	2.0
	Rubrax (Russia) [a]	2.5
	Colophony	2.0
	Carbon black (ASTM №220)	55.0
	Stabilizer	1.0
	Oil PN-6Sh (Russia) [b]	4.0
4	Synthetic butyl rubber BK-1675N (Russia)	100.0
	Stabilizer	0.2-0.4
	Stearic acid	3.0
	2-Mercaptobenzothiazole	1.3
	Thiuram TMTD	0.65
	Zinc oxide	4.0
	Carbon black (ASTM №220)	50.0
	Sulfur	2.0

Table 4.38. (Continued).

Formulation number	Ingredient	Content (wt%)
5	Synthetic *cis*-isoprene rubber SKI-3 (Russia)	100.0
	Stabilizer	0.6
	Stearic acid	1.0
	1,3-Diphenylguanidine	3.0
	Zinc oxide	5.0
	2,2'-Benzothiazyl disulfide	1.0
	Carbon black (ASTM №220)	40.0
	Sulfur	1.0
6	Synthetic *cis*-isoprene rubber SKI-3 (Russia)	56.0
	Synthetic *cis*-butadiene rubber SKD (Russia)	22.0
	Synthetic butadiene-α-methylstyrene rubber SKMS (Russia)	22.0
	Carbon black (ASTM №220)	57.0
	Zinc oxide	5.2
	Paralight-17 (Russia) [c]	2.0
	Colophony	1.0
	Koresin	2.0
	Stearic acid	2.0
	Oil PN-6Sh (Russia) [b]	14.2
	Phthalic anhydride	0.5
	2-(Cyclohexylaminothio)benzothiazole	1.5
	Sulfur	1.9
	Santoflex R	2.0
	Stabilizer	1.0
7	Synthetic butadiene-nitrile rubber BNKS-26 (Russia)	100.0
	Zinc oxide	3.0
	2-(Cyclohexylaminothio)benzothiazole	0.7
	Stearic acid	1.0
	Carbon black K-354 (Russia) [d]	40.0
	Sulfur	1.5
	Stabilizer	2.0

[a] plasticizing agent, high melting asphaltic bitumen exposed to caustic treatment;

[b] petroleum plasticizing agent, density is 0.96-0.98, 10-12% of paraffine-naphthenic hydrocarbons, 73-85% of aromatic hydrocarbons, 4% of resins;

[c] protective mineral wax – mixture of paraffins, ceresin and technical stearin, melting temperature ≈339 K;

[d] channel carbon black, active, obtained in diffusion flame during thermal-oxidative destruction of natural or associated gas, conditional geometric surface is 90-100 m^2/g, specific absorption surface area is no more than 150 m^2/g, dibutylphthalate adsorption is 70 sm^3/100g, ash percentage is no more than 0.5%.

REFERENCES

[1] Starinie i stabilizatsiya polimerov (Ageing and Stabilization of Polymers), Ed. by Neiman, M.B. Moscow: *Nauka;* 1964.

[2] Emanuel, N.M. and Buchachenko, A.L. Khimicheskaya physika stareniya i stabilizatsii polimerov (Chemical Physics of Ageing and Stabilization of Polymers). Moscow: *Nauka;* 1988.

[3] Scott, G. Atmosfernoe okislenie i antioksidanty (Atmospheric Oxidation and Antioxidants). Amsterdam: *Elsevier;* 1965.

[4] Kirillova, E.I. and Shul'gina, E.S. Starenie i stabilizatsiya termoplastov (Ageing and Stabilization of Thermoplasts). Leningrad: *Khimiya;* 1988.

[5] Minsker, K.S., Fedoseeva, G.T. Destruktsiya i stabilizatsiya polivinilkhlorida (Destruction and Stability of Polyvinylchloride). Moscow: *Khimiya;* 1972.

[6] Shlyapnikov, Yu.A. Kiryushkin, S.L., Mar'in, A.N. Antiokislitelnaya stabilizatsiya polimerov (Antioxidative Stabilization of Polymers). Moscow: *Khimiya;* 1986.

[7] Shlyapintokh, V.Ya. Fotokhimicheskie prevratscheniya i stabilizatsiya polimerov (Photochemical Transformations and Stabilization of Polymers). Moscow: *Khimiya;* 1979, p. 315-317.

[8] Piotrovskii, K.B., Tarasova, Z.N. Starenie i stabilizatsiya sinteticheskikh kauchukov i vulkanizatov (Ageing and Stabilization of Synthetic Rubbers and Vulkanizates). Moscow: *Khimiya;* 1980.

[9] Gladyshev, C.P., Nepalov, V.F. – *Usp. Khim.,* 1975, vol.44, №10, p. 1830-1850.

[10] Karpukhina, G.V., Emanuel, N.M. – *Dokl. AN SSSR,* 1984, vol.276, №5, p. 1163-1167.

[11] Emanuel, N.M. – *Usp. Khim.,* 1979, vol.48, №12, p. 2113-2158.

[12] Levin, P.I., Mikhailov, V.V., Medvedev, A.I. Ingibirovanie protsesov okisleniya polimerov smesyami stabilizatorov (Inhibition of Processes of Polymers Oxidation by Mixtures of Stabilizers). Moskow: *NIITEKhim;* 1970.

[13] Lugova, L.I. – Abstracts of Papers. The IX Conference "Destruction and Stabilization of Polymers", Moscow, Russia, 2001, p.11-112.

[14] Kirpichnikov, P.A., Kolyubakina, N.C., Mukmeneva, N.A. et al. – *Vysokomol. Soed.,* 1970, vol. B12, №3, p. 189-192.

[15] Kirpichnikov, P.A. Mukmeneva, N.A., Pobedimskii, D.G. – *Usp. Khim.,* 1983, vol.52, №11, p. 1831-1851.

[16] Pobedimskii, D.G., Mukmeneva, N.A., Kirpichnikov, P.A. In: Developments in Polymer Stabilization, Ed. by Scott, G. London: *Appl. Science Publ.*; 1980, p. 125-184.

[17] Gabutdinov, M.C. et al. Russia Patent 2140938 (1998); *Bull. Izobret.* (Rus), 1999, №31.

[18] Borisova, M.V., Fazlieva, L.K., Fokkho, Zh. Et al. – *Zh. Prikl. Khim.*, 2007, vol.74, №9, p. 1500-1504.

[19] Mukmeneva, N.A. et al. The USSR Author Certificate 2156263 (1999); *Bull. Izobret.* (Rus), 2000, №26.

[20] Mukmeneva, N.A., Akhmadullina, A.G., Kolubakina, N.S. et al. – *Vysokomol. Soed.,* 1974, vol. A16, №12, p. 370-375.

[21] Mukmeneva, N.A., Akhmadullina, A.G., Sabirova, Z.Kh. et al. – *Vysokomol. Soed.,* 1976, vol. B18, №2, p. 108-111.

[22] Kurashov, V.I., Pobedimskii, D.G., Kirpichnikov, P.A. – *Dokl. AN SSSR,* 1978, vol.28, №6, p. 1407-1410.

[23] Kurashov, V.I., Kubusheva, N.I., Pobedimskii, D.G. – *Vysokomol. Soed.,* 1980, vol. B22, №1, p. 219-222.

[24] Kurashov, V.I., Pobedimskii, D.G., Krivileva L.G. – *Zh. Phys. Khim.,* 1979, vol.53, №2, p. 501-502.

[25] Pobedimskii, D.G., Kurashov, V.I., Kirpichnikov, P.A. – *J. Polym. Sci., Polym. Chem. Ed.,* 1983, vol.21, №1, p. 55-58.

[26] Pobedimskii, D.G. – *Neftekhim.,* 1978, vol.18, №5, p. 701-707.

[27] Beresnev, V.V., Kirpichnikov, P.A. – *Kauch. Resina,* 1970, №10, p. 3-6.

[28] Beresnev, V.V., Kirpichnikov, P.A. – *Vysokomol. Soed.,* 1967, vol.B9, №11, p. 809-812.

[29] Minsker, K.S., Abdullin, M.I. – *Dokl. AN SSSR,* 1982, vol.263, №1, p. 140-143.

[30] Minsker, K.S., Abdullin, M.I., Kabashnikov, V.G. et al. – *Vysokomol. Soed.,* 1980, vol.A22, №9, p. 2131-2135.

[31] Minsker, K.S., Abdullin, M.I., Zueva, P.P. et al. – *Plast. Massy,* 1981, №9, p. 33-34.

[32] Stabilizatsiya plyonochnykh materialov I iskusstvennykh kozh (Stabilization of Filmy Materials and Synthetic Leathers), Ed. by Kutyanin. Moscow: *NIITEKhim;* 1974, p. 105-132.

[33] Voigt, I. Die Stabilisierung der Kunsfstoffe Gegen Licht und Wärme. Berlin: *Springer-Verlag;* 1966.

[34] Mukmeneva, N.A., Cherezova, E.N., Tuzova, N.G. et al. – *Zh. Prikl. Khim.,* 1994, vol.67, №4, p. 633-635.

[35] Serova, V.N., Vasil'ev, A.A., Koryagina, E.L. et al. – Abstract of Papers. The Conference "Advanced Solid-State Lazers", Washington, USA, 1993, vol.2, p. 173-175.

[36] Makhlis, F.A. Radiatsionnaya fizika i khimiya polimerov (Radiation Physics and Chemistry of Polymers). Moscow: *Atomizdat;* 1972.

[37] Barashkov, N.N., Sakhno, T.V. Opticheski prozrachnye polimery i meterialy na ikh osnove (Optically Transparent Polymers and Materials from them). Moscow: *Khimiya;* 1992.

[38] Dyumaev, K.M., Manenkov, A.A., Maslyukova, A.P. – *Izv. AN SSSR, Ser. Phyz.,* 1987, vol.51, №8, p. 1387-1395.

[39] Renbi, B., Rabek, Ya. Fotodestruktsya, fotookislenie i fotostabilizatsiya polimerov (Photodestruction, Photooxidation and Photostabilization of Polymers). Moscow: *Mir;* 1978.

[40] Vol'fson, S.I. et al. Russia Patent 2067103 (1996); *Bull. Izobret.* (Rus), 1996, №27.

[41] Krichevskii, G.E. Fotokhimicheskie prevratscheniya krasitelei i svetostabilizatsiya okrashennykh materialov (Photochemical Transformations of Dye-stuffs and Light Stabilization of Dyed Materials). Moscow: *Khimiya;* 1986.

[42] Moiseev, Yu.V., Zaikov, G.E. Khimicheskaya sto'kost polimerov v agressivnykh sredax (Kinetic Stability of Polymers in Hostile Environments). Moscow: *Khimiya;* 1979.

[43] Ershov, V.V., Nikiforov, G.A., Volod'kin, A.A. Prostranstvenno-zatrudnennye fenoly (Sterically Hindered Phenols). Moscow: *Khimiya;* 1972.

[44] Roginskii, V.A. Fenol'nye antioxidanty: reaktsionnaya sposobnost i effektivnost (Phenol Antioxidants: Reactivity and Efficiency). Moscow: *Nauka;* 1988.

[45] Nugumanova, G.N., Zhukova, R.S., Cherezova, E.N. et al. – *Zh. Prikl. Khim.*, 1997, vol.70, №9, p. 1520-1524.

[46] Mukmeneva, N.A., Cherezova, E.N., Rusina, I.F. et al. In: Ageing of Polymers, Polymer Blends and Polymer Composites – 2, Ed. by Zaikov, G.E., Buchachenko, A.L., Ivanov, V.B.. New-York: Nova Sci. Publ., 2002, vol.2, p. 197-205.

[47] Gol'dberg, V.M., Kolesnikova, N.N., Paverman, N.G. – Abstract of Papers. The IX Conference "Destruction and Stabilization of Polymers", Moscow, Russia, 2001, p. 49.

[48] Materialy resinovogo proizvodstva: spravochnik rezintschika (Materials of Rubber Productions: Reference-book for the Rubber Specialist), Ed. by Skuba, I.A. Moscow: *Khimiya;* 1971. Redactor

[49] Bukharov, S.V., Nugumanova, G.N., Teregulova, E.A. et al. – *Zh. Prikl. Khim.*, 2003, vol.76, №11, p. 1918-1920.

[50] Mukmeneva, N.A., Bukharov, S.V., Nugumanova, G.N. et al. – *Polym. Science, Ser. B,* 1998, vol.40, №9-10, p. 302-305.

CONCLUSION

In this monograph the most significant results on synthesis and study of element(N,P,S)-containing compounds, which are promising as polyfunctional stabilizers for polymers, are generalized .

It is shown that heteroatoms N,P,S, which may present in molecules of polyfunctional stabilizers both individually (phosphorous acid esters) and in the combination (phosphorus dithioacids, amides of phosphorous acids, thiocarbamide derivatives), condition nucleophilic properties and thereby high reactivity of polyfunctional stabilizers during the stabilization of polymers.

Possible ways are revealed to increase antioxidative activity of polyfunctional stabilizers by their additional functionalization that is by insertion of additional reactive groups (e.g. fragments of sterically hindered phenols) into their molecules, by co-ordination with variable valency metals etc.

The purposeful analysis of investigation results of polyfunctional stabilizers made possible to determine the correlation between structure, reactivity and inhibiting efficiency.

Polyfunctional stabilizers were found to have a number of peculiarities distinguishing them from traditional antioxidants.

The most important consequences of action polyfunctionality of the discussed stabilizers were revealed. Thus, among reactions of polyfunctional stabilizers there are such ones as reactions with free peroxide radicals ($RO_2^{\bullet}$), leading to the linear termination of oxidation chains; and reactions of polyfunctional stabilizers with hydroperoxides, preventing from degenerated branching of the chain process; such dual reactivity of polyfunctional stabilizers may result in either co-ordination or competition of reaction mechanisms acting simultaneously. Reactions of the "non-chain" inhibition (that is, reactions with active molecular products of polymers oxidation) are closely related to the problems of stabilization of polymer color.

The inhibition of polymer coloration under the conditions of their processing and ageing is one of the important peculiarities of the action of polyfunctional stabilizers. The most detailed study of polymer coloration has been carried out, organic phosphates being an example. The obtained results made it possible to determine that the necessary condition for the stabilization of polymer color by phosphites is their participation in a number of reactions, such as the blocking of anomalous fragments in macromolecules (at the initiation stage of destructive processes); interaction with colored products of oxidation, formed in the course of degradation of macromolecules (at the stage of development of destructive processes);

chemical deactivation of admixtures and colored products of conversion of additives in polymers etc.

The ability of polyfunctional stabilizers to react by many reaction mechanisms during stabilization of polymers creates the basis for the realization of effects of the intramolecular synergism, stipulating high inhibiting effects in polymers by this.

Moreover, the polyfunctional stabilizers in the mixed compositions with other stabilizers are capable to exhibit additional synergetic effects. The phenomenon of double synergetic effects opens up new possibilities for the efficient stabilization of practice-useful properties of polymers.

As a whole, the diverse information submitted in this monograph, on the one hand, makes possible to imagine and appreciate the significance of the discussed investigations in the total problem of stabilization of polymers; and on the other hand, indicates clearly to the expediency of further research in the field of element-containing polyfunctional stabilizers taking into account their potentially high efficiency and practical value.

INDEX

D

E

F

G

H

I

J

K

L

M

N

O

P

Q

R

S

T

U

V

W

X

Y

Z